CHRONIC
PAIN

New Molecular Insights into **Pain** and **Treatment**

CHRONIC PAIN

New Molecular Insights into Pain and Treatment

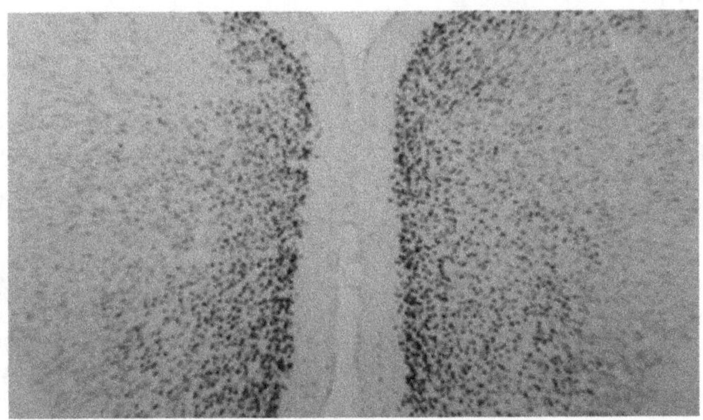

MIN ZHUO

University of Toronto, Canada

W♦ **World Scientific**

NEW JERSEY · LONDON · SINGAPORE · BEIJING · SHANGHAI · HONG KONG · TAIPEI · CHENNAI · TOKYO

Published by

World Scientific Publishing Co. Pte. Ltd.

5 Toh Tuck Link, Singapore 596224

USA office: 27 Warren Street, Suite 401-402, Hackensack, NJ 07601

UK office: 57 Shelton Street, Covent Garden, London WC2H 9HE

British Library Cataloguing-in-Publication Data
A catalogue record for this book is available from the British Library.

CHRONIC PAIN
New Molecular Insights into Pain and Treatment

ISBN 978-981-4324-52-6 (hardcover)
ISBN 978-981-4324-53-3 (ebook for institutions)

For any available supplementary material, please visit
https://www.worldscientific.com/worldscibooks/10.1142/7931#t=suppl

Typeset by Stallion Press
Email: enquiries@stallionpress.com

Key Abbreviations

1	5-HT	5-hydroxytryptamine
2	AA	arachidonic acid
3	AC	adenylyl cyclase
4	ACC	anterior cingulate cortex
5	ACSF	artificial cerebro-spinal fluid
6	AKAP	A-kinase anchoring protein
7	AM251	N-(piperidin-1-yl)-5-(4-iodophenyl)-1-(2,4-dichlorophenyl)-4-methyl-1H-pyrazole-3-carboxamide
8	AMPA	α-amino-3-hydroxy-5-methyl-4-isoxazole-propionic acid
9	AMPAR	α-amino-3-hydroxy-5-methyl-4-isoxazolepropionic acid receptor
10	AP-5	D-2-amino-5-phosphonopentanoic acid
11	ATP	adenosine triphosphate
12	BAPTA	1,2-bis(o-aminophenoxy)ethane-N,N,N′,N′-tetraacetic acid
13	BDNF	brain-derived neurotrophic factor
14	CaMKII	calcium/calmodulin-dependent protein kinase type II
15	CaMKIV	calcium/calmodulin-dependent protein kinase type IV

16	cAMP	cyclic adenosine monophosphate
17	CB1R	cannabinoid 1 receptor
18	CFA	complete Freund's adjuvant
19	ChR2	channelRhodopsin2
20	CNQX	6-cyano-7-nitroquinoxaline-2,3-dione
21	CNS	central nervous system
22	CO	carbonic oxide
23	CP-AMPARs	Ca^{2+}-permeable AMPARs
24	CPCCOEt	7-(hydroxyimino)cyclopropa[b]chromen-1a-carboxylate ethyl ester
25	CPN	common peroneal nerve
26	CPP	3-(R-2-carboxypiperazin-4-yl)-propyl-1-phosphonic acid
27	CREB	cAMP response element-binding protein
28	DHPG	(RS)-3,5-dihydroxyphenyl-glycine
29	DRG	dorsal root ganglion
30	DREAM	downstream regulatory element antagonist modulator
31	E-LTP	early-phase long-term potentiation
32	EM	electron microscopic
33	EPSC	excitatory postsynaptic current
34	EPSP	excitatory postsynaptic potential
35	Erk	extracellular signal-regulated kinases,
36	fEPSP	field excitatory postsynaptic potential
37	FMRP	fragile X mental retardation protein
38	GABA	gamma-aminobutyric acid
39	GluR1	GluA1
40	GluR2/3	GluA2/3
41	GRIP	Glutamate receptor-interacting protein
42	GRP	Gastrin-releasing peptide
43	HCN	hyperpolarization activated cyclic nucleotide-gated

44	HP	hot plate
45	IBS	irritable bowel syndrome
46	IC	insular cortex
47	KA	kainate acid
48	LFS	low-frequency stimulation
49	L-LTP	late-phase long-term potentiation
50	LTD	long-term depression
51	LTP	long-term potentiation
52	L-VGCCs	L-type voltage-gated calcium channels
53	MAPK	mitogen-activated protein kinase
54	mEPSC	miniature excitatory post-synaptic current
55	mGluRs	metabotropic glutamate receptors
56	MPEP	2-methyl-6-(phenylethynyl)-pyridine
57	NAPSM	1-naphthyl acetyl spermine N-[3-[[4-[(3-Aminopropyl)amino]butyl]amino]propyl]-1-naphthaleneacetamide trihydrochloride
58	NKA	neurokinin A
59	NMDA	N-methyl-D-aspartate
60	NMDAR	N-methyl-D-aspartate receptors
61	NR2A	GluN2A
62	NR2B	GluN2B
63	NO	Nitric Oxide
64	NSF	N-ethylmaleimide-sensitive factor
65	NVP-AAM077	[(R)-[(S)-1-(4-bromo-phenyl)-ethylamino]-(2,3-dioxo-1,2,3,4-tetrahydroquinoxalin-5-yl)-methyl]-phosphonic acid
66	PAG	periaqueductal gray
67	PBN	parabrachial nucleus
68	PBS	phosphate buffer solution
69	PDZ	postsynaptic density-95/Discs large/zona occludens-1
70	PFC	prefrontal cortex

71	PhTx	philanthotoxin-74
72	PICK1	protein interacting with C kinase 1
73	PKA	Protein kinase A
74	PKC	protein kinase C
75	PKCε	protein kinase C isoform epsilon
76	PKMζ	protein kinase Mζ
77	PP1	protein phosphatase 1
78	PPF	paired-pulse facilitation
79	PPR	paired-pulse ratio
80	pre-LTP	pre-synaptic LTP
81	post-LTP	post-synaptic LTP
82	PGE2	prostaglandin E2
83	RVM	rostral ventromedial medulla
84	S1	primary somatosensory cortex
85	S2	secondary somatosensory cortex
86	sEPSC	spontaneous excitatory post-synaptic current
87	SP	substance AP
88	STD	short-term depression
89	TBS	theta burst stimulation
90	TF test	tail-flick test
91	TRP	transient receptor potential
92	ZIP	ζ-pseudosubstrate inhibitory peptide; zeta inhibitory peptide

Contents

Introduction

I was contacted by Marina with a proposal to write a book on chronic pain, especially covering recent research progress in the field. Considering I had just finished editing a textbook on *Molecular Pain*, I feel that different approaches are needed in this new book. There are two major reasons that a textbook on this subject is important right now. First, pain research is undergoing dramatic changes due to the rapid progress made in the field of neurobiology, and many of the basic mechanisms for pain transmission, modulation, and plasticity have recently been discovered. Second, cross-field research has been facilitated by the development of new experimental agents and approaches, as well as transgenic/gene knockout mice.

Before accepting the invitation, I decided to screen current books available on chronic pain. Although there have been several books written about chronic pain, I feel that there is a significant gap in our understanding of the basic mechanisms of chronic pain. This book aims to address that gap by taking two major new approaches. First, I will focus on studies based on basic neurosciences. I believe these studies form the major milestones for future pain research. One problem with recent progress made in the genetic studies of mice is that many researchers tend to use behavioral reflexive studies to "replace" pain phenotypes. This is in part due to the limitation of electrophysiological approaches in mouse research (perhaps due to size). Recording from sensory neurons *in vitro*, or *in vivo*, preparations should be used as a "classic" measurement to assure that the changes observed are indeed occurring in the sensory neurons and not in the motor neurons! More importantly, knowledge of synapse functions and plasticity in pain pathways is essential for us to have a comprehensive understanding of pain. The second approach will lay out each chapter as a seminar-type text. Original figures and the reasons for experiments will be used to argue for the key mechanisms for chronic pain.

This book will focus primarily on recent progress (since 2002) made in the field. Previous works are cited for context and only when unavoidable. One critical basic neuroscience research area is understanding the synaptic mechanisms for memory and pain. In human history, the relationship between pain and memory has always been closely related to each other. In ancient oriental cultures, body punishment was found to be an effective (although brutal) tool to train child disciples. In ancient China, scholars often used body pain to achieve high levels of memory ability in order to pass the highly competitive national examinations. Painful experiences often create long-lasting memories in adult human brains. Fear memory, a classic memory test done by administering painful electric shocks, is one of the most commonly used animal tests. Being trained in a top-tier pain lab (Professor GF Gebhart), memory lab (ER Kandel lab), and calcium channel lab (RW Tsien lab), I believe that I may bring a unique perspective to these topics.

Despite the many recent advances in the studies of chronic, and pathological, pain, many general readers often confuse physiological pain with pathological pain. Physiological pain, or acute pain, can lead to the formation of a long-lasting fear memory that allows the organism to link specific environmental signals to an appropriate motor response. The expectation of acute pain can also cause anxiety. Whereas fear memory that is associated with acute pain is long-lasting, the unpleasantness associated with it rapidly disappears. Unlike acute pain, chronic pain, defined in humans as any pain that lasts several weeks or longer, serves no obvious beneficial function. In addition to its unpleasantness, chronic pain is also associated with fear, anxiety, and impairment of cognitive function. The cloning of a receptor important for physiological pain such as heat or cold pain does not necessarily mean that you have a target for treating chronic pain. What is exciting about chronic pain is that it indeed shares some of the common mechanisms with learning and memory. Drugs that enhance cognitive functions often lead to enhancement of chronic pain and/or fear. Many drugs that show efficacy in chronic pain are found to affect memory.

Thus, it becomes a new challenge for neuroscientists to identify any selective signaling molecules involved in one process but not the other.

Do we still need to perform basic experiments to understand pain and chronic pain? The failures of the discovery of novel analgesic drugs are not simply due to the restriction of animal models that do not apply to human patients. One key issue is that many of the so-called analgesic compounds or drugs are purely screened from behavioral assay without any direct neurobiological mechanisms. The behavioral studies can be affected by many non-sensory factors. For example, one drug candidate may affect the motivation to move or side effects on motor neurons to lead to inhibition of the reflex. However, it may be interpretated as 'analgesic' in such behavioral tests. Animal models are essential for our understanding of human pain. One purpose of this book is to reinforce the basic mechanisms for chronic pain in the view of neurobiology.

The organization of this book intends to explore the possible cellular and molecular mechanisms for chronic pain, along with pain transmission and perception pathways. Endogenous modulation will be also explored. Much of the original data that supports the argument/conclusion is used to enforce the critical roles of original discoveries. This book will focus on studies using integrative approaches. Studies that have only used one or two experimental approaches will be excluded for the most part. These selected topics are in the areas where I believe many new discoveries will be made in near future. It will lay out the basic neuronal mechanisms for chronic pain. I hope that many readers will find this book helpful, thereby triggering more collaborative efforts in our understanding of pain mechanisms.

Chapter 1 will introduce some of the most important discoveries in recent years, especially those that impact how we look at the problem of chronic pain. Chapter 2 will examine recent studies of peripheral nociceptors and the key mechanisms for peripheral sensitization will be discussed. Chapter 3 will focus on the excitatory sensory transmission in the spinal cord dorsal horn, and the synaptic mechanisms for pain gate control will be discussed. Chapter 4 will focus on silent synapses in the spinal cord dorsal horn and its contribution to spinal plasticity and long-term facilitation. Chapter 5 will move to cortical areas, especially the anterior cingulate cortex (ACC), and examine the cellular and molecular mechanisms for ACC LTP. Chapter 6 will focus on LTD — the opposite form of plasticity to LTP in the ACC. Chapter 7 will look at long-term plastic changes in the ACC, and other cortical areas, and compare its mechanisms with that of synaptic LTP. Chapter 8 will focus on the NMDA receptor NR2B subtype in the ACC and its upregulation, and candidacy as a potential analgesic target, in chronic pain conditions. Chapter 9 will explore recent animal studies in order to demonstrate how chronic pain may interfere with cognitive processes at higher levels of the CNS. The synaptic mechanisms for such interactions will be discussed. Chapter 10 will discuss the intersection of pain and emotional anxiety, an important area for our understanding of how pain can affect our mood. Chapter 11 will focus on endogenous pain modulatory systems and the top-down system from the supraspinal structures, including those areas involved in pain perception, fear, and anxiety, that facilitate spinal pain transmission. This positive loop is important for controlling chronic pain. Finally, in Chapter 12, we will discuss the basic neuronal mechanisms of current analgesic drugs and explore potential novel protein targets for chronic pain based on basic animal studies.

I would like to thank the following foundations for supporting my research over the years. Without their support, this would not be possible. I thank the National Institutes of Health (USA), the Canadian Institutes for Health Research (CIHR), the EJLB-Michael Smith Chair and Canada Research Chair, the International Spinal Cord Research Fund, the Canadian Fragile X Foundation, the Chang Jiang Scholar Program (China), and the World Class University Program (South Korea). The support from these organizations allows me to focus on the questions of pain and chronic pain without hinderance for over a decade. The research works from my laboratory, that are summarized in this book, were carried out by a group of hard-working and talented post-doctoral fellows and graduate students. Without their tireless efforts and teamwork, it would have been impossible to obtain these discoveries. I also like to thank Emily England and Jenny Wang for editorial help with this book. Their hard work and patience make this book possible.

Support from my daughters Morgan and Danielle is most critical, as their patience, independence, and generosity have allowed me to focus my energy on my research work and take on this book project.

Chapter 1

Basic Neurosciences and Pain

News and Views

十年寒窗无人问，一举成名天下知

不经一番寒彻骨，怎得梅花扑鼻香

Acute pain, also called physiological pain, is highly functional as it helps humans and animals learn about environmental dangers and avoid potential harm. Pain also activates the brain in a way that makes us more conscious and attentive. Fear memory model uses acute pain as an effective signal which generates long-lasting memory. In ancient China, the government employed a national examination to select candidates who would serve the government in its official duties. These positions were highly coveted as they were well respected, highly paid, and could even result in being married to the royal family. Competition was so fierce that students were known to employ extremely painful stimuli to keep themselves awaked and learning (see the following example)! There are many Chinese poems that describe the hard work and great rewards of these students.

With the exception of a few painful chronic conditions (allodynia and hyperalgesia, which prevent further injury), there is no physiological benefit to chronic pain. The famous self-portrait by artist Freda Kahlo vividly illustrates her experience of suffering from chronic pain. Needle-like pain spread throughout her body, including her face.

Summary

Pain research is often seen as a clinical problem when, in fact, pain is a brain problem. Future pain treatment depends on our understanding of the brain mechanisms of receiving, processing, and storing painful information. Cumulative research has mapped the pathways for pain transmission, modulation, and perception in the brain. This information helps us to locate and understand where pain is processed in the brain. Recent advances in the study of the neurobiology of pain reveal many new targets for pain and chronic pain, including new ion channels, new activity-dependent genes, and new synaptic mechanisms. We are in an exciting time for carrying out integrative experiments to understand the basic mechanisms of pain from genes to behaviors.

Keywords: CNS; plasticity; ACC; LTP; LTD; gene; AC; descending modulation; mice

Introduction

Basic investigations of pain have benefited greatly from the progress made in neuroscience. However, the relationship between neuroscience and pain is usually unidirectional, in that pain research often benefits from the progress made in basic neuroscience, whereas basic neuroscience does not often benefit from progress made in pain research. For example, analgesic compounds/drugs are likely to be tested in numerous models of chronic pain. Negative outcomes of these tests are likely to be discontinued as it is contrary to the interest/funding of this research. In this way, progress in pain research can be limited by its closed connection to clinic. This chapter will review some of the recent progress made in neuroscience and discuss how these basic concepts have helped us to understand the potential mechanisms for pain transmission, modulation, and more importantly pain plasticity.

Basic principle of anatomy of central nervous system (CNS): From molecules to behaviors

One major approach in neuroscience (or biology in general) is to identify the molecular basis of biological behaviors, including pathological conditions. One of the most important questions in neuroscience is as follows: What is the basic mechanism for learning and memory? Studies have been carried out at various levels of organization and proteins (see Figure 1) in an attempt to answer this question. Many new tools have been developed, and introduced, to address the basic cellular and molecular mechanisms of

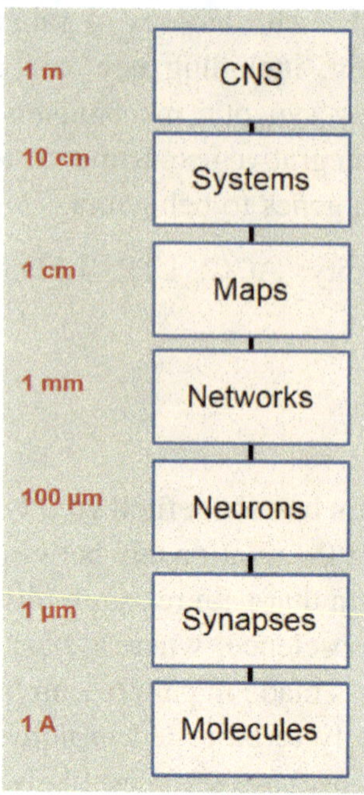

Figure 1.   From a molecule to brain circuit. Modern neuroscience researchers have investigated brain functions such as pain and memory at different levels. While rapid progresses have been achieved at each level, especially at the molecular level and synaptic level, putting this information together to explain the overall functions of brain has proved to be a rather difficult task. Simply adding up all signaling molecules will not form a functional synapse and summing up all synapses does not generate a functional neuron. Future integrative research works are clearly needed to link each signaling molecule to behavioral functions in an intact CNS system.

learning and memory, such as genetic manipulations, studies of receptor proteins (mainly glutamate receptors), and modern imaging approaches. Although we have made rapid progress in dissecting the whole organism into different levels of reduced preparations (from brain slices to cultured cells), it has become very clear that it is not a simple job to put the pieces back into the understanding of the physiological behavior of animals. Very often, reduced preparations are used in order to identify such detailed biological processes. However, along this path, many factors are often lost. It becomes challenging to try to put together the whole picture of what is likely to happen under *in vivo* conditions. Recent studies have

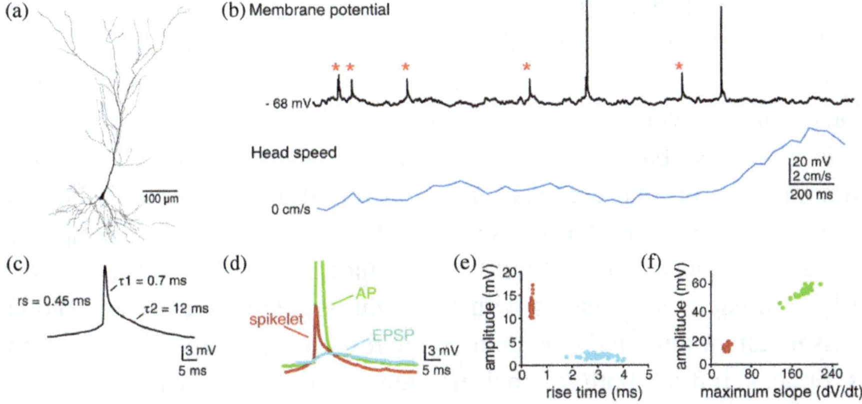

Figure 2.   Whole-cell patch recording from freely moving animals. Spikelets from an identified hippocampal CA1 pyramidal cell in a freely moving rat. (a) Reconstruction of the morphology of the recorded CA1 pyramidal neuron filled with biocytin during recording. (b) (Top) Membrane potential trace recorded during spatial exploration. Fast events of small amplitude (red stars) are present in addition to action potentials (APs). (Bottom) Corresponding speed of the animal's head. (c) Averaged spikelet ($n = 35$). Mean spikelet has a fast rise time (rs) and a decay time best fitted by the sum of two exponentials. (d) Superimposition of averaged APs (green), spikelets (red), and excitatory postsynaptic potentials (EPSPs, blue). (e) Scatterplot of EPSP (blue) and spikelet (red) amplitude versus rise time. Spikelets represent an independent population of events with faster rise times and larger amplitudes than EPSPs. (f) Scatterplot of AP (green) and spikelet (red) amplitude versus maximum rising slope (dV/dt). Spikelets represent an independent population of events with slower rising slopes and smaller amplitudes than APs (adapted from Epsztein *et al.* [1]).

made some progress in this arena, however, they are very technically difficult, such as performing whole-cell patch-clamp recordings in freely moving animals (Figure 2) [1].

Animal models for the study of pain

It is common knowledge in the medical community that pain consists of both a physical and an emotional component. The International Association for the Study of Pain (IASP) has adopted the following definition of pain: "An unpleasant sensory and emotional experience associated with actual, or potential, tissue damage or described in terms of such damage". It is unfortunate that this definition has led to some confusion in the research field and the suggestion that animal 'pain' is not pain, since animals failed to tell the experimenters whether they felt pain. Such a narrow interpretation excludes a wide range of people including deaf and hearing impaired, mute or non-verbal, infants, and foreign language speakers. It is this author's position that animals have pain sensations on par with human beings. A recent report even demonstrated that there is pain empathy between animals (Figure 3)! Indeed, rats, mice, and humans respond similarly to noxious stimulation with vocalization, escape behaviors, blood pressure alterations, and other autonomic responses. Analgesic drugs such as opiates inhibit both animal nocifensive behaviors and subjective responses in humans elicited by noxious stimulation. Furthermore, comparative neuroanatomy has established that similar pathways allow for the processing of pain in laboratory animals, such as rodents and cats, as well as in humans. Thus, similar anatomical, cellular, and molecular components appear to form the basis of human and animal pain.

Glutamate-mediated excitatory transmission

One major discovery is that the excitatory transmission of sensory pain transmission is mediated by glutamate receptors. From the spinal dorsal horn to the anterior cingulate cortex (ACC), fast excitatory transmission is mediated by postsynaptic glutamate receptors. The development of various selective agonists and antagonists against a subtype of glutamate receptors has greatly helped our understanding of the basic functions of glutamatergic systems in the CNS.

Glutamate binds to four major classes of receptors: three classes of ionotropic receptors favoring AMPA, KA, or NMDA as agonists and a family of at least eight G protein metabotropic receptors (Figure 4). AMPA receptors are heteromeric complexes of the subunits GluA1, 2, 3, and 4, each of which is expressed in neurons. Functionally, most of the synaptic transmissions are mediated by AMPA receptors with minor contributions of postsynaptic KA receptors in certain synapses. For example, in spinal dorsal horn synapses, AMPA receptors mediated more than 80% of EPSCs. However, in the hippocampal CA1, the same synaptic currents

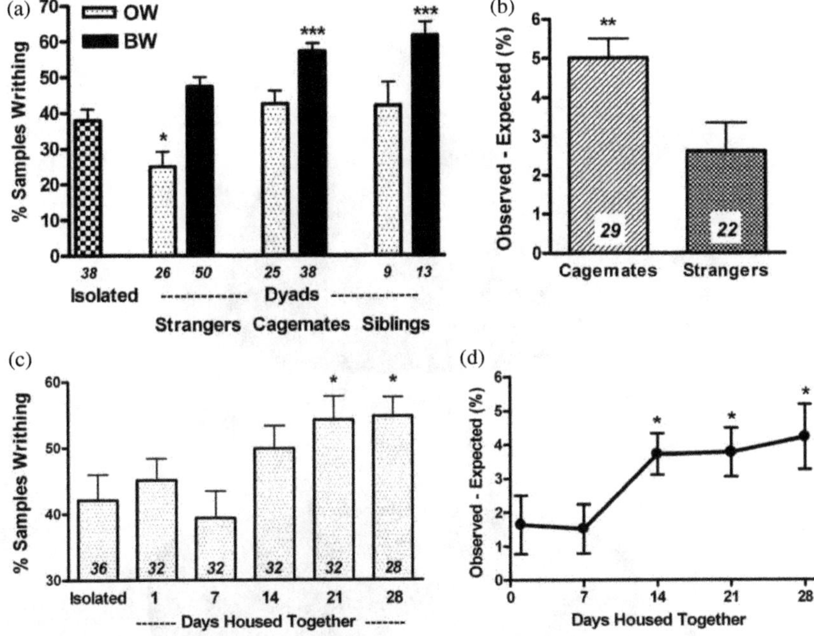

Figure 3.   Empathy in animals. (a–d): Mice injected with 0.9% acetic acid in the presence of similarly injected cagemates display higher levels of pain behavior, which co-occurs in time. In all graphs, group sample sizes are indicated in italics. (a) Mice were tested in isolation (isolated) or in dyads where either one mouse (One Writhing, OW) or both mice (Both Writhing, BW) received acetic acid injections. Bars represent the mean ± SEM percentage of sampled intervals showing writhing behavior (% samples writhing). $*P < 0.05$, $***P < 0.005$ by Dunnett two-way case-control comparison *post hoc* test compared to isolated mice. (b) Statistically significant co-occurrence in writhing behavior in the cagemate and stranger conditions (sign test, $P < 0.05$ in both cases); the co-occurrence was significantly higher in cagemates. Using data from (a), the expected number of samples with writhing in both mice of the dyad was calculated as a joint probability. Bars represent the mean ± SEM excess of observed samples with joint writhing above the expected value as a percentage. $**P < 0.01$ compared to strangers (Student's *t*-test). (c) Data from a separate experiment using naïve mice housed together for 1, 7, 14, 21, or 28 days and tested in BW dyads. Isolated mice were taken from the 28-day group but were tested alone. Bars are as in (a). Data in (d) were calculated from subjects shown in (c); symbols represent the mean ± SEM excess of observed samples with joint writhing above the expected value as a percentage (adapted from Langford *et al.* [48]).

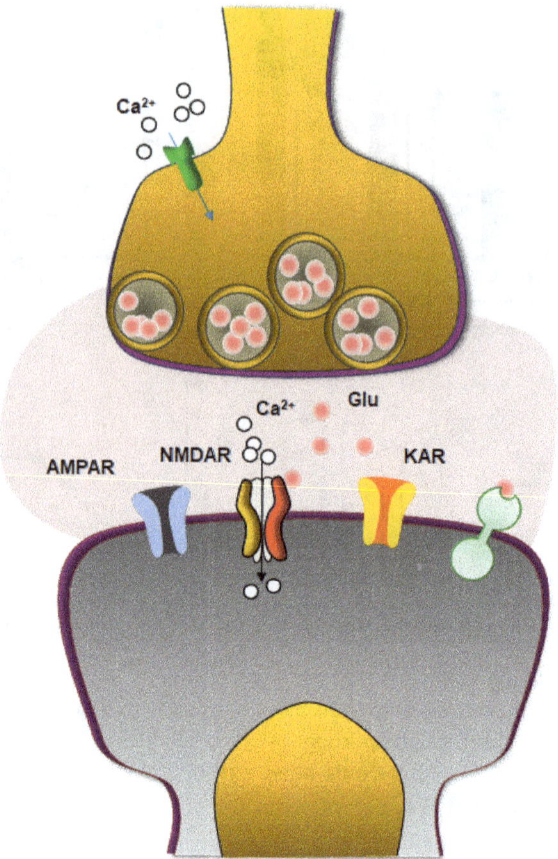

Figure 4.   Glutamate and its action in synapses. Glutamatergic synapses are the major excitatory in the CNS. Glutamate releases into synaptic cleft upon coming action potentials. The opening of voltage-gated calcium channels at presynaptic terminal is important for influx of calcium that is critical for triggering vesicle releases. At postsynaptic sites, glutamate AMPA receptor mediates most of the excitatory synaptic transmission, while in some of the synapses, glutamate kainate receptors, also contribute to postsynaptic transmission. Under the resting condition, postsynaptic NMDA receptors are mostly inactive. However, in freely moving animals, there is evidence that postsynaptic NMDA receptors may also contribute to synaptic transmission.

are strictly mediated by AMPA receptors: the KA receptor subunits GluA5, 6, and 7 and KA1 and 2. Some KA receptors contribute to fast synaptic transmission, while some KA receptors are only activated by an extra amount of glutamate, presumably by the activation of extrasynaptic receptors. In addition to postsynaptic location, there are reports of the presynaptic KA receptors at both the excitatory and inhibitory receptors that can be divided into subtypes, and activation of these KA receptors can also modulate the release of glutamate or GABA or the NMDA receptor subunits GluN1 and GluN2A-D. The expression of GluN2 subunit is regionally dependent and also undergoes changes during early development and adulthood. NMDA receptor plays a key role in central plasticity (see Chapter 5), including the synapses involved in sensory, learning, memory, emotion, and drug additions. The metabotropic receptors (mGluRs) contain at least eight subtypes. They are widely distributed in the brain including through the sensory-related spinal cord and cortex. In most cases, these receptors are not contributing to fast synaptic transmission, and they are likely activated in an activity-dependent manner, such as in case of learning and injury. The involvement of mGluRs in pain at the level of spinal cord has been investigated.

Adult glutamatergic synapses are highly plastic

It is well known that central synapses are highly plastic, and long-term changes in synaptic transmission contributes to different functions of the brain throughout the lifespan [2,3]. Two major forms of synaptic plasticity have been widely investigated: long-term potentiation (or called LTP) and long-term depression (or called LTD) (Figure 5). While LTP can enhance synaptic functions in certain regions of the brain, LTD attenuates or reduces the efficacy of synaptic transmission. Such biphasic synaptic plasticity is not limited at excitatory, glutamatergic synapses. Both LTP and LTD have also been reported in inhibitory synapses, and the underlying cellular and molecular mechanisms are different. Recent studies using different induction protocols reveal that the mechanisms for central LTP are likely to be different, depending on the induction protocols, regions of the CNS, input fibers, and postsynaptic neurons recorded. There is no doubt that many different molecular targets will be continuously revealed in the future; one urgent task is to verify the physiological or pathological relevance of synaptic LTP/LTD induced by experimental induction protocols. Furthermore, new forms of LTP and LTD remain to be discovered to mimic physiological and/or pathological changes *under in vivo* conditions

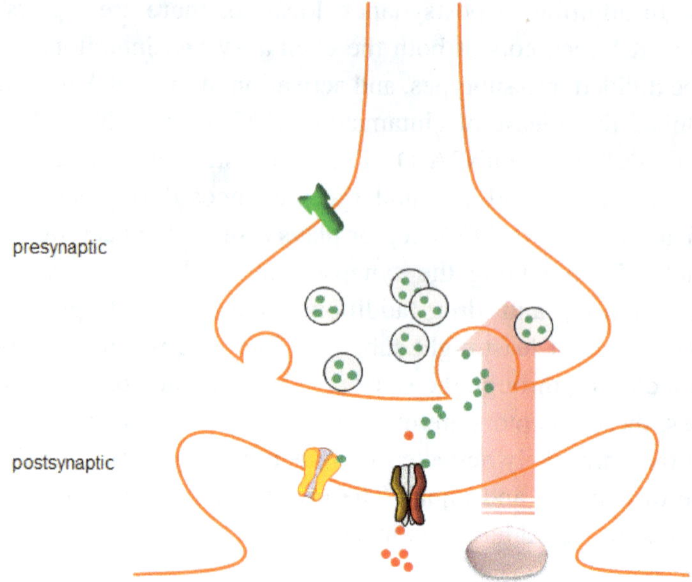

Figure 5.   LTP of glutamate synapses. LTP is a mostly investigated form of synaptic plasticity in the brain. Recent studies indicate that LTP can be recorded from most of the excitatory synapses in the adult brains. Depending on the regions of the brain, the experimental protocol for inducing LTP may be different, and the molecular mechanisms for the induction and expression for LTP are different from region to region. At least two major synaptic mechanisms may be contributed to most forms of LTP. First is the presynaptic enhancement of glutamate release after the activation of presynaptic induction mechanisms. In this case, there is no involvement of postsynaptic sites. Second is the presynaptic enhancement of glutamate after activation of postsynaptic receptors. Because the induction of LTP is postsynaptic, several diffusible messengers have been proposed to contribute to presynaptic expression of LTP, including NO, CO, and BDNF. Finally, both the induction and expression of LTP are all at postsynaptic sites.

(e.g., presynaptic enhancement of neurotransmitter release after tissue injury in the ACC).

What has been recognized about the potential functions of LTP is its contribution to many key brain functions in addition to learning and memory. At the spinal cord dorsal horn level — where the first sensory synapses are located — LTP of sensory synaptic transmission can be induced by different experimental protocols or peripheral injury.

Potentiated excitatory synaptic transmission is believed to contribute to spinal sensitization that, at least in part, attributes to behavioral hyperalgesia and allodynia during chronic pain. In the basolateral amygdala, LTP can be induced between thalamic/cortical inputs and postsynaptic principal neurons or fear conditions. Such enhanced responses are important for encoding fearful information. In the hippocampal CA1 region where most of the LTP studies have been reported, LTP can be induced and reliably detected, even with field recording electrodes. However, despite a huge amount of literature on hippocampal LTP, it remains to be demonstrated whether a simple spatial training trial may induce LTP in certain population of CA1 neurons.

Finally, in the prefrontal cortical (PFC) neurons including the ACC, LTP is induced by the pairing, spike-timing, and theta burst stimulation (TBS) protocols as well as peripheral injury. It has been proposed that the injury that causes the synaptic potentiation contributes to chronic pain and pain-related high brain functions including fear and emotion. Therefore, it is clear that studying central LTP provides us with a better understanding of the fundamental mechanisms for brain functions — from pain transmission to fear and chronic pain.

Some of the AMPA receptors are undergoing trafficking event

The study of AMPA receptor trafficking at postsynaptic sites was in part triggered by the discovery of silent glutamatergic synapses in the hippocampus. In these 'silent' synapses, no functional AMPA receptors were detected at resting membrane potentials. However, after the induction of LTP, active AMPA receptors appeared in tetanized/conditioned synapses. While different possible mechanisms may contribute to the recruitment of silent synapses, one direct possibility is the recruitment of postsynaptic AMPA receptors. Subsequent studies reveal that AMPA receptors in fact interact with many postsynaptic proteins, and some of these proteins are actively involved in the maintenance or plastic changes of AMPA subtype receptors, including removing (depression) or insertion of receptors (Figure 6). One major conclusion from these studies is

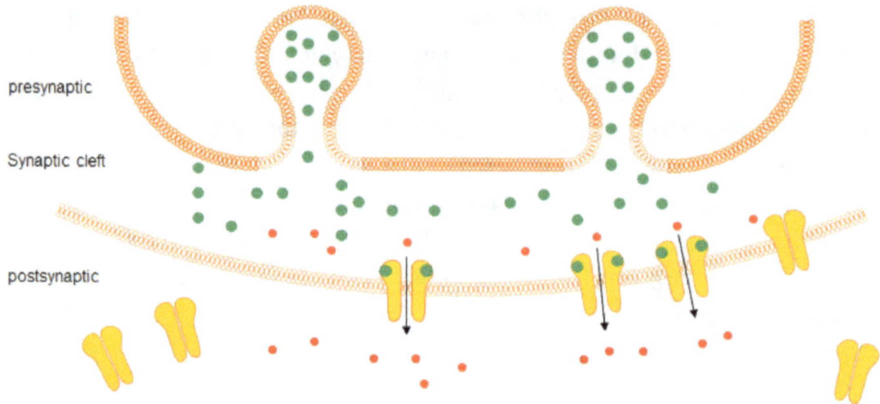

Figure 6. AMPA receptor trafficking and LTP. Recent neurobiological studies have provided evidence that postsynaptic glutamate AMPA receptors are mobile, and trafficking of AMPA receptors in and out of postsynaptic membrane contributes to certain forms of LTP in the CNS.

that postsynaptic receptor functions can be dynamically regulated. This is similar to events happening at the presynaptic terminals. These presynaptic and postsynaptic events strongly support that glutamatergic synapses — even those in adult synapses — are highly plastic and under fine, biphasic regulation.

Gene regulation and protein synthesis

One major cellular mechanism for plasticity is the importance of gene regulation and translation. Studies from long-term memory and late-phase synaptic plasticity consistently suggest that gene regulation is required for learning-related synaptic plasticity. The initial discovery comes from classic pharmacological evidence that reported that long-term memory was inhibited by transcriptional inhibitors. Subsequent molecular studies have shown that LTP-inducing protocol and behavioral training do indeed activate many activity-dependent immediate early genes in the neurons that are undergoing plastic changes (Figures 7 and 8). Recent studies using genetic and pharmacological approaches further confirm that both translation and transcriptional events are required for synaptic plasticity as well as related behavioral learning and memory.

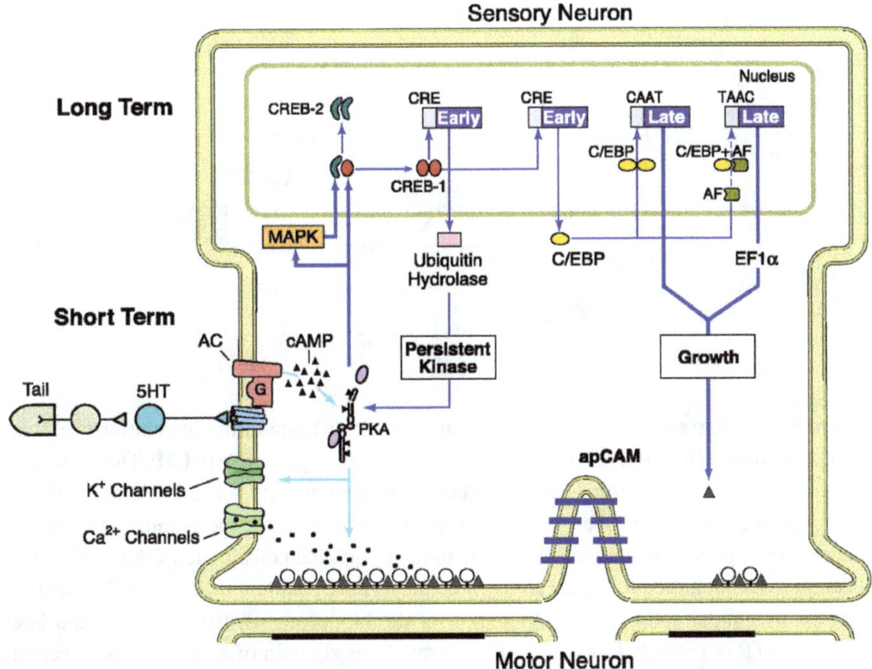

Figure 7.   Aplysia learning. Effects of short- and long-term sensitization on the monosynaptic component of the gill-withdrawal reflex of Aplysia. In short-term sensitization (lasting minutes to hours), a single tail shock causes a transient release of serotonin that leads to covalent modification of pre-existing proteins. The serotonin acts on a transmembrane serotonin receptor to activate the enzyme adenylyl cyclase (AC), which converts ATP to the second messenger cyclic AMP. In turn, cAMP recruits the cAMP-dependent protein kinase A (PKA) by binding to the regulatory subunits (spindles), causing them to dissociate from and free the catalytic subunits (ovals). These subunits can then phosphorylate substrates (channels and exocytosis machinery) in the presynaptic terminals, leading to enhanced transmitter availability and release. In long-term sensitization, repeated stimulation causes the level of cAMP to rise and persist for several minutes. The catalytic subunits can then translocate to the nucleus and recruit the mitogen-activated protein kinase (MAPK). In the nucleus, PKA and MAPK phosphorylate and activate the cAMP response element-binding (CREB) protein and remove the repressive action of CREB-2, an inhibitor of CREB-1. CREB-1 in turn activates several immediate-response genes, including a ubiquitin hydrolase necessary for regulated proteolysis of the regulatory subunit of PKA. Cleavage of the (inhibitory) regulatory subunit results in persistent activity of PKA, leading to persistent phosphorylation of the substrate proteins of PKA. A second immediate-response gene activated by CREB-1 is C/EBP, which acts both as a homodimer and as a heterodimer with activating factor (AF) to activate downstream genes [including elongation factor 1 (EF1)] that lead to the growth of new synaptic connections (adapted from Kandel [3]).

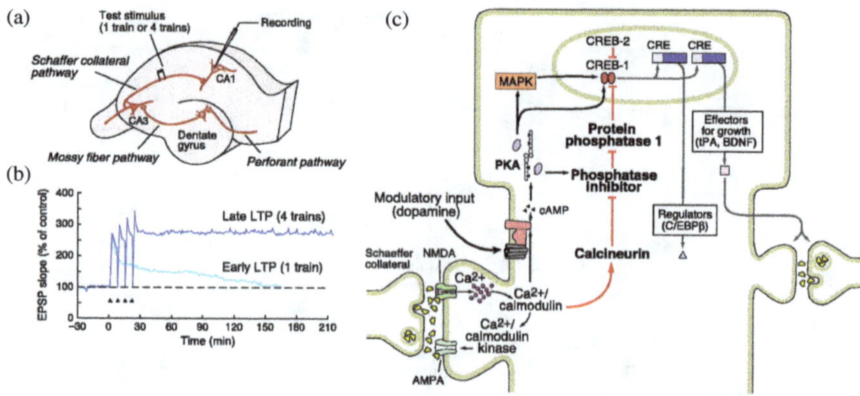

Figure 8.   Protein synthesis in learning-related memory. Long-term potentiation (LTP) in the hippocampus. (a) Three major pathways, each of which gives rise to LTP. The perforant pathway from the subiculum forms excitatory connections with the granule cells of the dentate gyrus. The mossy fiber pathway, formed by the axons of the granule cells of the dentate gyrus, connects the granule cells with the pyramidal cells in area CA3 of the hippocampus. The Schaffer collateral pathway connects the pyramidal cells of the CA3 region with the pyramidal cells in the CA1 region of the hippocampus. (b) The early and late phases of LTP in the Schaffer collateral pathway. A single train of stimuli for one second at 100 Hz elicits an early LTP, and four trains at 10-minute intervals elicit the late phase of LTP. The early LTP lasts about 2 hours, and the late LTP more than 24 hours. (c) A model for the late phase of LTP in the Schaffer collateral pathway. A single train of action potentials initiates early LTP by activating NMDA receptors, Ca^{2+} influx into the postsynaptic cell, and the activation of a set of second messengers. With repeated trains of action potentials (illustrated here), the Ca^{2+} influx also recruits an adenylyl cyclase (AC), which activates the cAMP-dependent protein kinase. The kinase is transported to the nucleus where it phosphorylates CREB. CREB in turn activates targets (C/EBPB, EPA, and BDNF) that are thought to lead to structural changes. Mutations in mice that block PKA or CREB reduce or eliminate the late phase of LTP. The adenylyl cyclase can also be modulated by dopamine signals and perhaps other modulatory inputs. In addition, there are constraints (in red) that inhibit L-LTP and memory storage. Removal of these constraints lowers the threshold for L-LTP and enhances memory storage (adapted from Kandel [3]).

Protein degradation: A new mechanism for learning

Traditional view of learning and memory is that the activation of new protein synthesis is involved in plastic changes. Recent studies, however, indicate that protein degradation is also involved in memory process (Figure 9) [4]. This finding strongly suggests that protein synthesis can be biphasically regulated for a long period of time.

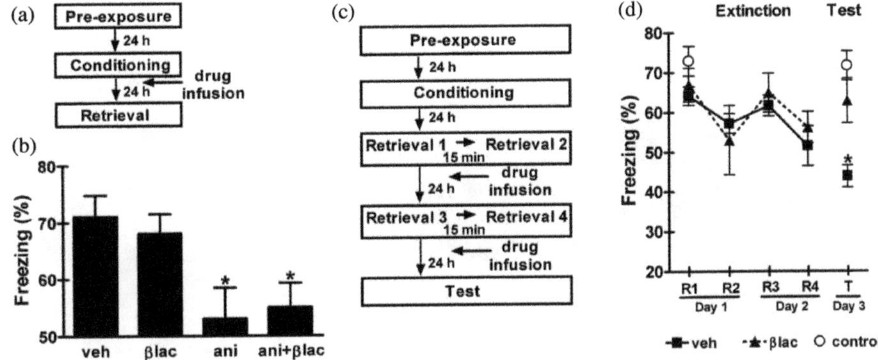

Figure 9. Protein degradation and memory. Protein degradation is not required for memory consolidation, whereas it is critical for memory extinction. (a) Schematic view showing the protocol of behavioral experiments to examine the effect of βlac on acquiring fear memory. (b) Bars represent the means ± SEM of the freezing level on the retrieval day. Single infusion of βlac ($n = 12$) had no effect on the formation of fear memory compared with the vehicle-infused control group ($n = 11$), whereas the anisomycin-infused group ($n = 14$) showed significant impairment of the fear memory. Co-infusion of βlac with anisomycin ($n = 12$) also had no effect on the anisomycin-induced amnesia during memory acquisition. (c) Schematic view of the behavioral experiments performed to examine the effect of βlac on the fear memory extinction. (d) Fear memory extinction is impaired in the βlac-infused group ($n = 7$) compared with the vehicle-infused group (veh, $n = 7$). Control group ($n = 8$) indicates the vehicle-infused animals without extinction training (adapted from Lee *et al.* [4]).

Neuronal network controls

Although the dorsal horn of the spinal cord is often regarded as a simple relay for sensory transmission, recent studies reveal that synaptic transmission in the dorsal horn of the spinal cord undergoes complicated, biphasic, and activity-dependent regulation. By doing so, sensory inputs from the periphery are appropriately coded and conveyed into the brain. Figure 10 shows models for some of the major forms of synaptic regulation in the spinal cord dorsal horn.

Postsynaptic regulation: DRG–dorsal horn synapses

Neurotransmitters or neuromodulators bind to their postsynaptic receptors at spinal dorsal horn neurons. Activation of these postsynaptic

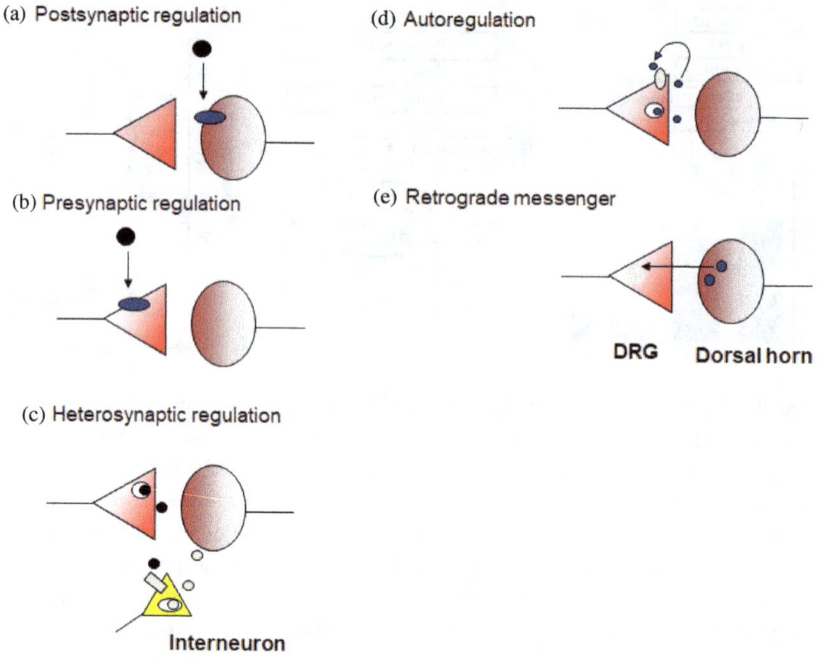

Figure 10. Feedback control as key mechanism. Models for spinal sensory transmission and modulation by postsynaptic regulation (a), presynaptic regulation (b), heterosynaptic regulation (c), autoregulation (d), and retrograde messenger (e). In case of heterosynaptic regulation, glutamate released from the central terminals of the primary afferent fibers may regulate spinal local inhibitory transmission through activation of presynaptic KA receptors.

receptors leads to changes in AMPA/KA receptor-mediated synaptic responses. They include acetylcholine, serotonin, opioids, norepinephrine, and oxytocin.

Presynaptic regulation: DRG–dorsal horn synapses

Sensory transmitters or neuromodulators bind to their target receptors on the central terminals of DRG cells in the spinal cord dorsal horn. Activation of these presynaptic receptors will lead to changes in the release of sensory transmitters in response to peripheral sensory stimulation. Many neurotransmitters and peptides have been reported to produce presynaptic regulatory effects in the DRG–dorsal horn synapses, such as ATP, serotonin, and opioids.

Heterosynaptic regulation: DRG–spinal inhibitory neurons

In the spinal cord dorsal horn, glutamate-containing sensory fiber terminals come into close proximity with the GABA- and glycine-containing boutons of local interneurons at synaptic glomeruli. In a recent study, we provided evidence that glutamate released from primary afferent sensory fibers can regulate spinal inhibitory transmission by activating KA receptors. These data suggest that heterosynaptic regulation of transmitter release by presynaptic ligand-gated ionic channels may be reciprocal between sensory fibers and dorsal horn interneurons. Because synaptically released glutamate suppressed evoked inhibitory transmission, it suggests that with sufficiently high levels of sensory input, inhibitory tone may be reduced, possibly facilitating the relay of sensory information to higher brain centers.

Autoregulation

In addition, neurotransmitters can act on their target receptors also expressed in the presynaptic terminals. These can be either excitatory glutamate or inhibitory GABA synapses. In case of glutamate synapses, glutamate may act on presynaptic KA receptor expressed on the central terminals of primary afferent fibers and regulates the release of glutamate. Similar autoregulation of GABA releases is also reported in the spinal cord.

Retrograde messengers

In central synapses, activation of postsynaptic receptors often leads to the production of diffusible messengers, such as nitric oxide (NO) and carbon monoxide (CO). In the spinal cord dorsal horn, enzymes that produce retrograde messengers are found in dorsal horn neurons. It is very likely that diffusible retrograde messengers affect presynaptic release of glutamate and/or neuropeptides.

Mice and humans

The use of mouse model for human studies is supported by the fact that human and mouse genes are similar as they share the same number of genes (97.5%), and the genome is organized similarly (Figure 11). Recent studies using genetically manipulated mice found that mutant genes that

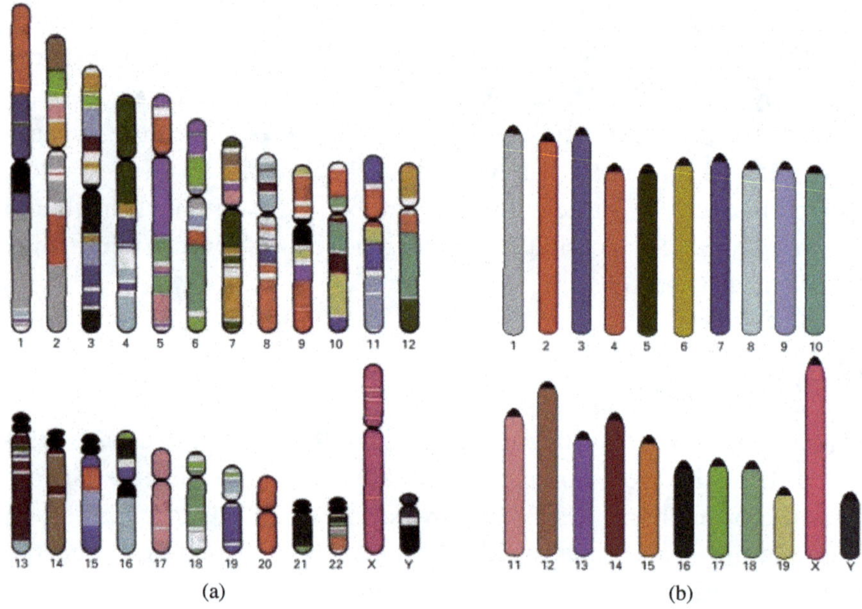

Figure 11. Conserved synteny between the human and mouse genomes. Regions from different mouse chromosomes (indicated by the colors of each mouse in b) show conserved synteny (gene order) with the indicated regions of the human genome (a). For example, the genes present in the upper portion of human chromosome 1 (orange) are present in the same order in a portion of mouse chromosome 4. Regions of human chromosomes that are composed primarily of short, repeated sequences are shown in black. Mouse centromeres (indicated in black in b) are located at the ends of chromosomes; no known genes lie beyond the centromere on any mouse chromosome. For the most part, human centromeres, indicated by constrictions, occupy more internal positions on chromosomes (adapted from Sinha and Meller [49]).

caused diseases in humans had similar effects in mice. The use of the mouse model for human diseases — including CNS diseases — has several advantages: the low cost of investigation, the short lifespan of mice, and the easy manipulation of genes/proteins before any selective drug is used.

Genetic approaches: A new light for molecular and cellular mechanism of pain

Genetically modified mice and other organisms are an essential tool in identifying the molecular pathways mediating pain transmission, modulation, and plasticity. Identification of the neurotransmitters and receptors involved has been possible using pharmacological techniques. However, most of the pharmacological agents have side effects, as well as unexpected interactions with molecules other than their targets, and the development of a selective pharmacological agent takes time. The possibility of genetically ablating or overexpressing a specific molecule has allowed for rapid progress in the identification of the molecular mechanisms of pain. Behavioral and electrophysiological examinations of mutant mice are necessary to identify specific deficits, or lack thereof, and their mechanisms. Therefore, a thorough study must include an analysis at the behavioral, electrophysiological, and molecular levels.

Ascending pain transmission

Noxious stimulation is detected by primary afferent fibers called nociceptors located in the skin and internal organs. Several types of afferent fibers are distinguished according to their conduction velocity. The large $A\delta$ fibers respond mostly to innocuous stimuli, whereas the slower $A\delta$ fibers and the thin unmyelinated C fibers primarily contribute to painful stimuli. We need to point out that the selective involvement of $A\delta$ and C fibers in pain transmission only holds up under normal physiological conditions. Recent studies indicate that non-nociceptive fibers such as $A\beta$ fibers may contribute to persistent pain after tissue or nerve injury.

Neurons in the spinal dorsal horn and related areas receive sensory inputs, including noxious stimuli, and convey them to supraspinal structures. Identifying molecules that are selectively involved in pain transmission is a major research focus and holds hope for the treatment of pain conditions, including persistent pain. Studies using pharmacological and behavioral approaches showed that glutamate and neuropeptides — including substance P (SP) — are likely transmitters of pain [5–8]. Electrophysiological investigations of sensory synaptic responses between primary afferent fibers and dorsal horn neurons provide evidence that glutamate was the principal fast excitatory transmitter and that synaptic responses were mediated by postsynaptic glutamate receptors. While AMPA receptors mediated most of the synaptic currents, KA receptors preferentially contribute to synaptic responses induced by higher or noxious intensities [6]. Consistent with this, receptor antagonists blocking KA and AMPA receptors yield greater analgesic effects in adult animals than AMPA receptor-selective antagonists. These novel findings suggest that a sensory modality may be coded by postsynaptic transmitter receptors.

Not all sensory synapses are functional, or effective, under normal conditions. In young animals, silent glutamate synapses containing only NMDA receptors were found in dorsal horn neurons and sensory afferent fibers [5]. Conversion of such 'silent' synapses contributes to the enhancement of synaptic responses by serotonin (5-HT), an important transmitter of descending projecting pathways. Furthermore, pure NMDA receptors were also reported in the spinal dorsal horns of adult animals. These NMDA synapses are functional due to possible distal dendrite locations of NMDA receptors and/or the insensitivity of NMDA receptors to magnesium blockade.

In addition to glutamate, several neuropeptides including SP are also thought to serve as sensory neurotransmitters. For many years, electrophysiological evidence for the monosynaptic nature of SP-mediated synaptic responses has been lacking, since SP-mediated responses have a very slow onset. Recent studies using whole-cell patch-clamp recordings revealed a rather fast SP- and neurokinin A (NKA)-mediated synaptic current at synapses between primary afferent fibers and dorsal horn neurons [7]. The currents summated upon repetitive stimulation at high frequencies. In sum, spinal sensory synapses are far more diverse and complicated than we previously believed.

Dorsal horn neurons project to the thalamus via the spinothalamic tract located in the anterolateral tract. Two pathways appear: a lateral one and a medial one. The lateral pathway consists of the lateral thalamic nuclei (ventroposterior lateral and medial) and their projection to the somatosensory cortices (S1 and 2). This pathway codes for the intensity of the stimulus from innocuous touch to noxious pinch. Receptive fields are small, allowing accurate localization of the stimulus. On the other hand, the ACC receives sensory input via the medial thalamic nuclei [9,10]. Human imaging experiments have shown that activity in the ACC correlates to the emotional, affective component of pain [11]. ACC responses to noxious stimulation are also recorded in other species [12].

Pain can be modulated

Sensory transmission within the dorsal horn is subject to biphasic modulation, including descending facilitatory and inhibitory regulation [13–20]. Alteration of spinal sensory transmission modulates both the behavioral responses to noxious stimulation and the information transmitted to supraspinal areas. Based on electrophysiological and pharmacological data, Melzack and Wall suggested the "gate control theory" in the 1960s [21]. Although A_β fibers do not contribute directly to pain processing, they can inhibit nociceptive transmission, as they synapse onto both projection neurons and local inhibitory interneurons. Recent studies have shown that glutamate released by primary afferent fibers can act on the presynaptic terminals of inhibitory interneurons to inhibit evoked GABA release. This effect is mediated by presynaptic glutamate KA receptors, suggesting that KA receptor antagonists would have an analgesic effect [22].

In addition to the interactions between different types of sensory afferent fibers, spinal sensory transmission, including pain transmission, is strongly modulated by inputs from various supraspinal areas. Dorsal horn neurons receive direct and indirect descending projections from numerous supraspinal areas, including the ACC, amygdala, hypothalamus, periaqueductal grey (PAG), and rostral ventral medulla (RVM). Among them, a major descending pathway consists of the PAG, RVM, and spinal cord connections (Figure 12). Many other central nuclei interact with this

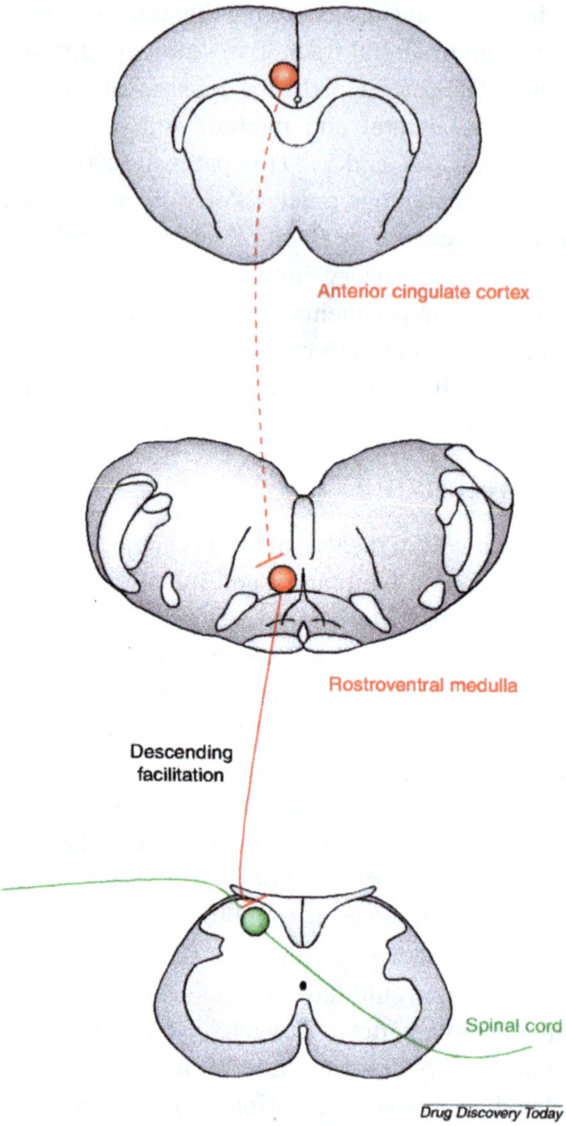

Anterior cingulate cortex

Rostroventral medulla

Descending
facilitation

Spinal cord

Drug Discovery Today

Figure 12.   Cortex-endogenous pain control. Neurons in the rostroventral medulla (RVM) project to the spinal dorsal horn and modulate sensory synaptic transmission in the spinal cord. Serotonin is the most likely transmitter for mediating this facilitatory effect. The facilitation induced by serotonin likely requires activation of specific subtypes of serotonin receptors and coactivation of cAMP signaling pathways to induce facilitation in adult

endogenous analgesia system and produced antinociceptive or analgesic effects. As the last step of relay nuclei, neurons in several nuclei in the brainstem play important roles in descending inhibition of spinal sensory transmission. In addition to descending inhibition, descending facilitatory influences from the brainstem and forebrain have also been characterized. Biphasic modulation of spinal sensory transmission affects not only inputs from somatosensory areas but also visceral organs. Biphasic modulation of spinal nociceptive transmission from the RVM, consistent with different types of neurons identified in this area, offers fine regulation of spinal sensory thresholds and responses. While descending inhibition is involved primarily in regulating the suprathreshold responses to noxious stimuli, descending facilitation reduces the neuronal threshold to nociceptive stimulation.

The ACC is a further source of descending facilitation (Figure 12). Electrical or chemical stimulation of the ACC facilitates the spinal nociceptive tail-flick (TF) reflex [23]. This effect is mediated by the RVM, as it is abolished following lidocaine inactivation of that area. This facilitation is clinically relevant because plastic changes occur in the ACC following injury. Many other more minor areas also contribute to modulation of pain. For instance, the hypothalamus projects directly to the dorsal horn and also releases hormones, which could alter nociceptive transmission. Relatively little is known regarding the exact roles of hormones and hypothalamic projections in sensory processing. Although more detailed data concerning the influence of gender on nociceptive and antinociceptive pathways are emerging, their mechanisms of action remain unknown. The use of knockout mice should shed some light on these mechanisms.

Figure 12. (*Continued*)   spinal dorsal horn neurons. Due to enhanced synaptic efficacy between primary afferent fibers and dorsal horn neurons, spike (action potential) responses to stimulation of afferent fibers were enhanced, as were behavioral nociceptive responses (e.g., decrease in response latencies). Stimulation of neurons in the ACC also activated descending facilitation, and activity within the RVM is required for mediating descending facilitation from the ACC to the spinal dorsal horn (adapted from Zhuo [49]).

Physiological pain versus pathological pain

Physiological pain is a very important physiological function for survival. Depending on the pain experience, animals and humans gain knowledge of potentially dangerous stimuli in their environments, and that pain-related unpleasantness helps form long-term avoidance memory in order to protect themselves long term [24]. Although animals have the capacity to enhance their sensitivity as well as their motor responses to subsequent noxious stimuli, animals' ability to distinguish pain from other sensations is intact or at least not permanently altered. Pathological pain happens only after injury (e.g., tissue or nerve injury) and is not the result of the repetitive application of physiological pain. Long-term changes are likely to occur after injury, both peripherally and centrally. Consequently, the injury and injury-related areas undergo long-term plastic changes, and pain sensations are significantly enhanced (hyperalgesia) or non-noxious stimuli cause pain (allodynia). It should be pointed out that allodynia is one of the major problems in pathological pain. Because it is induced by non-noxious stimuli, it is most likely that central plastic changes play important roles.

Central plasticity is most likely mechanism for pathological pain

Pathological pain is likely the result of long-term plastic changes along somatosensory pathways from the periphery to the cortex. Due to long-term plastic changes in the central regions, pain specificity is lost in the somatosensory pathway, at least in areas where allodynia was reported [25–30]. Thus, drugs developed based on physiological pain mechanisms may not be used for treating pathological pain. Understanding pathological pain requires an understanding of plastic changes in somatosensory pathways, mainly the CNS.

Pain plasticity: Key mechanisms underlying persistent pain

Prolonged nociceptive stimulation following tissue or nerve injury induces long-lasting changes at most levels of the nociceptive pathways. In the dorsal horn of the spinal cord, neurons exhibit increased responses to noxious stimulation. As mentioned previously, the selective activation of A_d and C fibers by nociceptive stimuli is selective in situations of 'normal' acute noxious stimulation. After tissue or nerve injury, such rules do not exist anymore. It is quite common that non-noxious stimuli such as a gentle touch or warm temperatures become very painful. Corresponding anatomical and chemical alterations have been reported. The synapses between the primary afferents and dorsal horn neurons further undergo plastic changes, which are partly dependent on NMDA receptors. Long-lasting changes of dorsal horn synaptic responses have been shown to occur *in vivo* following altered nerve activation.

Descending modulatory systems are also altered in persistent pain, leading to modification of spinal sensory transmission [20,31]. Both descending facilitatory and inhibitory modulation from the RVM may participate in the development and maintenance of hyperalgesia following inflammation and tissue injury. Following inflammation, reversible spinal inactivation produced greater increases in dorsal horn neuronal activity, receptive field, and response to noxious stimuli. Lesions of the dorsolateral funiculus, which mediates descending inhibition from the RVM, potentiated inflammation-induced hyperalgesia. This indicates that descending inhibition of dorsal horn nociceptive transmission from supraspinal structures is increased during inflammation. RVM lesions furthermore revealed that increased descending facilitation could contribute to the development and maintenance of secondary hyperalgesia. Long-term changes in RVM neuronal activity have been observed in polyarthritis. Studies using microinjections of excitatory amino acids or opiate receptor agonists suggest that RVM neuronal excitability and sensitivity to opiates are altered following persistent hind paw inflammation. Our recent study shows that changes in RVM neuronal activity and descending modulation occur rapidly (within 15 min) following formalin-induced injury [32]. One potential synaptic mechanism for descending facilitation during persistent pain is the recruitment of silent spinal synapses [20,33,34].

Plasticity can occur in cortical areas, even in adults. It has been proposed that use-dependent changes in synaptic strength may serve as key synaptic mechanisms of such cortical changes, although more direct evidence is needed. Cortical and subcortical reorganization occurs after limb or digit amputation [35]. Most human amputees experience phantom limb sensations or what we call 'phantom pain'. However, cellular and molecular mechanisms contributing to the plastic changes in the neocortex after amputation remain to be investigated. In addition to the re-organization of the somatosensory cortex, plasticity also occurs in the ACC. *In vitro*, in ACC slices, synaptic depression induced by low-frequency stimulation was abolished after the amputation of a single digit in rats. Activation of various immediate early genes was noted in the ACC, further suggesting that rapid plastic changes occur within the ACC after amputation [36]. Furthermore, electrophysiological studies in intact animals revealed long-term potentiation of sensory responses in the ACC to noxious hind paw stimulation after the amputation. Although more studies are needed to determine signaling molecules contributing to plastic changes, these studies provide strong evidence that long-lasting plastic changes occur in the ACC after the injury. Our recent data using mutant mice further supported this possibility. Mice lacking both Ca^{2+}-stimulated adenylate cyclase subtypes 1 and 8 (AC1 and AC8) exhibited reduced allodynia following inflammation induced by complete Freund's adjuvant (CFA), while behavioral responses to acute noxious stimuli were normal. Forskolin injection into the ACC, which activates the remaining ACs, rescued the phenotype [37,38]. This suggests that Ca^{2+}-stimulated AC activity in the ACC participates in the development of hyperalgesia following amputation [39].

Chronic pain is likely coded in multiple sites

It has become clear that chronic pain is likely coded at multiple sites along the sensory pathways for pain transmission, modulation, and plasticity [9,10,24]. The ultimate summary of sensitization in the periphery, spinal cord, and cortex sites contributes to the enhancement of pain in chronic pain conditions. It has been reported that inhibiting the plasticity of AMPA receptors can be analgesic at different levels of the CNS, and it is still unclear if chronic pain important is mainly stored at synaptic responses or by the other forms of plasticity, such as changes in action potentials.

Glial cells and chronic pain

Microglia are the principal immune-response cells in the CNS. In physiological conditions, they are found in a "resting" state — typically exhibiting ramified processes with high motility. Under pathological conditions, these cells are transformed from the resting condition to an activated condition, exhibiting phagocytoxic, chemotaxis, and secretory reactions. A growing body of literature indicates that spinal microglia can be activated after nerve injury, suggesting the possibility that neuronal activity may contribute to microglia activation [40,41]. This possibility is further supported by studies showing that several neurotransmitter receptors can be expressed on cultured microglia cells, including NMDA, GABA, opioid, and adrenergic receptors. However, in a recent study using the brain slice preparation for adult mice, we found that microglia did not respond to either a glutamate or GABA application, or activity-dependent LTP [42]. In addition to these findings, we found that nerve injury did not cause any activation of microglial cells in supraspinal central nuclei such as the ACC where excitatory synaptic transmission was significantly enhanced after nerve injury. In support of previous reports on the spinal cord, we also found that microglial cells were activated in the spinal cord dorsal horn after the nerve injury [43,44]. One possible explanation is that spinal microglia may be more sensitive to abnormal neuronal activity than those in higher brain regions.

NR2B: Smart gene for chronic pain treatment

The NMDA receptor is critical for learning-related LTP and behavioral memory [2]. Genetic overexpression of NMDA receptor 2B (NR2B) (also called GluN2B) in the forebrains of transgenic mice leads to enhanced activation of NMDA receptors, facilitating synaptic potentiation in the hippocampus [45]. These mice also exhibit superior ability in learning and memory in various behavioral tasks. Memory and pain are often seen as two distinct physiological functions, however, more and more evidence suggests that they may also be linked together. In transgenic NR2B mice, enhanced responsiveness to peripheral injection of two inflammatory stimuli, formalin and complete Freund's adjuvant (CFA), was found [30]. These results suggest that genetic modification of forebrain NMDA receptors can therefore influence pain perception, which suggests that forebrain-selective NMDA receptor antagonists, including NR2B-selective agents, may be useful analgesics for persistent pain.

PKMζ: memory kinase to maintain the pain

Multiple protein kinases are thought to contribute to the induction of LTP and initial consolidation of information storage. Among them, only protein kinase M zeta (PKMζ) maintains persistent synaptic changes [46]. In the hippocampus, LTP induction triggers the synthesis of PKMζ, and activation of PKMζ is critical for late-phase LTP (L-LTP) and memory consolidation. Interestingly, PKMζ was found to maintain pain-induced persistent changes in the mouse ACC [47]. Peripheral nerve injury caused the activation of PKMζ in the ACC, and inhibiting PKMζ by a selective inhibitor, ζ-pseudosubstrate inhibitory peptide (ZIP), erased synaptic potentiation. Microinjection of ZIP into the ACC blocked behavioral sensitization. These results suggest that PKMζ in the ACC acts to maintain neuropathic pain.

Conclusions

In summary, progress in basic neurosciences will continue to affect the way we study the mechanisms for physiological, pathological, and chronic pain. Our understanding of pain will improve with the development of new technology and the advancement of our molecular understanding of biological systems. At the systemic level, we are far behind in our understanding of how brains code sensory information and how consciousness, attention, and emotion are processed, and generated, at the molecular level. Without this information, we may never be able to understand and 'cure' pain.

References

[1] Epsztein, J. *et al.* (2010) Impact of spikelets on hippocampal CA1 pyramidal cell activity during spatial exploration. *Science* 327, 474–477. 10.1126/science.1182773.

[2] Bliss, T.V. and Collingridge, G.L. (1993) A synaptic model of memory: Long-term potentiation in the hippocampus. *Nature* 361, 31–39. 10.1038/361031a0.

[3] Kandel, E.R. (2001) The molecular biology of memory storage: A dialogue between genes and synapses. *Science* 294, 1030–1038. 10.1126/science.1067020.

[4] Lee, S.H. *et al.* (2008) Synaptic protein degradation underlies destabilization of retrieved fear memory. *Science* 319, 1253–1256. 10.1126/science.1150541.

[5] Li, P. and Zhuo, M. (1998) Silent glutamatergic synapses and nociception in mammalian spinal cord. *Nature* 393, 695–698. 10.1038/31496.

[6] Li, P. *et al.* (1999) Kainate-receptor-mediated sensory synaptic transmission in mammalian spinal cord. *Nature* 397, 161–164. 10.1038/16469.

[7] Li, P. and Zhuo, M. (2001) Substance P and neurokinin A mediate sensory synaptic transmission in young rat dorsal horn neurons. *Brain Res Bull* 55, 521–531. 10.1016/s0361-9230(01)00553-6.

[8] Yoshimura, M. and Jessell, T. (1990) Amino acid-mediated EPSPs at primary afferent synapses with substantia gelatinosa neurones in the rat spinal cord. *J Physiol* 430, 315–335. 10.1113/jphysiol.1990.sp018293.

[9] Zhuo, M. (2008) Cortical excitation and chronic pain. *Trends Neurosci* 31, 199–207. 10.1016/j.tins.2008.01.003.

[10] Zhuo, M. (2016) Contribution of synaptic plasticity in the insular cortex to chronic pain. *Neuroscience* 338, 220–229. 10.1016/j.neuroscience.2016.08.014.

[11] Craig, A.D. *et al.* (1996) Functional imaging of an illusion of pain. *Nature* 384, 258–260. 10.1038/384258a0.

[12] Koga, K. *et al.* (2010) In vivo whole-cell patch-clamp recording of sensory synaptic responses of cingulate pyramidal neurons to noxious mechanical stimuli in adult mice. *Mol Pain* 6, 62. 10.1186/1744-8069-6-62.

[13] Zhuo, M. and Gebhart, G.F. (1990) Characterization of descending inhibition and facilitation from the nuclei reticularis gigantocellularis and gigantocellularis pars alpha in the rat. *Pain* 42, 337–350. 10.1016/0304-3959(90)91147-B.

[14] Zhuo, M. and Gebhart, G.F. (1992) Characterization of descending facilitation and inhibition of spinal nociceptive transmission from the nuclei reticularis gigantocellularis and gigantocellularis pars alpha in the rat. *J Neurophysiol* 67, 1599–1614. 10.1152/jn.1992.67.6.1599.

[15] Zhuo, M. and Gebhart, G.F. (1997) Biphasic modulation of spinal nociceptive transmission from the medullary raphe nuclei in the rat. *J Neurophysiol* 78, 746–758. 10.1152/jn.1997.78.2.746.

[16] Zhuo, M. and Gebhart, G.F. (2002) Facilitation and attenuation of a visceral nociceptive reflex from the rostroventral medulla in the rat. *Gastroenterology* 122, 1007–1019. 10.1053/gast.2002.32389.

[17] Zhuo, M. *et al.* (2002) Biphasic modulation of spinal visceral nociceptive transmission from the rostroventral medial medulla in the rat. *J Neurophysiol* 87, 2225–2236. 10.1152/jn.2002.87.5.2225.

[18] Millan, M.J. (2002) Descending control of pain. *Prog Neurobiol* 66, 355–474. 10.1016/s0301-0082(02)00009-6.

[19] Porreca, F. *et al.* (2002) Chronic pain and medullary descending facilitation. *Trends Neurosci* 25, 319–325. 10.1016/s0166-2236(02)02157-4.

[20] Zhuo, M. (2017) Descending facilitation. *Mol Pain* 13, 1744806917699212. 10.1177/1744806917699212.

[21] Melzack, R. and Wall, P.D. (1965) Pain mechanisms: A new theory. *Science* 150, 971–979. 10.1126/science.150.3699.971.

[22] Kerchner, G.A. *et al.* (2001) Direct presynaptic regulation of GABA/glycine release by kainate receptors in the dorsal horn: An ionotropic mechanism. *Neuron* 32, 477–488. 10.1016/s0896-6273(01)00479-2.

[23] Calejesan, A.A. *et al.* (2000) Descending facilitatory modulation of a behavioral nociceptive response by stimulation in the adult rat anterior cingulate cortex. *Eur J Pain* 4, 83–96. 10.1053/eujp.1999.0158.

[24] Bliss, T.V. *et al.* (2016) Synaptic plasticity in the anterior cingulate cortex in acute and chronic pain. *Nat Rev Neurosci* 17, 485–496. 10.1038/nrn.2016.68.

[25] Woolf, C.J. and Salter, M.W. (2000) Neuronal plasticity: Increasing the gain in pain. *Science* 288, 1765–1769. 10.1126/science.288.5472.1765.

[26] Levine, J.D. (1998) New directions in pain research: Molecules to maladies. *Neuron* 20, 649–654. 10.1016/s0896-6273(00)81005-3.

[27] Sandkuhler, J. (2007) Understanding LTP in pain pathways. *Mol Pain* 3, 9. 10.1186/1744-8069-3-9.

[28] Zhao, M.G. *et al.* (2005) Roles of NMDA NR2B subtype receptor in prefrontal long-term potentiation and contextual fear memory. *Neuron* 47, 859–872. 10.1016/j.neuron.2005.08.014.

[29] Xu, H. *et al.* (2008) Presynaptic and postsynaptic amplifications of neuropathic pain in the anterior cingulate cortex. *J Neurosci* 28, 7445–7453. 10.1523/JNEUROSCI.1812-08.2008.

[30] Wei, F. *et al.* (2001) Genetic enhancement of inflammatory pain by forebrain NR2B overexpression. *Nat Neurosci* 4, 164–169. 10.1038/83993.

[31] Chen, T. *et al.* (2018) Top-down descending facilitation of spinal sensory excitatory transmission from the anterior cingulate cortex. *Nat Commun* 9, 1886. 10.1038/s41467-018-04309-2.

[32] Robinson, D. *et al.* (2002) Long-lasting changes in rostral ventral medulla neuronal activity after inflammation. *J Pain* 3, 292–300. 10.1054/jpai. 2002.125183.

[33] Fields, H. (2004) State-dependent opioid control of pain. *Nat Rev Neurosci* 5, 565–575. 10.1038/nrn1431.

[34] Gebhart, G.F. (2004) Descending modulation of pain. *Neurosci Biobehav Rev* 27, 729–737. 10.1016/j.neubiorev.2003.11.008.

[35] Kaas, J.H. *et al.* (1999) Subcortical contributions to massive cortical reorganizations. *Neuron* 22, 657–660. 10.1016/s0896-6273(00)80725-4.

[36] Wei, F. and Zhuo, M. (2001) Potentiation of sensory responses in the anterior cingulate cortex following digit amputation in the anaesthetised rat. *J Physiol* 532, 823–833. 10.1111/j.1469-7793.2001.0823e.x.

[37] Liauw, J. *et al.* (2005) Calcium-stimulated adenylyl cyclases required for long-term potentiation in the anterior cingulate cortex. *J Neurophysiol* 94, 878–882. 10.1152/jn.01205.2004.

[38] Wei, F. *et al.* (2002) Genetic elimination of behavioral sensitization in mice lacking calmodulin-stimulated adenylyl cyclases. *Neuron* 36, 713–726. 10.1016/s0896-6273(02)01019-x.

[39] Wang, H. *et al.* (2011) Identification of an adenylyl cyclase inhibitor for treating neuropathic and inflammatory pain. *Sci Transl Med* 3, 65ra63. 10.1126/scitranslmed.3001269.

[40] Coull, J.A. *et al.* (2005) BDNF from microglia causes the shift in neuronal anion gradient underlying neuropathic pain. *Nature* 438, 1017–1021. 10.1038/nature04223.

[41] Zhuo, M. (2017) New mechanisms for pain: From neurons to glia; from spinal cord to cortex. *J Neurochem* 141, 484–485. 10.1111/jnc.13985.

[42] Wu, L.J. and Zhuo, M. (2008) Resting microglial motility is independent of synaptic plasticity in mammalian brain. *J Neurophysiol* 99, 2026–2032. 10.1152/jn.01210.2007.

[43] Chen, T. *et al.* (2010) Spinal microglial motility is independent of neuronal activity and plasticity in adult mice. *Mol Pain* 6, 19. 10.1186/1744-8069-6-19.

[44] Zhang, F. *et al.* (2008) Selective activation of microglia in spinal cord but not higher cortical regions following nerve injury in adult mouse. *Mol Pain* 4, 15. 10.1186/1744-8069-4-15.

[45] Tang, Y.P. *et al.* (1999) Genetic enhancement of learning and memory in mice. *Nature* 401, 63–69. 10.1038/43432.

[46] Sacktor, T.C. (2008) PKMzeta, LTP maintenance, and the dynamic molecular biology of memory storage. *Prog Brain Res* 169, 27–40. 10.1016/S0079-6123(07)00002-7.

[47] Li, X.Y. *et al.* (2010) Alleviating neuropathic pain hypersensitivity by inhibiting PKMzeta in the anterior cingulate cortex. *Science* 330, 1400–1404. 10.1126/science.1191792.

[48] Langford, D.J. *et al.* (2006) Social modulation of pain as evidence for empathy in mice. *Science* 312, 1967–1970. 10.1126/science.1128322.

[49] Zhuo, M. (2002) Glutamate receptors and persistent pain: Targeting forebrain NR2B subunits. *Drug Discov Today* 7, 259–267. 10.1016/s1359-6446(01)02138-9.

Chapter 2

Peripheral Nociceptors and Sensitization

News and Views

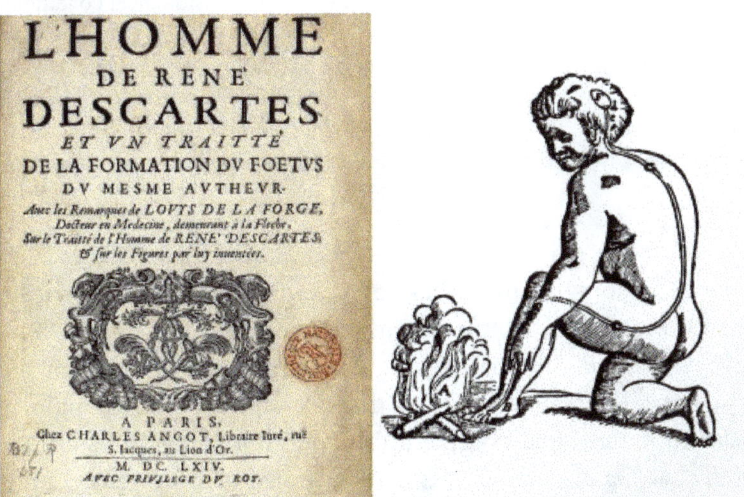

Note: Illustration depicting how the philosopher René Descartes imagined how heat sends mechanical signals to the brain.

In the 17th century, philosopher René Descartes described a theory in which he envisioned threads connecting different parts of the skin with the brain. In this way, a foot touching an open flame would send a mechanical

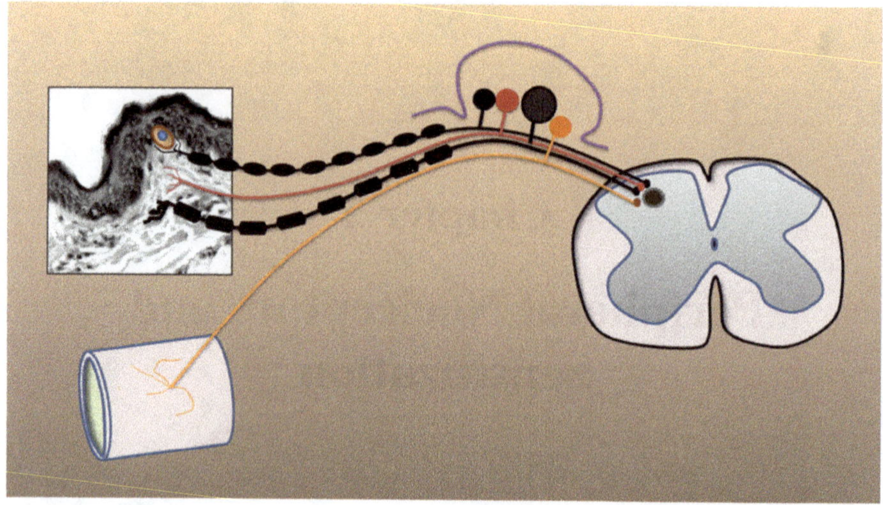

Figure 1.   Peripheral nociceptors and different fibers. In physiological condition, different sensory inputs are conveyed through various sensory fibers. For example, a gentle touch is carried out by myelinated afferent fibers, while noxious heat is conducted by small myelinated (Aδ) and unmyelinated fibers (C). Spinal dorsal horn sensory neurons receive different sensory inputs, including both non-noxious as well as noxious inputs. That is, some dorsal horn neurons can respond to a gentle touch as well as noxious heat.

signal to the brain (Figure 1). Discoveries later revealed the existence of specialized sensory neurons that register changes in our environment. Joseph Erlanger and Herbert Gasser received the Nobel Prize in Physiology or Medicine in 1944 for their discovery of different types of sensory nerve fibers that react to distinct stimuli, for example, in the responses to painful and non-painful touch. Since then, it has been demonstrated that nerve cells are highly specialized for detecting and transducing differing types of stimuli, allowing a nuanced perception of our surroundings, including our capacity to feel differences in the texture of surfaces through our fingertips or our ability to discern both pleasing warmth and painful heat.

The 2021 Nobel Prize in Physiology or Medicine was awarded jointly to David Julius and Ardem Patapoutian for their discoveries of receptors for temperature and touch. David Julius utilized capsaicin, a pungent compound from chili peppers that induces a burning sensation, to identify a sensor in the nerve endings of the skin that respond to heat. Ardem Patapoutian used pressure-sensitive cells to discover a novel class of sensors that respond to mechanical stimuli in the skin and internal organs. These breakthrough discoveries launched intense research activities leading to a rapid increase in our understanding of how our nervous system senses heat, cold, and mechanical stimuli.

Summary

Peripheral noxious stimuli such as heat and cold are converted into neuronal action potentials by nociceptors. Peripheral sensitization of nociceptors is thought to be important for many forms of chronic pain. Abnormal and ongoing activities from the affected peripheral area trigger and/or maintain long-term changes in the central nervous system, including the spinal cord and supraspinal structures. While some forms of chronic pain may be mainly driven by ongoing activities from peripheral nociceptors, other forms of chronic pain may be caused by central plasticity induced by previous peripheral inputs. Thus, investigation of the molecular mechanisms that contribute to peripheral sensitization is important not only for the understanding of peripheral-dependent chronic pain but also for understanding how peripheral molecular mechanisms may trigger CNS plasticity, making chronic pain resistant to any manipulation at the periphery.

Keywords: DRG; itch; cold; heat; peripheral sensitization; silent nociceptor; TRPV1; hyperalgesia

Introduction

One major hypothesis for pain transmission is the labeled theory. The labeled theory proposes that pain information is conducted by selective proteins and molecules in the central nervous system. Early identification of nociceptive fibers, and spinal nociceptive neurons, provides strong evidence for selective labeled lines. Recent research into the molecular identification of pain-related receptors provides strong evidence in support of this theory. However, there is also evidence to the contrary, suggesting that such distinct lines of pain pathways are unlikely, especially in case of pathological pain or central pain. Thus, it becomes critical to distinguish physiological pain from pathological pain. While it is true that under physiological conditions, some proteins are selective in pain transmission, chronic pain is likely mediated by a network.

In addition to selective lines of involvement, the contribution of peripheral sensitized activity and central plasticity to chronic pain may be simple additive or liner (see Figure 1). For certain types of pain, the central plasticity may be driven, and even maintained, by peripheral sensitized inputs. Thus, the removal or inhibition of abnormal inputs will lead to a significant reduction in chronic pain. However, in other cases, while peripheral inputs may still play minor roles, central potentiation may take over and play a major role in ongoing chronic pain. In such cases, inhibiting peripheral inputs may be insufficient, and in some cases, it is difficult to clearly identify which peripheral inputs are responsible for chronic pain.

Peripheral nerves and DRG cells

There are two major classes of peripheral nociceptors: unmyelinated C fibers and small-myelinated Aδ fibers. C fibers are small-diameter unmyelinated fibers with small-diameter cell bodies. Whereas, Aδ fibers have medium-diameter unmyelinated fibers. Each nociceptor can be divided into different subgroups according to their responses to different forms of sensory stimuli (thermal, mechanical, and chemical) or the chemicals they express. The cell bodies of nociceptors are generally located in the DRG. For the facial sensory nociceptors, they are located in trigeminal and nodose ganglia. Although the sizes of DRGs have often been used to identify nociceptors, it may not be appropriate to be used in case of visceral pain. Many somas of visceral nociceptors are larger than the DRG somas of somatic nociceptors. Performing electrophysiological recordings from nerve fibers and cells is the most reliable way to study the function of peripheral nociceptors.

Peripheral noxious stimuli activate peripheral nociceptive transducer receptors and/or ion channels causing membrane depolarization in sensory DRG cells. Recent studies indicate that not all nociceptors have a response threshold as high as previously believed. Some nociceptors have lower thresholds, just like non-nociceptors. Recent studies indicate that transducer proteins include a family of proteins: TRPV1-4, TRPM8, and different subtypes of ATP receptors. It becomes clear that no single protein or gene is responsible for a specific sensory process, such as heat pain or cold sensation (see Table 1).

Under physiological conditions, noxious stimuli stimulate both non-noxious and nociceptive fibers. It is almost impossible to deliver a selective noxious stimulus without activating some form of non-nociceptive receptors. In pathological pain conditions, typical allodynia — triggered by non-noxious stimulation — is also unlikely due to the selective activation of nociceptive fibers (Figure 1). Two voltage-gated sodium channel alpha subunits, Nav1.7 and Nav1.8, are expressed at high levels in nociceptor terminals. Deleting Nav1.7 has no effect on the development of neuropathic pain, and double knockouts of both Nav1.7 and Nav1.8 also develop normal levels of neuropathic pain. Inflammatory pain and mechanical and thermal acute pain were altered [1]. A recent study by Peirs *et al.* provides strong evidence that the neural circuits conveying mechanical allodynia in the dorsal horn differ by the nature of the injury [2]. Calretinin neurons in lamina II inner convey mechanical allodynia induced by inflammatory injuries, while protein kinase C gamma (PKCγ) neurons at the lamina II/III border convey mechanical allodynia induced by neuropathic injuries. It is safe to say that each sensory modality, or sensation, is a function of a specially organized neuronal circuit and network, from the periphery to the cortex with some of the key proteins playing major roles.

Table 1. Peripheral nociceptors for spinal transmitters for nociceptive pain and itch

Sensory modality	Peripheral receptors	Primary afferent fibers
Nociceptive heat	TRPV1, TRPV2-4,	A_δ; C fibers
Nociceptive cold	TRPM8, TRPA1, TRAAK/TREK-1	A_δ; C fibers
Nociceptive mechanical	ASIC1-3; Cav3,2; P2X3, TRAAK	A_δ; C fibers
Itch	H1,4; interlukin-31 receptors; PAR2	C fibers
Visceral pain	ASIC1-3; TRPV1; TRPV4	Visceral afferent fibers

Molecular biology of sensory receptors

Identification of heat receptor TRPV1 has greatly facilitated the discovery of many other sensory receptors [3]. Research into the expression of these genes in cell lines has provided the first direct evidence of their sensitivity to sensory stimuli. Development of selective inhibitors and gene knockout mice have further helped us to understand the physiological studies of these sensory receptors, although some of the results are complicated and different from the expectation (e.g., cold receptor knockout and mouse behavioral responses to cold). At the cellular level, the thermal threshold of TRPV1 is found to be similar to that of human thermal threshold for heat pain. In addition to the heat, TRPV1 is also activated by capsaicin and acids (protons). Knockout mice lacking the TRPV1 receptors not only avoid water with capsaicin in it but also have a diminished response to heat, providing strong evidence for its roles in sensory process in behavioral animals [4].

In addition to TRPV1, there are other types of heat receptors in the skin. They are all transmembrane proteins in the plasma membrane that are permeable to calcium ions and sodium ions when they are open. Between them, they cover a range of temperatures: TRPV4 Warm (~27–34°C), TRPV3 Warmer (~31–39°C), and TRPV2 Painfully hot (>52°C). Since the discovery of TRPVs, there has been a great interest in developing drugs that inhibit these proteins. The hypothesis is that by applying these receptor blockers, we may reduce the pain in patients with chronic pain. While it is true that some thermal hyperalgesia may be reduced, these inhibitors will also reduce normal thermal sensation in normal skin areas and thus may cause secondary damage in daily life due to the changed threshold.

Cold-sensitive thermoreceptors give rise to the sensations of cooling, cold, and freshness [5–7]. There are two candidate receptors for this: One, designated TRPM8, is a channel that admits Ca^{2+} and Na^+ in response to moderate cold (<26°C) or menthol (the ingredient that gives mint its "cool" touch and taste). However, knockout mice lacking the gene encoding the TRPM8 receptor do not avoid cold places as normal mice do, suggesting that its function can be compensated by other sensory proteins *in vivo*. A second candidate, designated TRPA1, responds to lower temperatures (<18°C). It also responds to several irritant chemicals eliciting signals that the brain interprets as pain. However, TRPA1 knockout mice respond normally to cold so the precise role of these receptors is still uncertain for those stimuli.

Recent physiological studies using gene knockout mice lacking TRPV1 or TRPM8 have nicely demonstrated that neither heat pain nor cold pain is mediated by a single protein/an ion channel. In mice lacking TRPV1, behavioral deficits in responses to noxious heat applying to the tail or hind paw are only partial. In mice lacking TRPM8, behavioral responses to noxious cold are partially affected or not affected at all. In some subpopulation of nociceptors, other ion channels such as voltage-gated K^+ channel may also contribute to cold responses. Thus, it is possible that a peripheral sensory protein may contribute to multiple sensory processes, such as heat, cold, itch, and touch. One common concern is that many reports only employed behavioral studies using gene-manipulated mice. Therefore, most of the conclusions are indirect and inconclusive. Electrophysiological experiments may be very difficult in some cases and are needed in future studies (Figure 2).

While the roles of TRPM8/TRPA1 in behavioral responses to physiological cold are inconclusive, recent studies found that they may contribute to cold stimuli after tissue inflammation or cold allodynia [8–12]. It is great of interest to confirm these findings at the cellular level (using electrophysiological recordings of fiber activity) and investigate the possible molecular mechanisms. Mechanical nociceptors respond to excess pressure or mechanical deformation. They also respond to incisions that break the skin's surface. The reaction to the stimulus is processed as pain by the cortex, just like chemical and thermal responses. Many times, these mechanical nociceptors have polymodal characteristics. Therefore, it is possible that some of the transducers for thermal stimuli are the same for mechanical stimuli. The same is true for chemical stimuli since TRPA1 appears to detect both mechanical and chemical changes.

Visceral nociceptors

Visceral receptors are less studied as compared with somatosensory nociceptors [13]. Recent studies from visceral nociceptors clearly show that it is unlikely that there is a common mechanism for nociceptors in

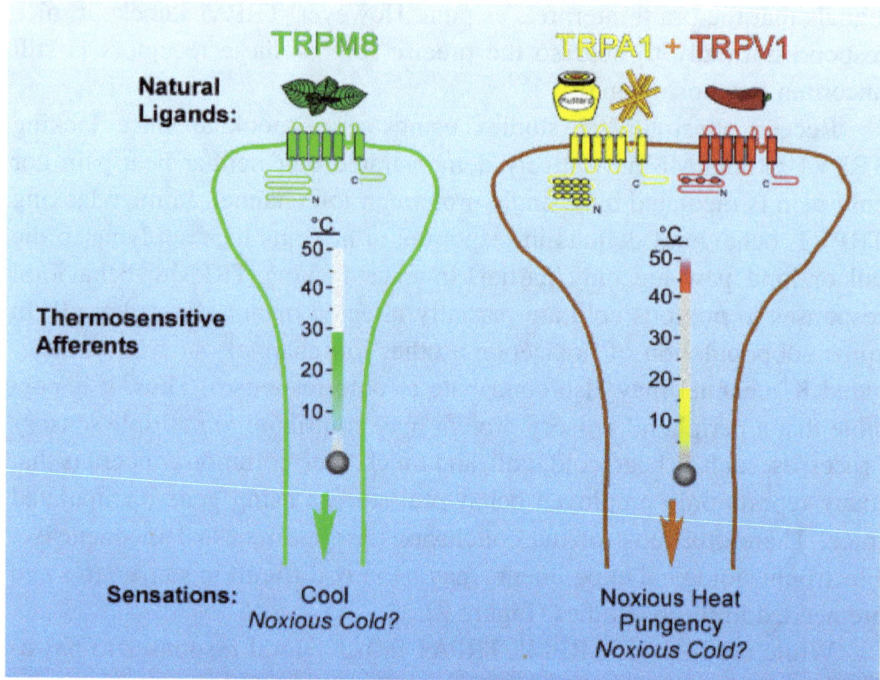

Figure 2. Molecular identity of thermosensitive afferents based upon the expression of TRPM8 and TRPA1. TRPM8 and TRPA1 are found in distinct and non-overlapping populations of sensory afferents, with TRPA1 expressed exclusively in some, but not all, neurons that express the heat-gated channel TRPV1. These thermosensitive TRP channels respond to a number of naturally occurring pungent compounds, such as menthol (mint), allyl isothiocyanate (mustard oil), cinnamaldehyde (cinnamon), and capsaicin ('hot' chili peppers), thus providing a molecular explanation for how these compounds provide distinct sensations of cold, heat, or spiciness. Based upon *in vitro* characterizations of these channels, along with their distinct expression patterns, thermal stimuli activating TRPM8-expressing afferents elicit the sensation of cool to potentially noxious cold, while TRPA1 afferents will merge both noxious cold and noxious heat, due to the expression of TRPV1 (adapted from McKemy [8]).

cutaneous tissues vs visceral organs. For example, some of the visceral nociceptors have large cell bodies of DRG as compared with typical somatic nociceptors. Unlike somatic nociceptors, visceral nociceptors are low-threshold mechanoceptors suggesting that they also respond to non-nociceptive stimuli. Recent studies reveal that different subtypes of ATP receptors are important for transducting visceral pain information [14,15].

Itching and pain: Two separate transmitter systems?

Research on itch has gained more interest recently. Similar to pain, one major theory is that itch may be mediated by selective, independent 'itch' transmitters and receptors, at least at the periphery [16,17]. The studies of mice lacking gastrin-releasing peptide (GRP) provide interesting findings that show selective reduction in 'itch' responses, while most of the pain responses remain intact. Considering the expression of GRP in selective DRG cells, it has been proposed that GRP may act as selective transmitter for itch. However, there is one key piece of evidence missing to support the theory of GRP as the neurotransmitter in the spinal cord for itch. It is essential to demonstrate that sensory-evoked responses between primary afferent fibers and GRP-responsive dorsal horn neurons are mediated by GRP itself. Without this direct evidence, it is impossible to distinguish GRP as a transmitter or neuromodulator for itch. The hypothesis of GRP as a selective itch transmitter is also questioned by recent studies using gene-mutant mice lacking VGLUT2 or BHLHB5 [18]. In the study of BHLHB5 mutant mice, loss of inhibitory interneurons in the spinal dorsal horn contributes to elevated itch responses in behavioral animals. In VGLUT2 conditional null mice — in which vesicle glutamate transport type 2-dependent synaptic release of glutamate was abolished — behavioral responses to itch stimuli were either significantly reduced or enhanced. Electrophysiological studies showed that while GRP activated certain dorsal horn neurons that received C fiber inputs, excitatory synaptic responses onto the same neurons are purely mediated by glutamatergic receptors, suggesting that glutamate is also the major transmitter for itching at the level of the spinal cord [19] (Figure 3). These conflicting results raise new questions about whether GRP serves as a selective transmitter of itch or GRP and/or glutamate may serve as transmitters in spinal itch synapses (Figure 4).

Peripheral sensitization

It is well known that nociceptor neuron sensitivity is modulated by a large variety of mediators in the extracellular space [20,21]. Peripheral sensitization represents a form of functional plasticity of the nociceptor. The nociceptor can simply change from a noxious stimulus detector to a

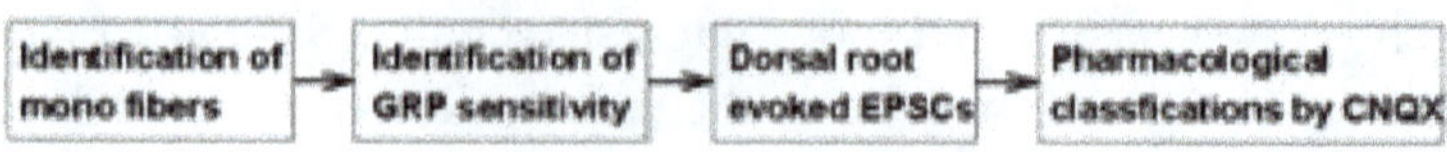

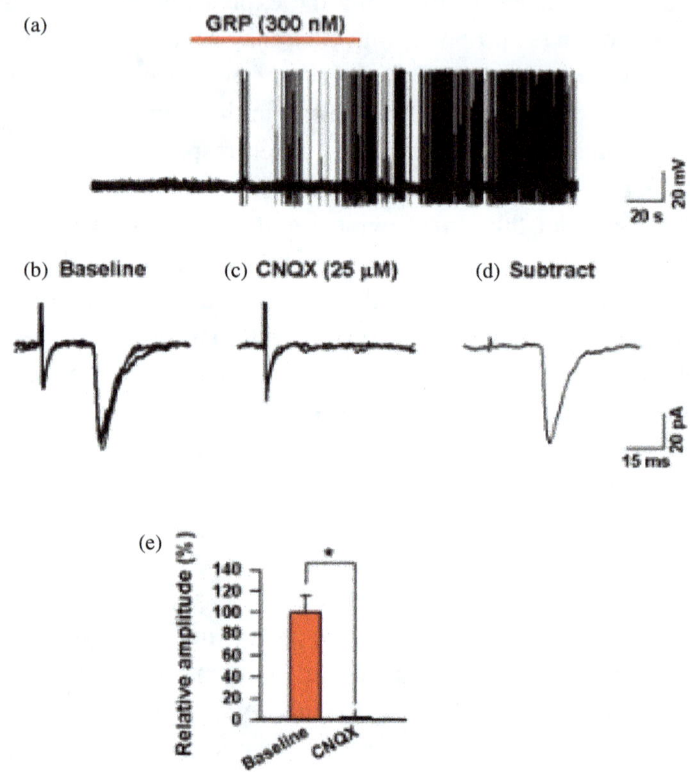

Figure 3. C fiber-evoked responses in GRP-positive neurons are blocked by glutamate antagonists. (A) The experimental procedure to identify the transmitter between C fibers and GRP-positive neurons in rats. (B) In a GRP-sensitive neuron (a), a monosynaptic C fiber-evoked EPSCs (b) was totally blocked by a bath application of CNQX (25 μM), an AMPA/KA receptor antagonist (c–e, $n = 4$) (adapted from Koga *et al.* [19]).

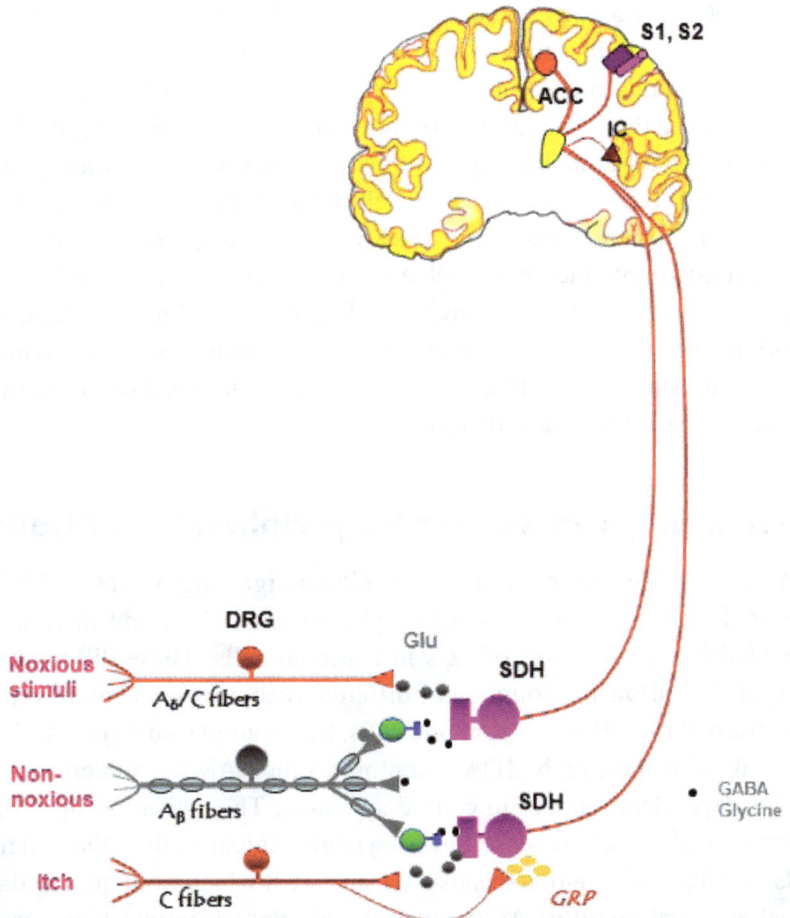

Figure 4. Itching and pain pathways. Under normal physiological conditions, noxious stimuli activate nociceptive afferent fibers (A$_\delta$ and C fibers). Incoming action potentials trigger a release of excitatory transmitter glutamate in the spinal dorsal horn. In addition, some neuropeptides are also released including substance P (SP) and neurokinin A (NKA). Glutamate and neuropeptides activate spinal dorsal horn neurons, including those that send projection terminals to supraspinal structures. Neurons in the thalamus play key roles in relaying these ascending inputs. Five major cortical areas, the ACC, IC, S1, S2, and PFC, are activated and contribute to different aspects of pain perception, including the unpleasantness of pain. Recent studies have suggested that glutamate may also serve as a key fast excitatory transmitter for itch, and a peptide GRP plays modulatory role in the transmission of itch.

detector of non-noxious stimuli. The result is that low-intensity stimulus from regular activity initiates a painful sensation. This is commonly known as hyperalgesia and allodynia. Inflammation is one common cause that results in the sensitization of nociceptors. Normally, hyperalgesia ceases when inflammation goes down, however, sometimes genetic defects and/or repeated injury can result in allodynia: where a completely non-noxious stimulus like light touch causes extreme pain. It becomes clear that no simple factor is involved in peripheral sensitization. Instead, different factors and cells are involved. It is believed that peripheral sensitization may contribute to behavioral hyperalgesia and/or allodynia in chronic pain state. It is unlikely that there is one universal mechanism for all kinds of peripheral sensitization.

cAMP as a key messenger for peripheral sensitization

cAMP is a key second messenger for cellular signaling process. cAMP is generated by the activation of adenylyl cyclases (ACs), and there are at least 11 different isoforms of ACs in neuronal cells. These different isoforms of ACs allow the coupling of different receptors/ion channels to the production of cAMP in postsynaptic cells. For example, subtypes AC1 and AC8 link activation of NMDA receptors to the cAMP production in an activity-dependent manner in central synapses. The activation of NMDA receptors leads to an increase in postsynaptic calcium, then that calcium binds to calmodulin and activates AC1 and AC8 which both are regulated by calcium calmodulin. At the peripheral, non-selective activators of ACs — forskolin as well as cAMP analogs — trigger peripheral nociceptor sensitization (i.e., causing behavioral hyperalgesia or allodynia). At least one major mechanism is enhancing responses of primary afferent fibers to sensory stimuli. Interestingly, opioid receptor agonists produced their analgesic effects at least in part by reducing the intracellular cAMP level by different types of G protein-related signaling pathways.

In addition to cAMP and PKA pathways, another major protein kinase PKC has also been found to be important for peripheral sensitization. Interestingly, among different isoforms of PKCs, only calcium-independent PKCε is mainly involved in peripheral sensitization. It can be activated by different types of inflammation-related mediators, such as bradykinin, epinephrine, tumor necrosis factor, and carrageen.

Sensitization of TRPV1 receptor

Among many different possible target proteins that may be sensitized or enhanced in peripheral sensitization, the modulation of TRPV1 receptor function is an attractive target (Figure 5) [22,23]. Due to its important role in processing thermal heat transmission, it is obvious that its sensitization will significantly enhance the responses of nociceptors to thermal stimuli and contribute to thermal hyperalgesia. It may also contribute to thermal allodynia if the responsive threshold is also reduced. The cAMP-PKA pathway that is known to be critical for peripheral sensitization has been reported to enhance the function of TRPV1 receptors in nociceptors [24]. Capsaicin responses in nociceptors are enhanced by the activation of PKA. One possible major mechanism is that PKA phosphorylates TRPV1 at Ser 116, and this phosphorylation reduces the desensitization of TRPV1.

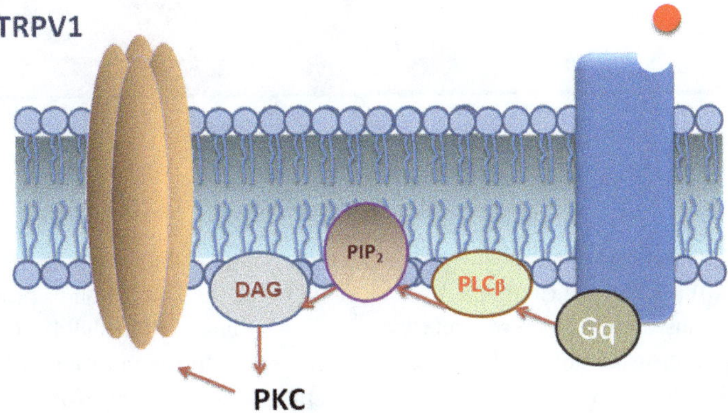

Figure 5.   Plasticity of TRPVR1 receptors. The capsaicin receptor, TRPVR1, is a sensory neuron-specific ion channel that serves as a polymodal detector of pain-producing chemical and physical stimuli. ATP potentiates the TRPVR1 currents evoked by capsaicin through metabotropic P2Y1 receptors in a protein kinase C (PKC)-dependent pathway.

Hyperalgesic priming and its related molecular mechanism

The peripheral sensitization mechanism for hyperalgesia has been greatly improved by the discovery of hyperalgesia priming in the periphery. In this form of peripheral nociceptor plasticity, acute inflammatory insult can trigger long-lasting hypersensitivity of nociceptors to inflammatory cytokines (Figure 6). This form of priming depends on one form of protein kinase C (PKCε). cAMP signaling pathway has been known to contribute

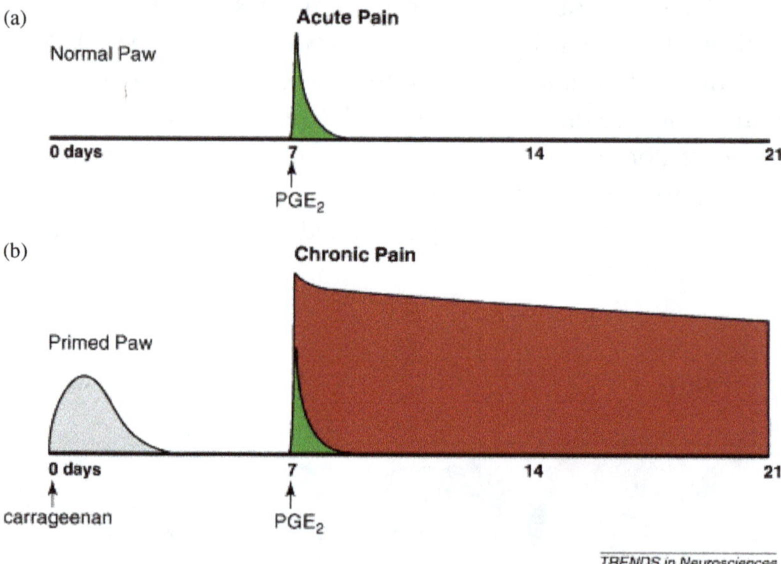

Figure 6. Chronic hyperalgesia associated with inflammation-induced hyperalgesic priming. (a) In the normal (unprimed) paw, a small intradermal injection of prostaglandin E2 (PGE2) causes an episode of acute hyperalgesia (decreased threshold for paw withdrawal from a pressure stimulus) lasting less than 4 h. (b) Injection of the inflammogen, carrageenan, causes an episode of hyperalgesia that lasts less than 4 days (gray-filled curve). After carrageenan-induced hyperalgesia is no longer present, the paw remains in a latent state of hyperalgesic priming. In this state, an injection of PGE2, which would cause only acute hyperalgesia in the normal (unprimed) paw, now induces an additional chronic hyperalgesia. In comparison to the unprimed paw, this hyperalgesia is greater in magnitude and is greatly prolonged, lasting at least 3 weeks (adapted from Reichling and Levine [27]).

to the priming. The peripheral activity of PKCε is required for both the initiation and maintenance of primed hyperalgesia.

Silent nociceptors

One interesting finding related to peripheral sensitization is the silent nociceptors [25]. In experimental conditions, no physiological stimuli are found to be able to activate these 'silent' receptors. However, after injury, the silent receptors become active. These silent receptors are mostly mechanosensitive receptors (or called MIAs, mechanically insensitive afferents) and have been found in tissues, such as joints, muscles, and visceral organs. Thus, the recruitment of silent nociceptors provides novel mechanisms for peripheral sensitization. It is likely that peripheral sensitization can occur via three different mechanisms: (1) enhanced responses to stimuli, (2) the recruitment of silent nociceptors that are previously inactive or silent, and (3) sprouting of nociceptors and recruiting new neuronal networks in the spinal cord (Figure 7). However, due to the poor access to direct recording from silent fibers, most of the intracellular molecular mechanisms that lead to re-activation of silent nociceptors remain unknown. Future advances in single-fiber electrophysiology and molecular biology interference techniques will facilitate research in this area.

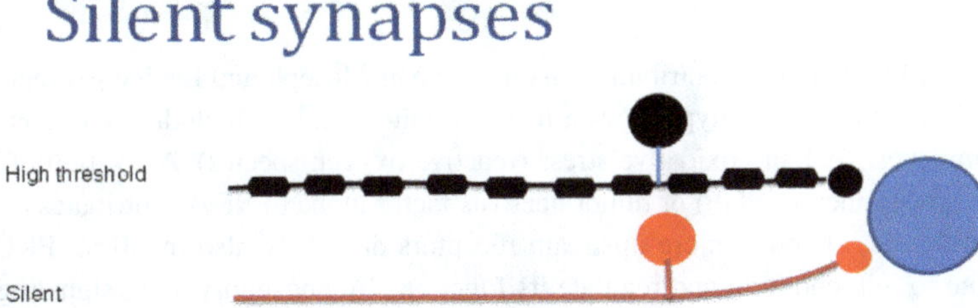

Figure 7.   Silent nociceptors. Spinal dorsal horn neurons receive different types of sensory fibers. Some of these fibers may be 'silent' due to the absence of functional postsynaptic receptors to respond to glutamate. These silent receptors may exist in adult synapses, although they may be difficult to detect by using the electrode placed in the soma of dorsal horn neurons.

Mitochondrial contribution to peripheral sensitization

Recent studies showed that the mitochondrion is a downstream element of the pathway through which PKCε generates mechanical hyperalgesia [26]. Inhibition of two closely related mitochondrial functions — electron transport (complexes I–V) and oxidative stress (reactive oxygen species) — selectively attenuated the mechanical hyperalgesia (Figure 8). The

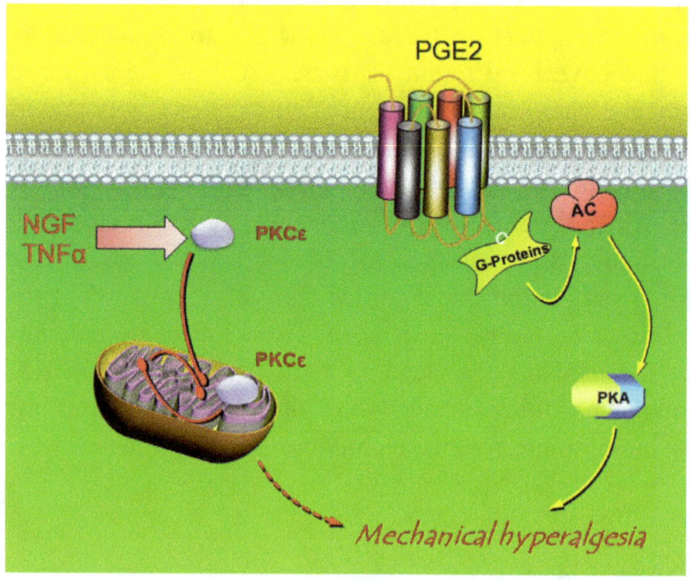

Figure 8.   Mitochondrial contribution to chronic pain. Joseph and Levine [3] report that PKCε-induced mechanical hyperalgesia involves mitochondria, including electron transport (complexes I–V) and oxidative stress (reactive oxygen species). Activation of PKCε by nerve growth factor (NGF) or tumor necrosis factor alpha (TNFα) contributes to priming hyperalgesia, although more upstream receptors are likely also involved. PKCε can translocate to mitochondria and regulate its functions by phosphorylating signaling molecules, such as respiratory chain proteins. The exact molecular mechanism connecting mitochondria to mechanical hyperalgesia remains to be studied. Short-lasting hyperalgesia induced by intradermal injection of PGE2 is likely mediated by a G protein-regulated cAMP signaling pathway and downstream activation of cAMP-dependent protein kinase (PKA). This PGE2-induced short-lasting hyperalgesia apparently does not involve a mitochondrial contribution (adapted from Zhuo [27]).

PKCε-dependent form of mechanical hyperalgesia induced by PGE2 was not attenuated by inhibition of mitochondrial function. These studies suggest that at least two downstream signaling pathways mediate the hyperalgesia induced by activating PKCε. Mitochondria apparently contribute selectively to the long-term hyperalgesia.

Conclusions

Integrative neurobiological approaches have provided us with new information for our understanding of pain biology. Many key proteins and ion channels that are critical for the physiological process of sensory stimuli have been identified and characterized. While the use of receptor antagonists for these receptors may cause side effects in patients, it has become clear that the study of the sensitization, or modification, of these sensory receptors may reveal new mechanisms for peripheral sensitization of nociceptors. Selectively inhibiting or preventing sensitization without affecting normal responses of nociceptors to sensory noxious stimuli will be a challenge for all sensory neuroscientists who are interested in developing any novel drug for treating chronic pain that is dependent on peripheral sensitization.

References

[1] Nassar, M.A. *et al.* (2005) Neuropathic pain develops normally in mice lacking both Na(v)1.7 and Na(v)1.8. *Mol Pain* 1, 24. 10.1186/1744-8069-1-24.

[2] Peirs, C. *et al.* (2021) Mechanical allodynia circuitry in the dorsal horn is defined by the nature of the Injury. *Neuron* 109, 73–90 e77. 10.1016/j.neuron.2020.10.027.

[3] Caterina, M.J. and Julius, D. (2001) The vanilloid receptor: A molecular gateway to the pain pathway. *Annu Rev Neurosci* 24, 487–517. 10.1146/annurev.neuro.24.1.487.

[4] Caterina, M.J. *et al.* (2000) Impaired nociception and pain sensation in mice lacking the capsaicin receptor. *Science* 288, 306–313. 10.1126/science.288.5464.306.

[5] Colburn, R.W. *et al.* (2007) Attenuated cold sensitivity in TRPM8 null mice. *Neuron* 54, 379–386. 10.1016/j.neuron.2007.04.017.

[6]   Peier, A.M. *et al.* (2002) A TRP channel that senses cold stimuli and menthol. *Cell* 108, 705–715. 10.1016/s0092-8674(02)00652-9.

[7]   Story, G.M. *et al.* (2003) ANKTM1, a TRP-like channel expressed in nociceptive neurons, is activated by cold temperatures. *Cell* 112, 819–829. 10.1016/s0092-8674(03)00158-2.

[8]   McKemy, D.D. (2005) How cold is it? TRPM8 and TRPA1 in the molecular logic of cold sensation. *Mol Pain* 1, 16. 10.1186/1744-8069-1-16.

[9]   Bautista, D.M. *et al.* (2007) The menthol receptor TRPM8 is the principal detector of environmental cold. *Nature* 448, 204–208. 10.1038/nature05910.

[10] Abrahamsen, B. *et al.* (2008) The cell and molecular basis of mechanical, cold, and inflammatory pain. *Science* 321, 702–705. 10.1126/science.1156916.

[11] Chung, M.K. and Caterina, M.J. (2007) TRP channel knockout mice lose their cool. *Neuron* 54, 345–347. 10.1016/j.neuron.2007.04.025.

[12] Dhaka, A. *et al.* (2007) TRPM8 is required for cold sensation in mice. *Neuron* 54, 371–378. 10.1016/j.neuron.2007.02.024.

[13] Gold, M.S. and Gebhart, G.F. (2010) Nociceptor sensitization in pain pathogenesis. *Nat Med* 16, 1248–1257. 10.1038/nm.2235.

[14] Chen, T. *et al.* (2010) Spinal microglial motility is independent of neuronal activity and plasticity in adult mice. *Mol Pain* 6, 19. 10.1186/1744-8069-6-19.

[15] Cockayne, D.A. *et al.* (2000) Urinary bladder hyporeflexia and reduced pain-related behaviour in P2X3-deficient mice. *Nature* 407, 1011–1015. 10.1038/35039519.

[16] Sun, Y.G. and Chen, Z.F. (2007) A gastrin-releasing peptide receptor mediates the itch sensation in the spinal cord. *Nature* 448, 700–703. 10.1038/nature06029.

[17] Sun, Y.G. *et al.* (2009) Cellular basis of itch sensation. *Science* 325, 1531–1534. 10.1126/science.1174868.

[18] Ross, S.E. *et al.* (2010) Loss of inhibitory interneurons in the dorsal spinal cord and elevated itch in Bhlhb5 mutant mice. *Neuron* 65, 886–898. 10.1016/j.neuron.2010.02.025.

[19] Koga, K. *et al.* (2011) Glutamate acts as a neurotransmitter for gastrin releasing peptide-sensitive and insensitive itch-related synaptic transmission in mammalian spinal cord. *Mol Pain* 7, 47. 10.1186/1744-8069-7-47.

[20] Bhave, G. and Gereau, R.W.t. (2004) Posttranslational mechanisms of peripheral sensitization. *J Neurobiol* 61, 88–106. 10.1002/neu.20083.

[21] Hucho, T. and Levine, J.D. (2007) Signaling pathways in sensitization: Toward a nociceptor cell biology. *Neuron* 55, 365–376. 10.1016/j.neuron.2007.07.008.

[22] Moriyama, T. *et al.* (2005) Sensitization of TRPV1 by EP1 and IP reveals peripheral nociceptive mechanism of prostaglandins. *Mol Pain* 1, 3. 10.1186/1744-8069-1-3.

[23] Zhang, X. *et al.* (2008) Proinflammatory mediators modulate the heat-activated ion channel TRPV1 via the scaffolding protein AKAP79/150. *Neuron* 59, 450–461. 10.1016/j.neuron.2008.05.015.

[24] Bhave, G. *et al.* (2002) cAMP-dependent protein kinase regulates desensitization of the capsaicin receptor (VR1) by direct phosphorylation. *Neuron* 35, 721–731. 10.1016/s0896-6273(02)00802-4.

[25] Wall, P.D. (1988) Recruitment of ineffective synapses after injury. *Adv Neurol* 47, 387–400.

[26] Zhuo, M. (2010) Mitochondrial connection in chronic pain. *Pain* 150, 1–2. 10.1016/j.pain.2010.04.019.

[27] Reichling, D.B. and Levine, J.D. (2009) Critical role of nociceptor plasticity in chronic pain. *Trends Neurosci* 32, 611–618. 10.1016/j.tins.2009.07.007.

Chapter 3

Spinal Dorsal Horn Synaptic Transmission and Gate Theory

News and Views

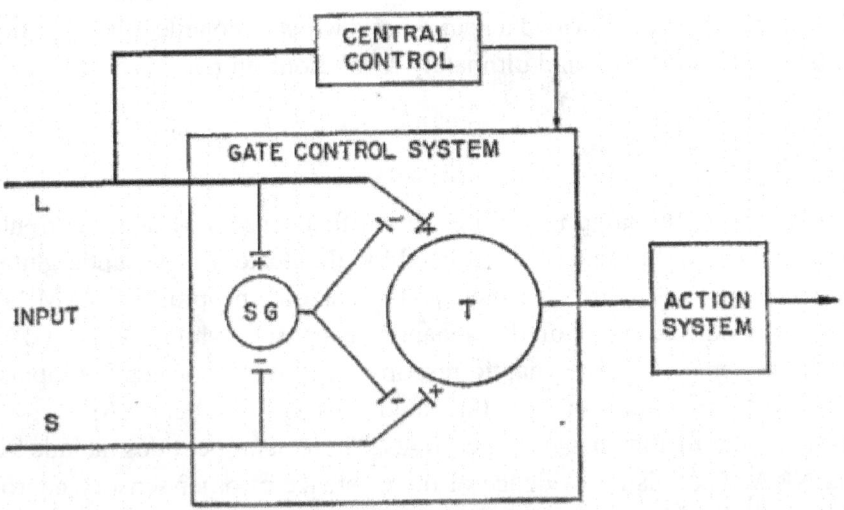

Fig. 4. Schematic diagram of the gate control theory of pain mechanisms: L, the large-diameter fibers; S, the small-diameter fibers. The fibers project to the substantia gelatinosa (SG) and first central transmission (T) cells. The inhibitory effect exerted by SG on the afferent fiber terminals is increased by activity in L fibers and decreased by activity in S fibers. The central control trigger is represented by a line running from the large-fiber system to the central control mechanisms; these mechanisms, in turn, project back to the gate control system. The T cells project to the entry cells of the action system. +, Excitation; −, inhibition (see text).

The gate control theory of pain describes how non-painful sensations can override, and even reduce, painful sensations. A painful, nociceptive stimulus stimulates primary afferent fibers and travels to the brain via transmission cells. Increasing the activity of the transmission cells results in increased perceived pain. Conversely, decreasing the activity of transmission cells reduces perceived pain. In the gate control theory, a closed "gate" describes when input to transmission cells is blocked, therefore reducing the sensation of pain. An open "gate" describes when input to transmission cells is permitted, therefore allowing the sensation of pain.

First proposed in 1965 by Ronald Melzack and Patrick Wall, the gate theory offers a physiological explanation for the previously observed effect of psychology on pain perception. Combining early concepts derived from the specificity theory, and the peripheral pattern theory, the gate control theory is considered to be one of the most influential theories of pain. This theory provided a neural basis which reconciled the specificity and pattern theories and ultimately revolutionized pain research.

Summary

Whole-cell patch-clamp recordings reveal that sensory synaptic currents in the spinal cord are mainly mediated by glutamate. Postsynaptic glutamate α-amino-3-hydroxy-5-methyl-4-isoxazole propionate (AMPA) receptor mediated most of the synaptic responses, while kainate (KA) receptors contributed to synaptic responses receiving nociceptive inputs. Repetitive stimulation of high-threshold nociceptive fibers also triggers neuropeptide-mediated synaptic currents. These neuropeptides include SP and NKA. There is no evidence of the existence of other sensory neurotransmitters. Synaptic excitatory transmission is biphasically modulated by G protein-coupled receptors; both presynaptic and postsynaptic mechanisms are likely involved. For postsynaptic regulation, G protein-coupled receptors — cholinergic, serotonergic, and adrenergic — are involved. Presynaptic regulation can be mediated by different receptors, including ATP P2X and KA receptors. Activation of KA receptor regulates spinal inhibitory transmission as well; both GABA and glycine had mediated responses. Inhibition or the reduction of these modulations can produce either analgesic or facilitation of behavioral nociceptive responses in freely moving animals.

Keywords: Kainate receptor; glutamate; SP; NKA; gate control; presynaptic regulation; mGluRs; spinal dorsal horn; nociception; whole-cell patch-clamp

Introduction

Spinal cord dorsal horn is the first synapse of the CNS. It has received much attention because it is functionally simpler than the synapses in the brain. It is believed that the spinal cord circuit is simple and the understanding of it will help us to develop analgesics for the treatment of chronic pain. However, the reality is that it is not as simple as we think. The neuronal information travels through the spinal cord in a complicated manner, and the subsequent neuronal activities quickly "leak" into the supraspinal structures, inducing long-term plastic changes. Even at the spinal cord level, synaptic transmission and modulation are very sophisticated, and the roles of "pain" transmitters are often biphasic. This chapter will review recent progress made in this area.

Glutamate is the major excitatory transmitter

Neurons in the spinal cord dorsal horn, and related areas, receive sensory inputs, including noxious information, and convey them to supraspinal structures. Studies using pharmacological and behavioral approaches show that glutamate and neuropeptides, including SP, are excitatory transmitters for pain [1–5]. Electrophysiological investigation of sensory synaptic responses between primary afferent fibers and dorsal horn neurons provides evidence that glutamate is the principal fast excitatory transmitter, and synaptic responses are mediated by postsynaptic glutamate receptors [6–9]. While AMPA receptors mediate the largest component of postsynaptic currents (Figure 1), KA receptors preferentially contribute to synaptic responses induced by higher (noxious) stimulation intensities (Figure 1). Consistent with this, antagonism of both KA and AMPA receptors yields greater analgesic effects in adult animals than AMPA receptor

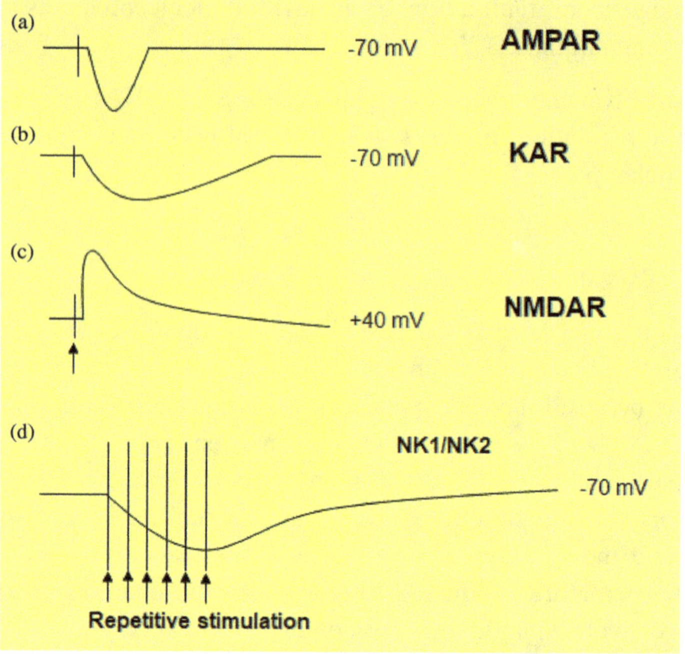

Figure 1.   Summary of spinal excitatory transmission mediated by glutamate and neuro-peptides. At spinal dorsal horn excitatory synapses, AMPA receptor mediated fast EPSCs (a), while KA receptor mediated relatively small and slow EPSCs (b). In some silent synapses, only postsynaptic NMDA receptors are available to detect the release of gluta-mate (c). Repetitive stimulation triggers slower EPSCs that are insensitive to the inhibition of glutamate receptors, and these slow EPSCs were mediated by neuropeptide receptors.

antagonism alone. These findings suggest that sensory modality may be coded, in part, by different postsynaptic neurotransmitter receptors.

In addition to glutamate, ATP is also proposed as another fast trans-mitter for nociception. In the spinal dorsal horn, several ATP receptor subtypes including P2X2,4,6 receptor subtypes are expressed in presynap-tic terminals of primary afferent fibers and postsynaptic dorsal horn neu-rons [7,10–15]. Although a presynaptic action of ATP P2X receptors on glutamate release was recently reported, the involvement of postsynaptic P2X receptors in synaptic transmission between primary afferent fibers and dorsal horn neurons is still unclear. ATP may serve as a neuromodula-tor to enhance glutamate-induced responses by a postsynaptic mechanism or induce postsynaptic currents in a subpopulation of dorsal horn neurons.

We used *in vitro* whole-cell patch-clamp recording techniques and *in vivo* behavioral tests to examine the role of P2X receptors in sensory synaptic transmission and behavioral nociception. To induce synaptic responses in the superficial dorsal horn, stimulating electrodes were placed at the dorsal root entry zone (DREZ) or attached dorsal root nerves. EPSCs were recorded from dorsal horn neurons in both cases. Direct stimulation of the dorsal root nerves induced monosynaptic EPSCs at low intensities; however, polysynaptic responses were often recorded when the intensity was increased. To stimulate a smaller population of afferent fibers, we placed stimulating electrodes near the DREZ. Short and constant latency fast EPSCs were induced. These EPSCs were monosynaptic because the response latency did not change with increasing intensities of electrical stimulation.

To test if ATP may also be involved in fast synaptic transmission, we carried out experiments with two types of glutamate receptor antagonists, 6-cyano-7-nitroquinoxaline-2,3-dione (CNQX; 10 μM) for AMPA/KA receptors and D(-)-2-amino-5-phosphonovaleric acid (AP5; 50 or 100 μM) for NMDA receptors, to block glutamate receptor-mediated synaptic responses. If there are other fast transmitters involved, we should have been able to detect some residual currents. In a total of 56 neurons tested, fast EPSCs were completely blocked by co-application of CNQX and AP-5; 28 neurons were successfully labeled. They were located in laminae I, II, and III. Similar results were obtained with direct stimulation of dorsal root nerves. The intensity of electrical stimulation used was strong enough to activate high-threshold nociceptive unmyelinated C fibers because at the same intensity slow, substance P/neurokinin A-mediated synaptic currents were induced when six pulses of stimuli were delivered at 25 Hz.

Postsynaptic KA receptors contribute to synaptic transmission

Previous studies have shown that there are two components of transmission at these synapses: a rapid component, blocked by the non-NMDA-receptor antagonist CNQX, and a slower component, mediated by AP5-sensitive NMDA receptors. Because CNQX blocks both AMPA and

KA receptors, this earlier work did not resolve the different contributions of these two receptor subtypes. To address this question, we obtained whole-cell patch-clamp recordings from neurons in the superficial dorsal horn of the spinal cord, which receives afferent inputs from primary sensory neurons in the periphery. Fast, monosynaptic EPSCs were induced by electrical stimulation delivered to the DREZ (Figure 2). We blocked NMDA receptors with the selective inhibitor AP5 (100 M) and tested the action of two selective, non-competitive AMPA-receptor antagonists, GYKI 53655 and SYM 2206. Each drug was used at 100 M, a concentration that produces maximal inhibition of AMPA receptors (half-maximal

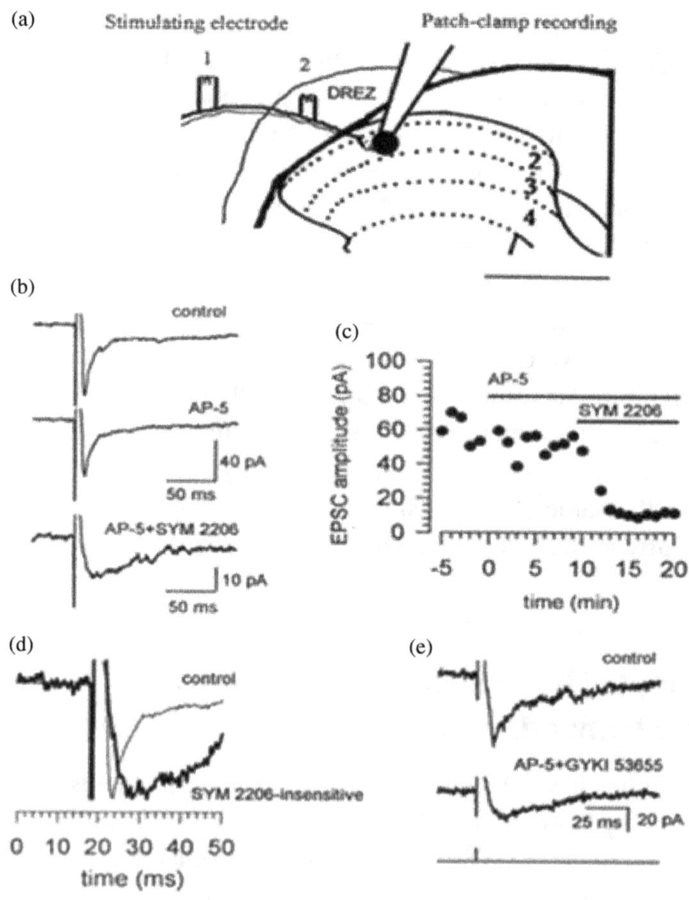

inhibitory concentration = 1–2 M) but less than 20–30% inhibition of KA receptors [8].

Superfusion of slices with AP5 plus GYKI 53655 or AP5 plus SYM 2206 reduced, but did not completely block, the EPSCs. In contrast, AP5 plus 10 M CNQX or 2,3-dihydroxy-6-nitro-7-sulphamoylbenzo quinoxaline (NBQX) always produced almost complete inhibition of EPSCs. For EPSCs studied before and after exposure to AMPA-receptor antagonists, the relative peak amplitude of the residual EPSC (the EPSC remaining after exposure to AMPA-receptor antagonists) was 31.3% of control. In two cases, the peak currents after superfusion of GYKI 53655 or SYM 2206 were small. The rise (10–90%) and decay time constant for residual EPSCs were significantly longer than that for AMPA-receptor-mediated currents. To confirm that the residual currents were mediated by KA receptors, we used SYM 2081, a selective KA-receptor-agonist that causes potent receptor desensitization. SYM 2081 (1 or 5 M) significantly decreased or abolished the residual EPSCs in the presence of AP5 and SYM 2206 but did not affect the fast component of control EPSCs

◄──

Figure 2 (*Continued*).   Excitatory monosynaptic responses in the superficial dorsal horn of the lumbar spinal cord are mediated by glutamatergic AMPA/KA receptors and NMDA receptors. (a) EPSCs induced electrical stimulation of the dorsal root nerves. Five responses induced by stimulation at a low intensity were displayed (with a 20-s interval). The time of stimulation is indicated by an arrow. (b) Evoked EPSCs induced electrical stimulation of the dorsal root nerves. In the same experiment as A, electrical stimulation at a high intensity elicited polysynaptic responses. Two responses induced at a 20-s interval were displayed. (c) EPSCs induced by electrical stimulation of dorsal root nerves were completely blocked by application of both an AMPA/KA receptor antagonist CNQX (10 sM) and an NMDA receptor antagonist AP-5 (50 μM). (d) Evoked monosynaptic EPSCs induced by placing a stimulating electrode at the DREZ in the dorsal horn. The time of stimulation is indicated by an arrow. Five responses induced at a 20-s interval were displayed. One response of reduced magnitude was induced by stimulation at a low intensity. No polysynaptic response was observed. (e) EPSCs induced by stimulation at the DREZ were completely blocked by application of both an AMPA/KA receptor antagonist CNQX (10 μM) and an NMDA receptor antagonist AP-5 (50 μM). (f) Distribution of cells, which were labeled and from which no residual current was recorded after the blockade of glutamatergic receptors, illustrated on a representative coronal spinal cord slice (Paxinos and Watson 1997). Twenty-eight of 56 cells were successfully labeled (adapted from Li *et al.* [8]).

recorded in AP5 alone. Cyclothiazide (10 M), which potentiates currents mediated by AMPA receptors but not by KA receptors, did not affect SYM-2206-insensitive EPSCs. These results indicate that the residual EPSCs were mediated by KA receptors.

In young rats, sensory neurons in the superficial dorsal horn of the spinal cord receive both low- and high-threshold afferents. To determine whether KA and AMPA receptors co-exist at these synapses, we used low-intensity stimulation, which activates a minimal number of presynaptic fibers. If KA and AMPA receptors always co-localize, then EPSCs evoked by minimal stimulation should include both AMPA- and KA-receptor-mediated components. However, only rapidly decaying EPSCs, typical of the AMPA-receptor-mediated component, were observed following low-intensity stimulation. The addition of GYKI 53655 or SYM 2206 blocked these EPSCs completely, even when a brief train of low-intensity stimulation was delivered (25 Hz, 6 pulses; Figure 3). These results raise the possibility that KA receptors might be displayed selectively at synapses formed by high-threshold afferents and not at synapses that receive low-threshold afferent inputs. Because accurate measurement of EPSC latency is difficult in thin-slice preparations, we studied the dependence of synaptic responses on stimulation intensity more systematically, using intensities that yield activation of different primary afferent fiber types in adult rats. The amplitude of SYM-2206-resistant EPSCs increased disproportionately with stimulus intensity (Figure 3). Thus, a significantly greater percentage of the EPSC was resistant to block by SYM 2206 during stimulation at high intensities, which were sufficient to activate A_δ and/or C fibers *in vivo*, than was the case for lower intensity stimulation (Figure 3).

KA receptor-mediated sensory responses in spinothalamic projection cells

To determine whether KA receptors contribute to sensory transmission in ascending pathways, we injected a retrogradely transported fluorescent tracer into the thalamus and, two or three days later, recorded from labeled cells in spinal slices [8]. Many cells in the superficial dorsal horn were labeled (Figure 4). Recordings were obtained from four labeled cells

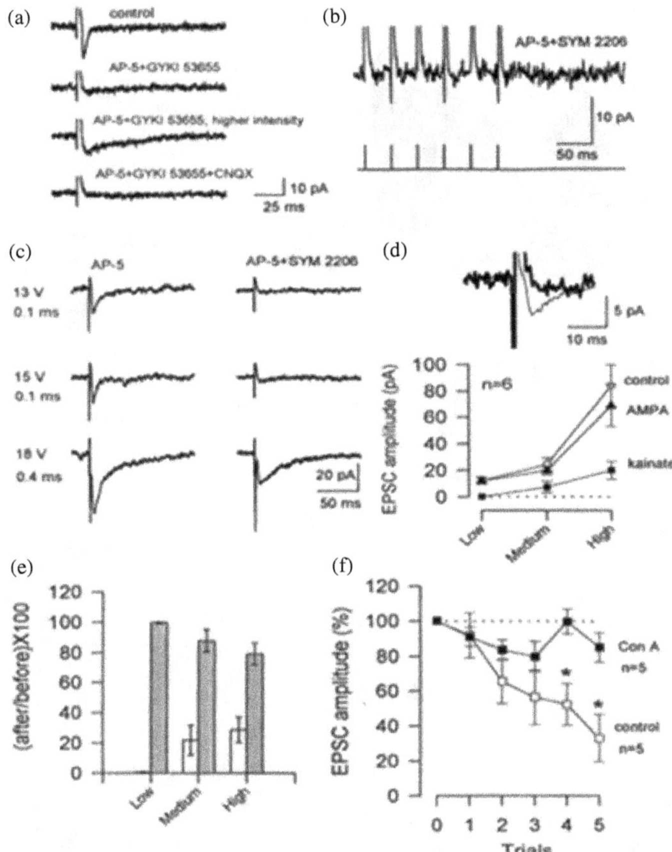

Figure 3. KA receptors reside at synapses receiving high-threshold afferents. (a) Stimulation at low intensities induced GYKI-53655-sensitive AMPA currents (top two traces). At a higher intensity, a GYKI-53655-resistant current was induced and could be blocked by 10 μM CNQX (bottom two traces). (b) Repetitive stimulation in the presence of AP5 and SYM 2206 did not elicit KA currents. (c) Single traces of EPSCs induced by stimulation at different intensities in 100 M AP5 or AP5 plus 100 μM SYM 2206. (d) (Bottom) Summary of within-cell data from c (AP5, $n = 6$, open squares; AP5 plus SYM 2206, $n = 6$, filled squares). (Inset) The control EPSC at low-stimulation intensity, scaled to 30% of its original amplitude (thin line) and the EPSC in SYM 2206 (thick line); single traces are shown. (e) SYM-2206-sensitive (AMPA; shaded bars) and -resistant (KA; open bars) EPSCs as a percentage of control EPSCs. Differences between the means within each set are significant ($P < 0.01$). Responses to low intensity are different from responses to medium and high intensities by pairwise comparison ($P < 0.05$). AMPA component is the control EPSCs–KA component. (f) Pretreatment with conA (2 μM) reduced the decay of KA currents during 1-Hz stimulation. Asterisk indicates $P < 0.05$ (adapted from Li. *et al.* [8]).

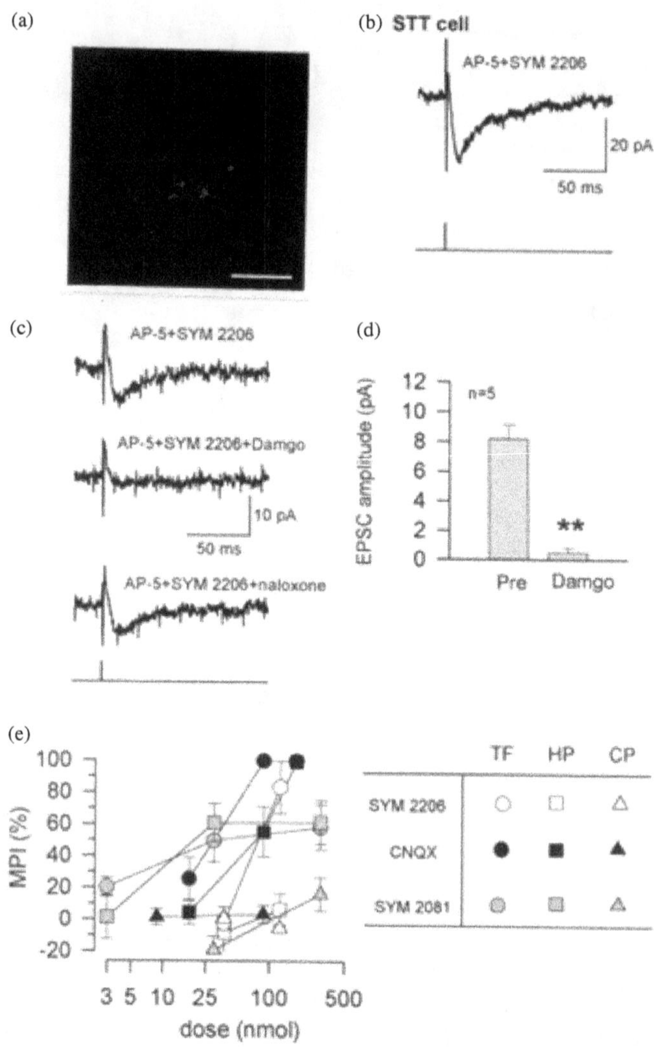

Figure 4. KA receptors contribute to ascending sensory transmission. (a) A fluorescence photomicrograph of a dorsal horn cell that was labeled by fluorescent latex microspheres (light areas) 2 days after thalamic injection. (b) An EPSC from a labeled spinothalamic tract cell in 100 M AP5 and 100 M SYM 2206. (c) EPSCs in AP5 plus SYM 2206, in AP5, SYM 2206, and Damgo (1 μM), and in AP5, SYM 2206, and naloxone (20 μM). (d) Mean amplitude (s.e.m.) of KA EPSCs before and after application of Damgo in the presence of AP5 and SYM 2206 ($n = 5$). Double asterisk indicates $P < 0.001$. (e) Antinociceptive effects of intrathecal administration of antagonists ($n = 3$–7 for each dose). TF, tail flick; HP, hot plate; CP, cold plate (adapted from Li *et al.* [8]).

(for 10–15 min), all of which exhibited KA-receptor-mediated EPSCs upon stimulation of the DREZ or dorsal root nerves (Figure 4).

Opiates produce antinociceptive effects by inhibiting sensory transmission in the dorsal horn. Damgo (1 μM), a selective-opiate-receptor agonist, blocked KA-receptor-mediated EPSCs (Figure 4). To study the role of KA receptors in nociception further, we carried out behavioral experiments (Figure 4). Intrathecal administration of CNQX produced dose-dependent antinociception in the TF and HP tests. SYM2206 had no effect in the HP test but inhibited the TF reflex at the highest dose. SYM2081 was antinociceptive in both the TF and the HP tests. Latency in the cold-plate test was not affected by any of the non-NMDA-receptor antagonists. The NMDA-receptor antagonists MK801 (7 nmol) and AP5 (25 nmol) increased latency in the cold-plate test, although motor side effects were also observed.

These findings indicate that spinal neurons express functional KA receptors which contribute to synaptic transmission between primary afferent fibers and dorsal horn neurons. In contrast to previous results obtained with hippocampal mossy-fiber synapses, substantial KA-receptor-mediated EPSCs were observed following individual, low-frequency, high-intensity stimuli. Low-threshold afferents appear to form synapses with few postsynaptic KA receptors, whereas synapses that include both AMPA and KA receptors are likely to receive inputs with a high activation threshold from, for example, C and/or A_δ fibers. Moreover, different subsets of postsynaptic glutamatergic receptors may be useful in promoting behavioral responses to noxious heat and cold. Although glutamatergic synapses mediate pain sensations at all intensities, recent studies indicate that SP and/or NKA is important in sensing intense pain. Thus, it appears that both intensity and modality of pain are coded by different subtypes of postsynaptic receptors and by the release of different transmitters from primary afferent fibers.

Neuropeptide-mediated synaptic responses induced by repetitive stimulation

In addition to glutamate, several neuropeptides including SP are thought to act as sensory transmitters. For many years, there was a lack of electrophysiological evidence to show that SP can mediate monosynaptic responses due to the fact that SP-mediated responses have a very slow onset. Recent studies using whole-cell patch-clamp recordings reveal relatively faster SP and neurokinin A-mediated synaptic currents in the synapses between primary afferent fibers and EPSCs. The amplitudes of the residual currents were small (approximately 50-fold smaller) as compared to the AMPA/KA receptor-mediated responses to single shock stimulation of primary afferent fibers. However, sensory synapses in the spinal dorsal horn are normally activated *in vivo* by a train of stimulation. *In vivo* electrophysiological recordings from primary afferent fibers showed that nociceptive fibers fire spikes at high frequencies in response to peripheral noxious stimulation (e.g., noxious heating of the skin at 49°C). Thus, in order to mimic this pattern of primary afferent fiber activity, we applied a short train of stimuli at different frequencies (15, 20, or 25 Hz; containing six pulses, delivered at 3-min intervals) in the presence of ionotropic glutamate receptor antagonists. As shown in Figure 5, a summation of individual EPSC responses was observed.

A total of 41 experiments were performed. While some responses (in 24 neurons) peaked at the end of the last stimulus, in 17 experiments, the summated response reached a peak between the second and sixth stimulation pulse [9]. The mean peak currents were 14 pA, significantly smaller than the EPSCs recorded before perfusion of CNQX and AP5 in the same cells (mean 49 pA). The summated, residual currents therefore represented approximately 28% of the total current synaptic current during 25-Hz stimulation. Because the residual currents had slow onset and decay phases, we also estimated the area of responses with a single pulse of stimulation before application of CNQX and AP5 as well as ones with six pulses of stimulation after CNQX and AP5. Interestingly, the total positive charge transfer in residual EPSCs (the estimated mean of 2670 ms.pA) was comparable to that in fast, glutamate-mediated EPSCs (2620 ms.pA).

SP produces its effects mainly by binding to NK_1 receptors in the spinal cord. Due to potential concerns over the selectivity and effectiveness of NK_1 receptor antagonists, we tested three different NK_1 receptor antagonists: GR82334 (5 μM), WIN-51,708 (10 μM), or L-733060 (300 nM). Application of these antagonists for 10 min effectively abolished residual

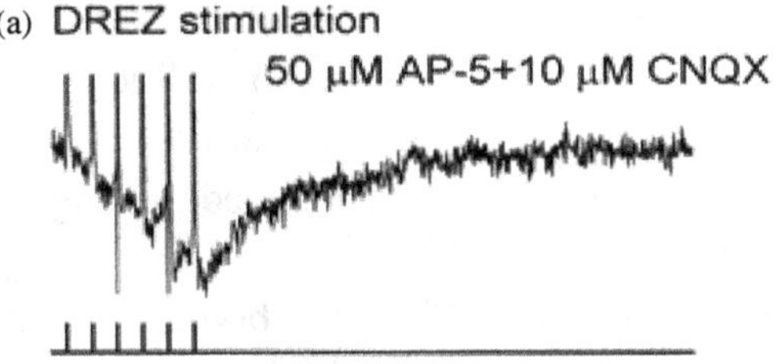

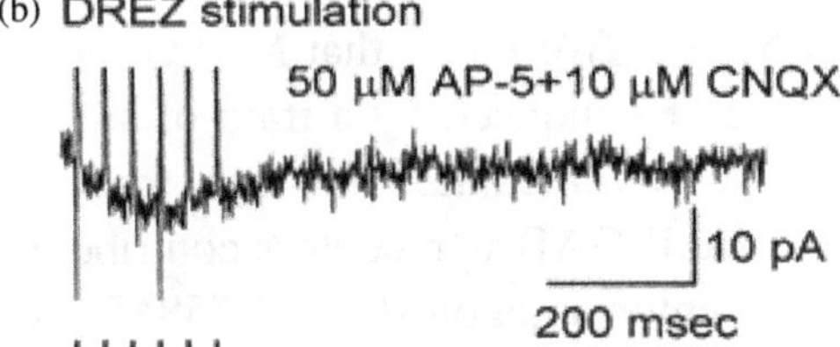

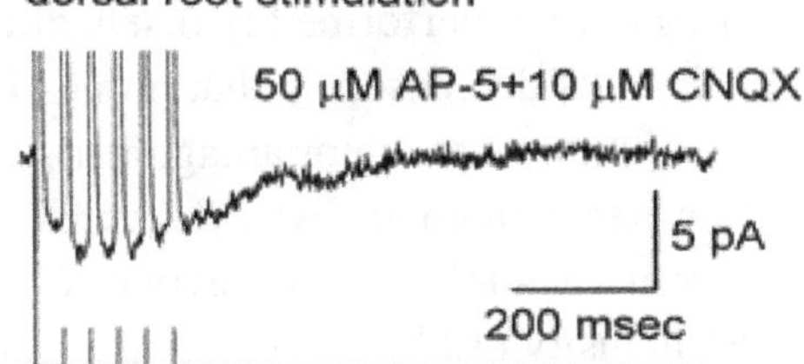

Figure 5. Synaptic transmission in the superficial dorsal horn of the spinal cord following a train of stimulation. (a,b) EPSCs were induced by 25-Hz stimulation (six pulses) after application of CNQX (10 μM) and AP-5 (50 μM) to abolish glutamatergic EPSCs; 25-Hz stimulation is meant to resemble the firing frequency of C fibers in response to peripheral noxious stimulation. In (b), the summated EPSC peaked at the third pulse. (c) An EPSC induced by 25-Hz stimulation in the dorsal root in the same conditions. Abbreviation: DREZ, dorsal root entry zone (adapted from Li and Zhuo [9]).

currents in 14 of 20 experiments. In the six experiments where NK_1 antagonists did not affect the residual currents, the NK_2 receptor antagonists MEN-10,736 (1 μM) or L-659,877 (6.5 μM) blocked responses. In five additional experiments, NK_2 receptor antagonists, applied before any NK_1 receptor antagonist, had no effect on residual currents. Subsequent application of NK_1 receptor antagonists blocked responses in each cell. In all experiments, no residual response was observed after blockade of either NK_1 or NK_2 receptors, suggesting that the residual EPSCs were mediated by either NK_1 or NK_2 receptors. In addition, we perform experiments to test if MEN-10,736 affects glutamate-mediated currents. We found that glutamate-mediated EPSCs induced by single shocks were not significantly affected by MEN-10,736.

The expression of several different subtypes of metabotropic glutamate receptors (mGluRs) has been reported in the spinal cord dorsal horn. To test the possible involvement of mGluRs, we tested the effect of the mGluRs antagonist MCPG. We found that MCPG (500 μM) did not significantly affect the EPSCs induced by a train of stimulation, indicating that these currents are not mediated through metabotropic glutamate receptors. We also tested if $GABA_B$ receptors contributed to EPSCs using a selective $GABA_B$ receptor antagonist CGP 55845 (10 μM). We found that the EPSCs were not affected in the presence of CGP 55845.

In order to determine the stimulus intensity threshold for inducing summated, residual EPSCs, we performed two sets of experiments. First, the stimulation intensity was decreased so that no fast EPSC was induced (in the absence of any glutamate receptor antagonist). Then, a train of (25 Hz) stimulation at the same intensity was delivered. No response was found (see Figure 6 for an example). In the second set of experiments, we slightly increased the intensity (12 V, 0.1 ms), and a small, fast, glutamate receptor-mediated EPSC was induced (Figure 6). After complete blockade of synaptic EPSCs with 10 μM CNQX (50 μM AP-5 was also included), a train of 25-Hz stimulation was applied at the same intensity. No response was detected. However, a summated EPSC was observed when the stimulation intensity was increased above the threshold for unmyelinated C fibers (18 V, 0.4 ms) (Figure 6).

Both NK_1 and NK_2 receptors are G protein-coupled receptors, and NK_1/NK_2 receptor-mediated effects depend on the activation of G protein signaling pathways. Postsynaptic application of a G protein inhibitor should abolish slow EPSCs if these EPSCs are mediated by NK_1/NK_2 receptors. As expected, GDP-β-S (2 mM) applied postsynaptically through the recording pipette eliminated residual EPSCs. Fast EPSCs induced by single-pulse stimuli in the same cells tested before application of CNQX (10 μM) and AP-5 (50 μM) were not significantly affected.

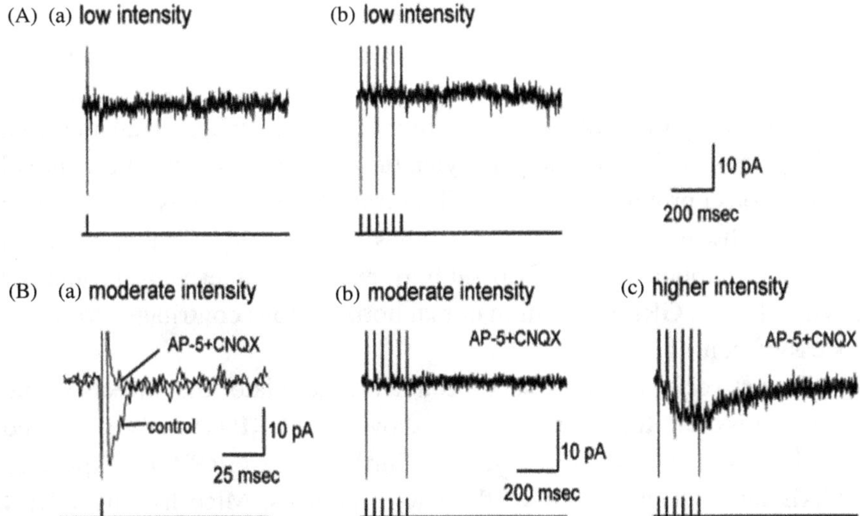

Figure 6.   The summated EPSCs induced by a stimulus train (6 pulses at 25 Hz) required greater intensities of stimulation than fast EPSCs induced by a single pulse of stimulation. (A) A single pulse of stimulation at a low intensity, which did not induce any fast EPSC and also did not induce any summated EPSCs when a train of 25 Hz stimulation was applied at the same intensity. (B) A single pulse of moderate-intensity stimulation induced a small fast EPSC which was completely blocked by the AMPA receptor antagonist CNQX. A train of 25 Hz stimulation at the same intensity did not induce any slow EPSC. However, a summated EPSC was induced in the same cell when the intensity was increased (adapted from Li and Zhuo [9]).

Taken together, these pharmacological data strengthen the hypothesis that the residual EPSCs are due to postsynaptic NK_1/NK_2 receptor activation.

Gastrin-releasing peptide (GRP) and itching

Recently, it has been reported that GRP may act as a transmitter for behavioral itch [16]. The evidence for GRP being an "itching" transmitter is mainly from anatomic studies. In these studies, GRP is expressed in some DRG cells, while GRP receptors (GRPR) are mainly expressed in spinal lamina I cells (presumably ascending projection cells). However, previous electrophysiological studies failed to observe any residual currents that may be mediated by non-SP/NKA-mediated currents [9,17]. For any spinal transmitter to be qualified as one, the critical criterion is to demonstrate the involvement of this transmitter in spinal synaptic transmission. If this proves to be true, spinal synaptic transmission may be a novel pathway for G protein-related signaling transduction in these itch-related specific pathways. In fact, recent studies indicate that GRP is released from spinal subpopulation of dorsal horn excitatory interneurons, and the activity of these GRP-containing dorsal horn neurons contributes to spinal process of itch.

Alternatively, GRP may serve as a neuromodulator that affects spinal itch transmission. Recent studies have shown that GRP is highly expressed in the lateral nucleus of the amygdala. Furthermore, GRPR are expressed in GABAergic interneurons of the lateral nucleus. Mice lacking GRPR show enhanced LTP and greater fear/anxiety responses. It remains to be determined whether GRP receptors may also be expressed in inhibitory neurons in the spinal cord and thus contribute to disinhibition of spinal sensory transmission or itch.

Presynaptic regulation by ATP receptor

Sensory transmitters or neuromodulators bind to their target receptors on the central terminals of DRG cells in the spinal cord dorsal horn [18]. Activation of these presynaptic receptors will lead to changes in the release of sensory transmitters in response to peripheral sensory stimulation. Many neurotransmitters and peptides have been reported to produce presynaptic regulatory effects in the DRG–dorsal horn synapses, such as ATP, serotonin, and opioids.

P2X receptors are highly expressed in DRG, and activation of presynaptic P2X receptors elicited glutamate release. Several drugs were used to block P2X receptors. Among them, pyridoxal-phosphate-6-azophenyl-2',4'-disulfonic acid (PPADS) is a selective antagonist for P2X receptors, and in this way, we investigated the possible presynaptic effects of PPADS. EPSCs to a paired-pulse stimulation at a 50-ms interval were recorded before and after application of 10 μM PPADS (PPADS at this dose did not affect other receptors) [7]. Interestingly, EPSCs were significantly decreased. This inhibition was accompanied by significant changes in the ratio of paired-pulse responses. PPADS produced its effect by blocking presynaptic P2X receptors. A previous study demonstrated that PPADS did not produce any antinociceptive effect in the TF test. We re-examined the effect of PPADS in the TF reflex and also carried out experiments with the HP test. Intrathecal administration of PPADS (0.6 or 5.9 μg) into the lumbar spinal cord produced no significant change in response latencies. The skin temperature of the tail was not significantly affected. At a higher dose (59 μg), PPADS induced abnormal aggressive behavior and made measurements impossible. Rats became hypersensitive to gentle touch and thus no more testing was able to be carried out. Intrathecal injection of vehicle saline (10 μl) affected neither the TF nor HP latency. Morphine, however, produced significant antinociceptive effects in the TF and HP tests when administered intrathecally at a similar quantity (10 μg).

Presynaptic regulation of excitatory transmission by KA receptor

Presynaptic KA receptors regulate the transmitter release from DRG cells onto spinal dorsal horn neurons [19]. Evoked transmission was inhibited by exposure to KA or ATPA in both dissociated cell cultures and intact slices (Figures 7 and 8). In addition, our results demonstrate a significant difference in the pharmacology of KA receptors expressed by DRG cells and spinal cord neurons. DRG cell KA receptors were sensitive to ATPA, whereas ATPA produced little or no response in dorsal horn neurons and was unable to cross-desensitize the currents evoked by KA. This difference provides one of the key pieces of evidence that KA receptor agonists act presynaptically to reduce evoked transmission. KA, but not ATPA, was effective at suppressing spontaneous action potential-dependent transmission at excitatory spinal synapses, whereas both agonists suppressed DRG spinal transmission. Thus, the selectivity of ATPA for DRG cell KA receptors strongly indicates a presynaptic locus of action in reducing evoked transmission at DRG spinal synapses (Figure 9) [20,21].

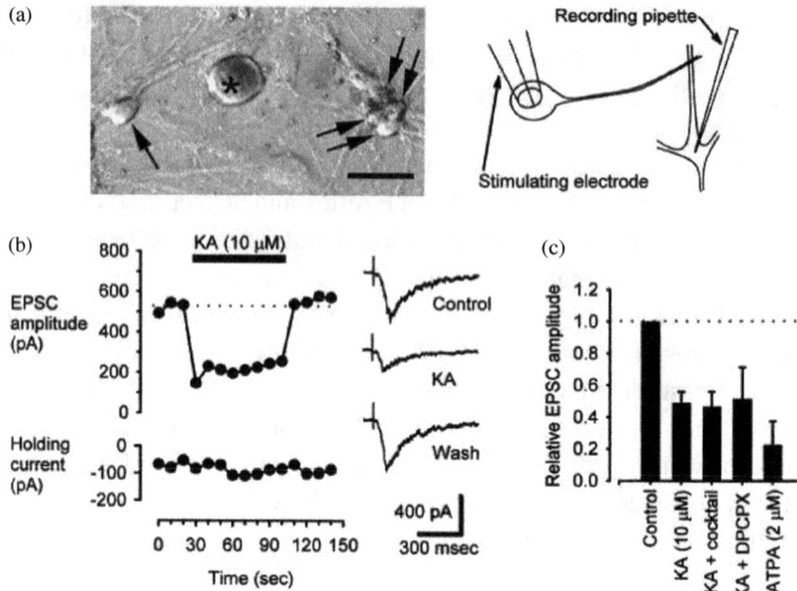

Figure 7. KA suppresses evoked excitatory transmission between DRG neurons and dorsal horn neurons in co-culture. (a) A photograph of co-cultured DRG (asterisk) and dorsal horn neurons (arrows) and a diagram showing placement of recording and stimulating electrodes illustrate the experimental system. Scale bar, 20 μm. (b) In a representative neuron, 10 μM KA reversibly reduced NMDA receptor-mediated EPSC amplitude. Traces at the right show the shapes of EPSCs before, during, and after KA treatment. This recording was from a neuron treated with the antagonist cocktail described in C. The dotted line indicates the baseline level. (c) Pooled data illustrate the relative amplitudes of NMDA receptor-mediated EPSCs in control conditions and after treatment with 10 μM KA (n = 8) or 2 μM ATPA (n = 6) alone; reduction of EPSC amplitude in both cases was statistically significant ($p < 0.05$; paired t-test comparing absolute EPSC amplitudes in control and test conditions) (adapted from Kerchner *et al.* [20]).

Postsynaptic regulation of sensory synaptic transmission

Neurotransmitters or neuromodulators bind to their receptors postsynaptically at spinal dorsal horn neurons. Activation of these postsynaptic receptors leads to changes in AMPA/KA receptor-mediated synaptic responses. They include acetylcholine, serotonin, opioids, norepinephrine, and

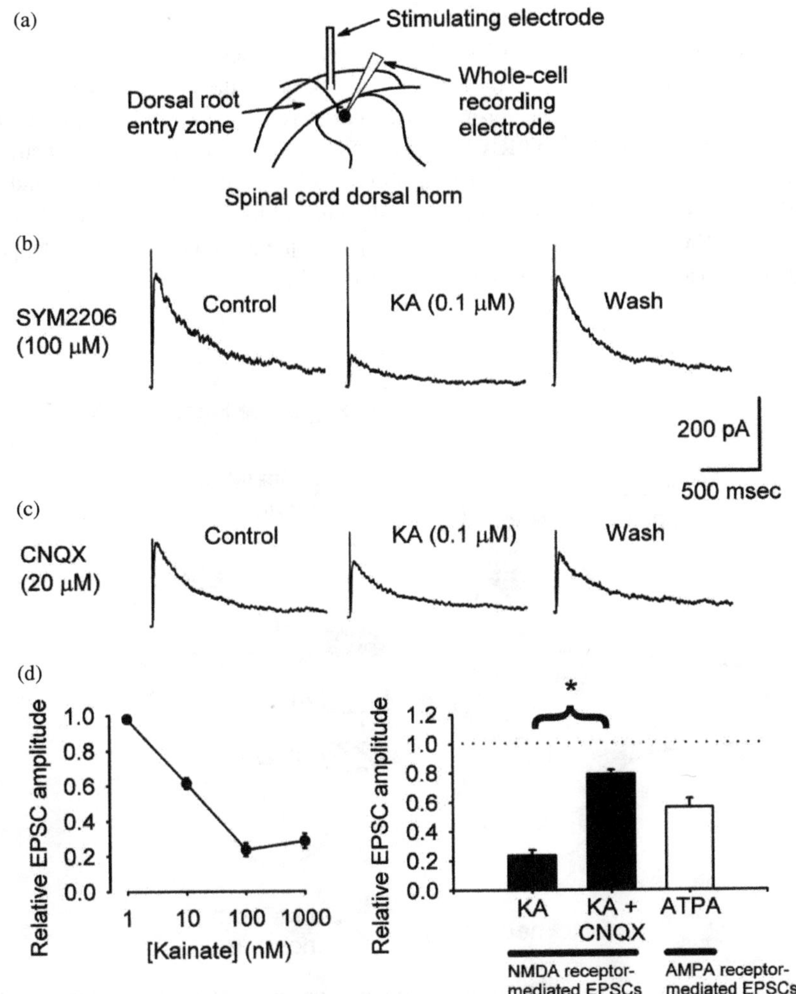

Figure 8. KA receptor suppresses primary afferent neurotransmission in spinal cord slices. (a) The diagram illustrates the placement of stimulating and recording electrodes in the dorsal horn of a spinal slice. (b) NMDA receptor-mediated EPSCs were isolated at a holding potential of +40 mV in the presence of 100 μM SYM2206. Traces from a representative neuron show reversible inhibition of EPSCs by 0.1 μM KA. (c) When 20 μM CNQX replaced SYM2206 in experiments, as in B, 0.1 μM KA had less of an effect. (d) KA exerted a dose-dependent inhibition of NMDA receptor-mediated EPSCs (n = 3–4 cells per concentration). (e) Summarized data showing inhibition of NMDA receptor-mediated EPSCs by 0.1 μM KA in the presence of 100 μM SYM2206 (KA; n = 3) that was

Figure 8 (*Continued*)   sensitive to 20 μM CNQX (KA + CNQX; $n = 3$). Treatment with 100 μM CNQX ($n = 3$) afforded no additional blockade of the effect of KA (data not shown). In addition, AMPA receptor-mediated EPSCs were suppressed by application of 2 μM ATPA ($n = 5$). ATPA- and KA-induced EPSC suppressions were both statistically significant ($p < 0.05$; paired t-test comparing absolute EPSC amplitudes in control and test conditions). *$p < 0.05$ indicates a significant difference in EPSC reduction by KA in the presence of SYM2206 versus CNQX; t-test comparing the percentage of EPSC suppression in each condition (adapted from Kerchner *et al.* [20]).

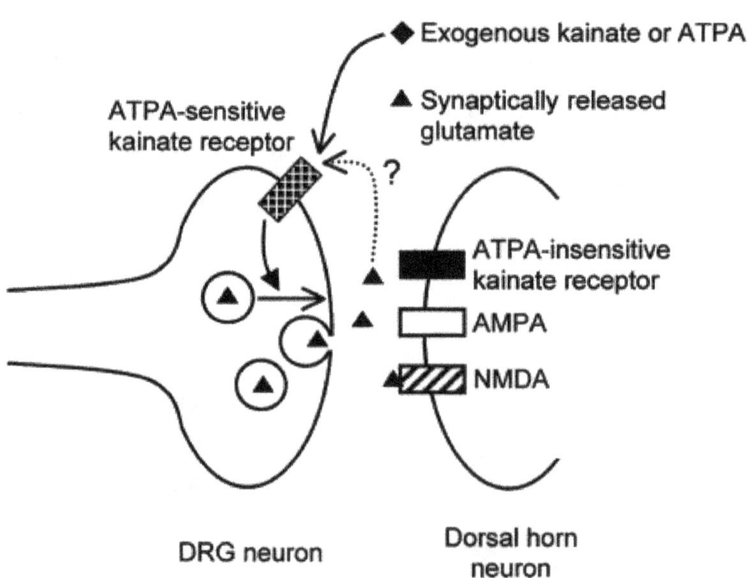

Figure 9.   KA autoreceptors at primary afferent synapses. The model illustrates the hypothesis that KA receptors, in addition to residing in the postsynaptic membrane of dorsal horn neurons, are also present on the presynaptic terminals of DRG neurons. Presynaptic and postsynaptic KA receptors could be distinguished in this study by their sensitivity to the GluR5-selective agonist ATPA, and activation of presynaptic KA receptors, at least by exogenous agonists, inhibited evoked release of glutamate. Unresolved is whether the presynaptic receptors can be activated by synaptically released glutamate (dotted line), and, if so, whether subsequent glutamate release would be inhibited or enhanced (adapted from Kerchner *et al.* [20]).

oxytocin. Spinal nociceptive transmission is modulated by an endogenous antinociceptive or analgesic system, consisting of the midbrain periaqueductal gray (PAG) and the rostral ventral medulla (RVM). The possible top-down modulatory systems from the cortex have been also reported recently. The RVM serves as an important relay for descending influences from the PAG to the spinal cord. Activation of neurons in the RVM inhibits spinal nociceptive transmission and behavioral nociceptive reflexes. The inhibitory effect is mediated directly by descending pathways projecting bilaterally in the dorsolateral funiculi and indirectly by descending activation of local spinal inhibitory neurons. In the spinal cord, muscarinic, noradrenergic, and serotonergic receptors are important for descending inhibition of behavioral nociceptive reflexes.

Electrophysiological studies using intracellular or whole-cell patch-clamp recordings of dorsal horn neurons allow for the investigation into the cellular mechanisms for the antinociceptive or analgesic effects induced by these transmitters. In anesthetized whole animals, electrical stimulation applied to sites within the nucleus raphe magnus or PAG produced inhibitory postsynaptic potentials (IPSPs) in dorsal horn neurons including ascending projection spinothalamic tract cells. More detailed pharmacological analysis came from studies using an *in vitro* brain/spinal cord slice preparation. In trigeminal nuclei, all three major transmitters, acetylcholine, serotonin, and norepinephrine, are reported to inhibit glutamatergic transmission. Consistent with their effects in trigeminal nuclei, application of the cholinergic receptor agonist carbachol produced a dose-dependent inhibition of glutamate-mediated EPSCs. Postsynaptic injection of two different types of G protein inhibitors, guanosine 5'-O-2-thiophosphate or guanosine 5'-O-3-thiotriphosphate, blocked the inhibition produced by carbachol. Clonidine, a selective alpha-adrenergic receptor agonist, also produced a dose-dependent inhibition of EPSC that was reduced by postsynaptic inhibition of G proteins. The inhibitory effect of serotonin was likewise mediated by postsynaptic G proteins. These results suggest that activation of postsynaptic neurotransmitter receptors plays a critical role in inhibition of glutamate-mediated sensory responses by acetylcholine, norepinephrine, and serotonin.

Presynaptic regulation of inhibitory transmission by KA

In the spinal cord dorsal horn, glutamate-containing sensory fiber terminals come into close proximity with the GABA- and glycine-containing boutons of local interneurons at synaptic glomeruli. In a recent study, we provide evidence that glutamate released from primary afferent sensory fibers can regulate spinal inhibitory transmission by activating KA receptors [22]. These data suggest that the heterosynaptic regulation of transmitter release by presynaptic ligand-gated ionic channels may be reciprocal between sensory fibers and dorsal horn interneurons. Because synaptically released glutamate suppressed evoked inhibitory transmission, it suggests that with sufficiently high levels of sensory input, inhibitory tone may be reduced, possibly facilitating the relay of sensory information to higher brain centers.

To test for the presence of presynaptic KA receptors at spinal dorsal horn inhibitory synapses, we examined the effect of KA receptor activation on spontaneous miniature inhibitory postsynaptic currents (mIPSCs). Recordings were obtained from dorsal horn neurons in a dissociated culture to allow for a more rapid and complete exchange of the extracellular medium than is possible during acute slice recordings. The speed of exchange is an important consideration because KA receptors exhibit prominent desensitization. Recordings were made in the presence of 500 nM tetrodotoxin (TTX), which blocked all voltage-gated Na^+ current, as well as SYM2206 (100 μM) and AP5 (25 μM) to block AMPA and NMDA receptors, respectively. Neurons were voltage-clamped at 0 mV and perfused intracellularly with a low-Cl^- pipette solution. In these conditions, mIPSCs appeared as outward currents at a background frequency of 3.8 ± 0.4 s^{-1}. Upon application of KA (10 μM), this frequency increased to 850% $\pm$ 100% of the control value during the first four seconds of the exposure, suggesting that KA acted at a presynaptic locus. A small but significant increase in mIPSC amplitude was also observed during KA exposure, but because of the prevalence of overlapping mIPSCs triggered by KA, it is likely that this small change in amplitude was biased by coordinated, multiquantal release events. Application of bicuculline (10 μM) plus strychnine (1 μM) eliminated mIPSCs, confirming that the postsynaptic events resulted from the release of GABA and/or glycine.

In addition, co-application of KA with the non-selective AMPA/KA receptor antagonist CNQX (100 μM) resulted in no significant change in mIPSC frequency compared to control.

KA receptors appeared to play a unique role in stimulating the release of inhibitory transmitters. AMPA receptor activation by AMPA (50–100 μM) or KA (10 μM) had no significant effect on mIPSC frequency when applied in the presence of TTX, AP5, and SYM2081 (1–10 μM), a selective KA receptor agonist that induces potent, complete receptor desensitization. Similarly, application of NMDA (10 μM) in the presence of TTX and SYM2206 also had no effect. Although presynaptic AMPA and NMDA receptors have been reported to modulate inhibitory transmission in other systems, these receptors do not appear to play such a role in the dorsal horn.

Glycine and GABA are probably co-packaged in and co-released from spinal interneurons [23,24]. Bicuculline and strychnine each blocked a portion of IPSCs evoked by extracellular stimulation (eIPSCs), and a combination of the two antagonists blocked eIPSCs completely, suggesting that both GABA and glycine were released from individual cultured dorsal horn neurons and contributed to postsynaptic currents. We predicted that GABAergic and glycinergic mIPSCs would be regulated similarly by presynaptic KA receptor activation. Indeed, in the presence of strychnine (500 nM), KA enhanced mIPSC frequency in all nine neurons tested. Similarly, in the presence of bicuculline (5 μM), KA increased mIPSC frequency in six of eight experiments. In two of the experiments, no change in frequency was noted. Interestingly, whereas bicuculline and strychnine each depressed mIPSC frequency when applied alone, neither antagonist produced any significant effect on mIPSC amplitude, suggesting that individual mIPSCs were typically mediated by GABA or glycine but not both. Such a phenomenon has been attributed to the differential clustering of the two receptor types at postsynaptic sites, not to the differential release of GABA and glycine.

Synaptically released glutamate activates presynaptic KA receptors

We next examined whether the effects of KA receptor activation could be reproduced in a more intact preparation using synaptic glutamate — instead of exogenously applied agonists — as the trigger. Intracellular recordings were achieved from superficial dorsal horn neurons in spinal cord slices from adult rats, and two stimulating electrodes were introduced: one locally in the dorsal horn and one in the attached dorsal root (Figure 10). In the presence of SYM2206 (100 μM) and AP5 (50–100 μM), local stimulation produced a hyperpolarizing inhibitory postsynaptic potential (IPSP) that was sensitive to picrotoxin (20 μM) and strychnine (1 μM), indicating that this stimulation activated synaptically released glutamate local GABA and glycinergic interneurons. By contrast, stimulation of excitatory sensory fibers in the dorsal root produced no response when the stimulus intensity was below the threshold needed to activate fibers eliciting EPSCs mediated by postsynaptic KA receptors.

Primary afferent sensory fibers release glutamate in the vicinity of GABA- and glycine-containing boutons in the superficial dorsal horn. With two types of synaptic glomeruli and their distribution in laminae I–III of the rat spinal cord, we hypothesized that dorsal root stimulation may activate presynaptic KA receptors on spinal interneurons and thereby modulate inhibitory transmission. Indeed, when a conditioning stimulus train (50 Hz, 20 pulses) was delivered to the dorsal root (Figure 10), both the amplitude and the slope of the rising phase of an IPSP generated 50 ms later by local stimulation were reduced relative to the baseline, unconditioned response. The conditioning train itself triggered no EPSP or IPSP. In the presence of CNQX (20 μM) instead of SYM2206, conditioning did not affect IPSPs, suggesting that KA receptor activation was necessary (Figure 10). Furthermore, conditioning produced no effect in the presence of CGP55845 (10 μM). Thus, in agreement with the effects of KA detailed above in cultured neurons, inhibitory transmission in spinal slices was downregulated by synaptic glutamate through a mechanism involving KA receptors and $GABA_B$ receptors.

A mechanism for presynaptic KA receptor-mediated regulation of transmitter release

A model of a synaptic glomerulus depicts the proposed function of presynaptic KA receptors at dorsal horn inhibitory synapses (Figure 11). These receptors — which can be activated by glutamate released from primary afferent sensory fibers — mediate Na^+ entry and terminal depolarization

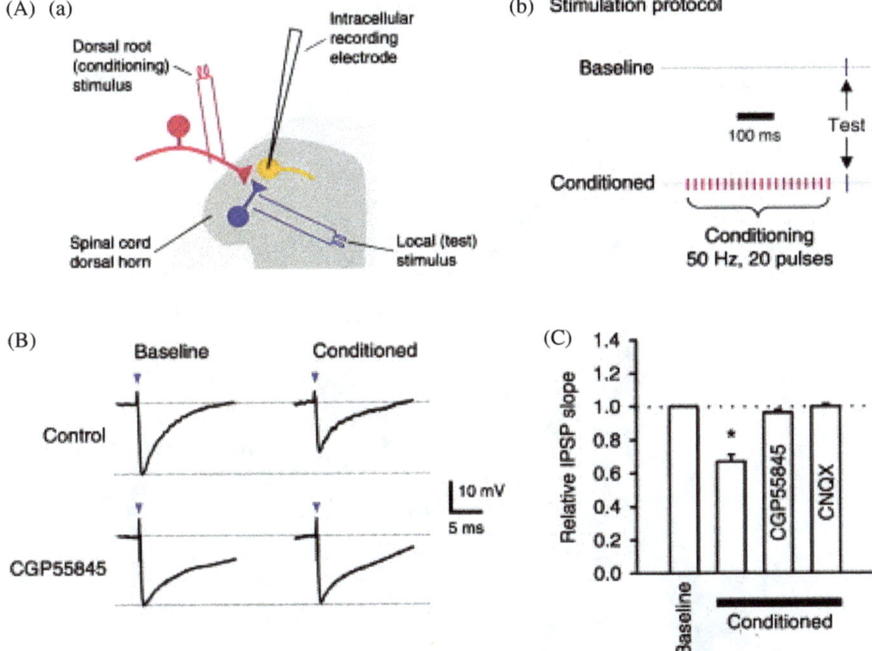

Figure 10.   Synaptic glutamate from sensory fibers suppresses inhibitory transmission in the dorsal horn through KA and $GABA_B$ receptor activation. (A) The diagram illustrates the placement of recording and stimulating electrodes in a transverse spinal cord slice. Intracellular recordings were performed from dorsal horn neurons (dark yellow), and stimulating electrodes were placed in the dorsal root to activate sensory fibers (red) and in the dorsal horn to activate inhibitory interneurons (blue). The stimulation protocols are illustrated for baseline and conditioned responses. (B) In representative recordings, averaged traces are shown of IPSPs evoked by local test stimulation before (baseline) and immediately after a train of dorsal root stimulation (conditioned). One cell was tested in the presence of SYM2206 and AP-5 alone (control), and another was tested in the additional presence of CGP55845 (10 μM). Arrowheads indicate the time of test stimulation. During conditioning stimulation, no synaptic response was observed in the recorded neuron (data not shown). (C) Pooled data illustrate the effects of conditioning on the slope of the rising phase of IPSPs evoked by test stimulation in control conditions ($n = 6$) or in the presence of CGP55845 (10 μM; $n = 5$) or CNQX (20 μM; $n = 5$). (*) represents a significant difference from baseline (adapted from Kerchner *et al.* [22]).

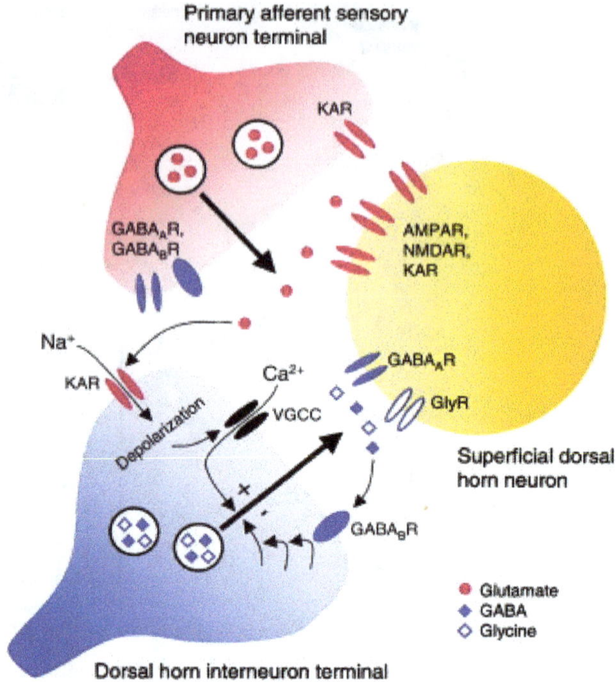

Figure 11.   Presynaptic KA receptors regulate spinal inhibitory transmission. The model of a synaptic glomerulus depicts the proposed function of presynaptic KA receptors at dorsal horn inhibitory synapses. These receptors, which can be activated by glutamate released from primary afferent sensory fibers, mediate Na$^+$ entry and terminal depolarization, triggering opening of VGCCs and Ca^{2+}-dependent vesicle fusion. GABA released in this manner may activate presynaptic GABA$_B$ autoreceptors, reducing action potential-dependent transmitter release. Previous work has shown that sensory neuron terminals contain GABA$_A$ and GABA$_B$ receptors, indicating that sensory neurons and dorsal horn interneurons engage in reciprocal heterosynaptic regulation of transmitter release. Other studies have documented additional roles for KA receptors in spinal sensory transmission. Along with AMPA and NMDA receptors, KA receptors mediate a component of the post-synaptic response of dorsal horn neurons to high-threshold sensory fiber stimulation. In addition, sensory fibers themselves express presynaptic KA receptors that regulate gluta-mate release (adapted from Kerchner *et al.* [22]).

thereby triggering the opening of voltage-gated Ca^{2+} channels (VGCCs) and Ca^{2+}-dependent vesicle fusion. GABA released in this manner may activate presynaptic GABA$_B$ autoreceptors, reducing action potential-dependent transmitter release. Previous work has shown that sensory neu-ron terminals contain GABA$_A$ and GABA$_B$ receptors, indicating that sensory neurons and dorsal horn interneurons engage in reciprocal heterosynaptic regulation of transmitter release. Other studies have documented additional roles for KA receptors in spinal sensory transmission. Along with AMPA and NMDA receptors, KA receptors mediate a component of the postsynaptic response of dorsal horn neurons to high-threshold sensory fiber stimulation. In addition, sensory fibers themselves express presynaptic KA receptors that regulate glutamate release.

The biphasic effects of presynaptic KA receptor activation on excitatory transmission have also been demonstrated at hippocampal mossy fiber-CA3 synapses. KA receptor activation caused a Cd^{2+}-sensitive increase in mEPSC frequency in CA3 neurons and enhanced the excitability of mossy fiber axons in a manner consistent with terminal depolarization. KA also suppressed mossy fiber-CA3 transmission and reduced action potential-evoked Ca^{2+} entry into mossy fiber axon terminals, possibly reflecting voltage-gated Na^+ or Ca^{2+} channel inactivation, current shunting, or stimulation of metabotropic glutamate autoreceptors, in analogy to the role of $GABA_B$ receptors in this study. Further work will be required to establish whether presynaptic KA receptors function similarly at other synapses in the central nervous system as they do at inhibitory synapses in the dorsal horn.

Pharmacological and genetic inhibition of KA receptor is analgesic

We have presented considerable evidence to show that pain associated with peripheral tissue or nerve injury involves KA receptor activation, highlighting the role of KA receptors in physiological and pathological pain. In accordance with this idea, systemic administration of KA acid produces persistent hyperalgesia in rodents. A myriad of studies report that KA receptor antagonists alleviate pain-related behaviors in animal models and clinical settings, although negative or conflicting results have also been reported (Table 1).

GluR5 is involved in both *in vitro* and *in vivo* nociceptive responses; however, intraperitoneal injection of either ATPA or LY382884 had no effect on HP responses. Moreover, HP and TF responses were unaltered in GluR5 and GluR6 knockout mice when compared to wild-type

Table 1. Fast and slow sensory synaptic transmission in spinal sensory synapses.

Type of transmission	Transmitter	Receptors
Fast sensory transmission	Glutamate	AMPA receptor KA receptor
Slow sensory transmission	Substance P Neurokinin A	NK1 receptor NK2 receptor

controls [25]. It is interesting to point out that other studies report a role for spinal KA receptors in acute nociception. This was first demonstrated by Li and colleagues, who found that intrathecal injection of the AMPA and KA receptor antagonist CNQX produced greater antinociceptive effects than an antagonist selective for AMPA receptors alone in the HP and TF, but not cold-plate, tests. The same study showed that the non-selective KA receptor antagonist SYM2081 alone could have analgesic effects. SYM2081 had a similar effect on acute pain induced by mechanical or thermal stimulation. In contrast, intrathecal injection of ATPA was reported to have antinociceptive effects in mechanical or thermal stimulation-induced acute pain.

A number of recent studies have focused on the role of KA receptors in chronic pain. Behavioral experiments using systemic drug administration showed that GluR5-containing KA receptors may play a role in formalin-induced inflammatory pain, while SYM 2081 was shown to attenuate mechanical allodynia and thermal hyperalgesia in a rat model of nerve injury. Similar analgesic effects were found by several groups using different chronic pain models and pharmacological tools, such as intraperitoneal injection of SYM2081 in a freeze injury neuropathic pain model, intrathecal injection of NS-102 (KA receptor antagonist) or LY382884 in the CFA inflammation model, intrathecal or intraperitoneal injection of SYM2081 in capsaicin-induced hyperalgesia or carrageenan-induced inflammation, intraperitoneal injection of NS1209 (AMPA/GluR5 antagonist) in the formalin test or chronic constriction nerve injury model, and oral administration of LY467711 or LY525327 (GluR5 antagonist) in capsaicin-induced hyperalgesia, formalin or carrageenan inflammation. Results from these studies, employing a wide variety of pain models and pharmacological tools, highlight the potential utility of targeting the GluR5 subunit when developing new drug therapies.

The selective role of GluR5 in mediating nociceptive responses was reported recently using GluR5 knockout mice. Ko *et al.* found that GluR5, but not GluR6, knockout mice showed reduced responses to capsaicin or formalin-induced inflammatory pain while acute nociceptive responses and CFA-induced allodynia were unchanged compared to controls (Figure 12). Furthermore, GluR6 mice showed a deficit in fear memory while remaining intact in GluR5 knockout mice. This study points out the subunit specificity of KA receptors to different behaviors

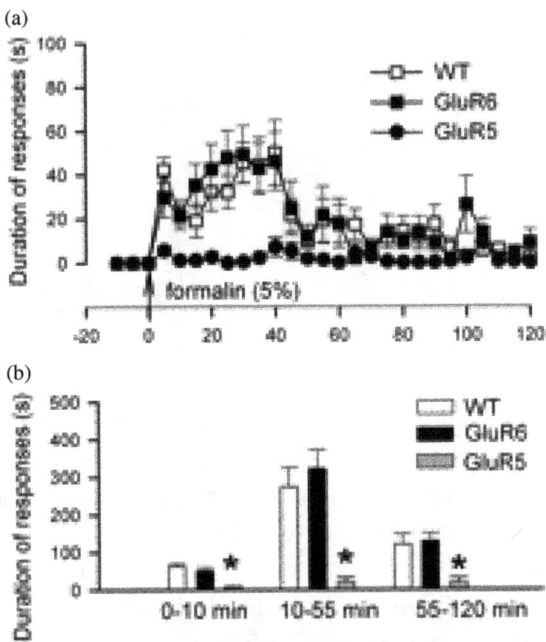

Figure 12.   Reduced response to formalin in GluR5 but not GluR6 knockout mice. (a) Behavioral responses to formalin injection, plotted in 5 min intervals, in wild-type (WT) mice compared with GluR5 and GluR6 knockout mice. (b) Data from A grouped into three phases. (c) Behavioral responses after CFA injection. Data are plotted as a percentage of positive responses to stimulation of the ipsilateral or contralateral hind paw in wild type, GluR5 knockout, and GluR6 knockout mice. (d) Hind paw edema was measured with a fine caliper in wild type, GluR5 knockout, and GluR6 knockout mice 3 days after CFA injection. *$p < 0.05$, significant difference from wild-type mice injected with formalin (adapted from Ko *et al.* [25]).

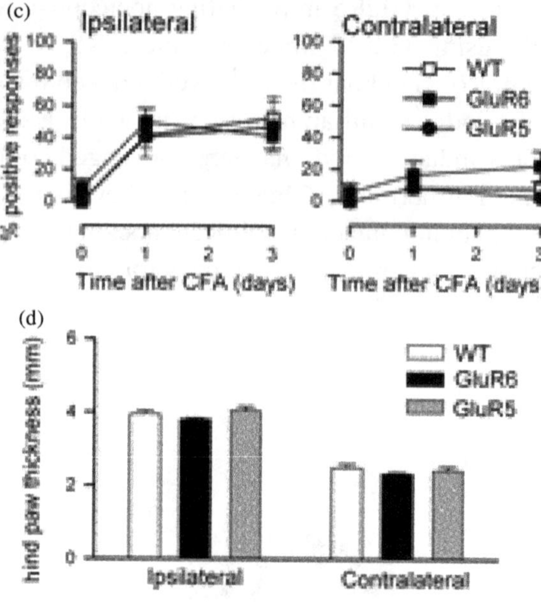

Figure 12.   (*Continued*)

and further suggests that the GluR5 subunit plays a selective role in nociception.

While these animal studies reinforce the role of KA receptors in nociception, clinical studies also provide some preliminary evidence for the role of KA receptors in pain. Sang *et al.* reported that intravenous infusion of LY293558 reduced capsaicin-evoked hyperalgesia in humans — reducing pain intensity, pain unpleasantness, and the receptive field. Moreover, a recent study reported that intravenous administration of LY293558 was efficacious in the treatment of acute migraines. Collectively, these behavioral studies indicate that pharmacological or genetic manipulation of KA receptors has the potential to affect acute or chronic nociceptive transmission. Although the exact contribution of each KA receptor subunit to acute nociception versus chronic pain remains unexplored, it is clear that in light of the potentially different distribution and function of KA receptor subunits in the pain pathway, selectively targeting different KA receptor subunits may provide a useful strategy for treating persistent pain [26].

Conclusions

Synaptic studies reveal that glutamate is a major excitatory transmitter in the spinal cord, and AMPA and KA receptors contribute to sensory synaptic transmission. Glutamate mediates excitatory transmission and undergoes presynaptic and postsynaptic regulations by different neurotransmitters and neuromodulators. Some of these transmitters can be released locally in the spinal cord, such as glutamate, acetylcholine, opioid peptides, and ATP, while some are released from descending modulatory projection systems, such as serotonin and norepinephrine. In addition to glutamate, neuropeptides such as substance P also contribute to synaptic transmission when the primary afferent fibers are stimulated repetitively. It is likely that spinal nociceptive transmission is carried out by the co-action of glutamate and neuropeptides.

References

[1] Hollmann, M. and Heinemann, S. (1994) Cloned glutamate receptors. *Annu Rev Neurosci* 17, 31–108. 10.1146/annurev.ne.17.030194.000335.

[2] Liu, H. *et al.* (1997) NMDA-receptor regulation of substance P release from primary afferent nociceptors. *Nature* 386, 721–724. 10.1038/386721a0.

[3] Lu, Y. and Perl, E.R. (2005) Modular organization of excitatory circuits between neurons of the spinal superficial dorsal horn (laminae I and II). *J Neurosci* 25, 3900–3907. 10.1523/JNEUROSCI.0102-05.2005.

[4] Popratiloff, A. *et al.* (1996) AMPA receptor subunits underlying terminals of fine-caliber primary afferent fibers. *J Neurosci* 16, 3363–3372.

[5] Tachibana, M. *et al.* (1994) Light and electron microscopic immunocytochemical localization of AMPA-selective glutamate receptors in the rat spinal cord. *J Comp Neurol* 344, 431–454. 10.1002/cne.903440307.

[6] Yoshimura, M. and Jessell, T. (1990) Amino acid-mediated EPSPs at primary afferent synapses with substantia gelatinosa neurones in the rat spinal cord. *J Physiol* 430, 315–335. 10.1113/jphysiol.1990.sp018293.

[7] Li, P. *et al.* (1998) ATP P2x receptors and sensory synaptic transmission between primary afferent fibers and spinal dorsal horn neurons in rats. *J Neurophysiol* 80, 3356–3360. 10.1152/jn.1998.80.6.3356.

[8] Li, P. *et al.* (1999) Kainate-receptor-mediated sensory synaptic transmission in mammalian spinal cord. *Nature* 397, 161–164. 10.1038/16469.

[9] Li, P. and Zhuo, M. (2001) Substance P and neurokinin A mediate sensory synaptic transmission in young rat dorsal horn neurons. *Brain Res Bull* 55, 521–531. 10.1016/s0361-9230(01)00553-6.

[10] Gu, J.G. and MacDermott, A.B. (1997) Activation of ATP P2X receptors elicits glutamate release from sensory neuron synapses. *Nature* 389, 749–753. 10.1038/39639.

[11] Bardoni, R. *et al.* (1997) ATP P2X receptors mediate fast synaptic transmission in the dorsal horn of the rat spinal cord. *J Neurosci* 17, 5297–5304.

[12] Nakatsuka, T. and Gu, J.G. (2001) ATP P2X receptor-mediated enhancement of glutamate release and evoked EPSCs in dorsal horn neurons of the rat spinal cord. *J Neurosci* 21, 6522–6531.

[13] Nakatsuka, T. *et al.* (2002) Activation of central terminal vanilloid receptor-1 receptors and alpha beta-methylene-ATP-sensitive P2X receptors reveals a converged synaptic activity onto the deep dorsal horn neurons of the spinal cord. *J Neurosci* 22, 1228–1237.

[14] Nakatsuka, T. *et al.* (2003) Distinct roles of P2X receptors in modulating glutamate release at different primary sensory synapses in rat spinal cord. *J Neurophysiol* 89, 3243–3252. 10.1152/jn.01172.2002.

[15] Shiokawa, H. *et al.* (2006) Direct excitation of deep dorsal horn neurones in the rat spinal cord by the activation of postsynaptic P2X receptors. *J Physiol* 573, 753–763. 10.1113/jphysiol.2006.108613.

[16] Sun, Y.G. and Chen, Z.F. (2007) A gastrin-releasing peptide receptor mediates the itch sensation in the spinal cord. *Nature* 448, 700–703. 10.1038/nature06029.

[17] Koga, K. *et al.* (2011) Glutamate acts as a neurotransmitter for gastrin releasing peptide-sensitive and insensitive itch-related synaptic transmission in mammalian spinal cord. *Mol Pain* 7, 47. 10.1186/1744-8069-7-47.

[18] Engelman, H.S. and MacDermott, A.B. (2004) Presynaptic ionotropic receptors and control of transmitter release. *Nat Rev Neurosci* 5, 135–145. 10.1038/nrn1297.

[19] Huettner, J.E. *et al.* (2002) Glutamate and the presynaptic control of spinal sensory transmission. *Neuroscientist* 8, 89–92. 10.1177/107385840200800204.

[20] Kerchner, G.A. *et al.* (2001) Presynaptic kainate receptors regulate spinal sensory transmission. *J Neurosci* 21, 59–66.

[21] Kerchner, G.A. *et al.* (2002) Kainate receptor subunits underlying presynaptic regulation of transmitter release in the dorsal horn. *J Neurosci* 22, 8010–8017.

[22] Kerchner, G.A. *et al.* (2001) Direct presynaptic regulation of GABA/glycine release by kainate receptors in the dorsal horn: An ionotropic mechanism. *Neuron* 32, 477–488. 10.1016/s0896-6273(01)00479-2.

[23] Todd, A.J. (1996) GABA and glycine in synaptic glomeruli of the rat spinal dorsal horn. *Eur J Neurosci* 8, 2492–2498. 10.1111/j.1460-9568.1996.tb01543.x.

[24] Yasaka, T. *et al.* (2007) Cell-type-specific excitatory and inhibitory circuits involving primary afferents in the substantia gelatinosa of the rat spinal dorsal horn in vitro. *J Physiol* 581, 603–618. 10.1113/jphysiol.2006.123919.

[25] Ko, S. *et al.* (2005) Altered behavioral responses to noxious stimuli and fear in glutamate receptor 5 (GluR5)- or GluR6-deficient mice. *J Neurosci* 25, 977–984. 10.1523/JNEUROSCI.4059-04.2005.

[26] Jane, D.E. *et al.* (2009) Kainate receptors: Pharmacology, function and therapeutic potential. *Neuropharmacology* 56, 90–113. 10.1016/j.neuropharm.2008.08.023.

Chapter 4

Silent Synapse

News and Views

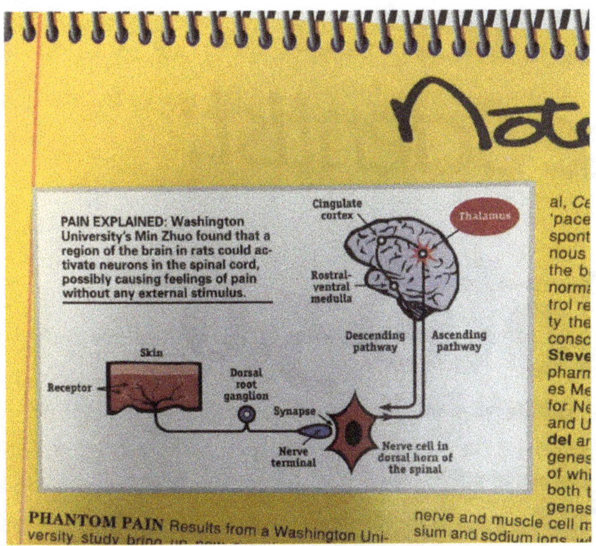

Research from Washington University on phantom pain raises new questions about your high-school gym teacher's proclamations that pain is "all in your head." The study found that chemical signals from the brain can condition nerve synapses in the spinal cord so that pain persists even in the absence of a stimulus. While investigators have known that the brain effectively reduces pain by inactivating synapses, few have considered the effect of the brain on enhancing signals of pain throughout the nervous system. Dr. Min Zhuo and his team found that a region in the brain of rats could activate neurons in the spinal cord, causing feelings of pain without an apparent source [2]. In the study, Zhuo, along with research assistant Ping Li, showed that the rostroventral medulla region of the brain sends a serotonin signal to the dorsal horn of the lumbar spinal cord. Here, the signal reactivates the so-called "silent," or inefficient, synapses at nerve junctions. This finding is significant because it provides insight as to why some patients continue to experience pain even when the stimulus is gone, Zhuo says. For example, cancer patients often have persisting pain even after tumors are removed. "Through the process we call 'long-term potentiation,' neurons in the spinal cord become efficient for transmitting pain," he says, "Once 'silent' synapses have been activated during tissue or nerve injury, they tend to remain functionally active and continue to send messages to the brain even when there is no stimulus to speak of." Zhuo says the research could have significant clinical implications because it suggests new targets and pathways for therapies aimed at treating chronic pain.

Reported by: *The Scientist* (1998).

Summary

Neurons in the superficial dorsal horn of the spinal cord are important for conveying sensory information from the periphery to the central nervous system. It has been proposed that some synapses between primary afferent fibers and spinal dorsal horn neurons are inefficient or silent. Ineffective sensory transmission could result from a small postsynaptic current that fails to depolarize the cell to the threshold for an action potential or a cell with a normal postsynaptic current but an increased threshold for action potentials. One possible mechanism for ineffective synapses may be due to silent glutamatergic synapses. In silent synapses, postsynaptic dorsal horn neurons lack functional AMPA/KA receptors, and no synaptic responses are detected even when glutamate is released from presynaptic sensory afferent fibers. Serotonin (5-HT), a major neurotransmitter of the raphe-spinal projecting pathway, transforms silent glutamatergic synapses into functional ones by recruiting postsynaptic functional AMPA receptors. AMPA receptor interaction with PDZ-containing proteins is critical for 5-HT-induced recruitment. Silent synapses, as well as their recruitment mechanisms, are developmentally regulated and can still be found in adult synapses. However, they are "functional" and contribute to some sensory transmission. The recruitment of AMPA receptors into pure NMDA receptors containing sensory synapses requires the coactivation of 5-HT and postsynaptic ACs. The recruitment of silent glutamatergic synapses provides a synaptic mechanism for spinal sensitization in pathological pain, including possible phenotype switching of dorsal horn neurons. These results suggest that the transformation of silent glutamatergic synapses may serve as a cellular mechanism for central plasticity in the spinal cord.

Keywords: Silent synapse; AMPA receptor; Spinal cord; serotonin; descending facilitation; cAMP; glutamate

Introduction

Primary afferent fibers form synapses with dorsal horn sensory neurons in the spinal cord. Some of these dorsal horn neurons send ascending projecting fibers that synapse with neurons located at supraspinal sites, such as the thalamic nuclei. These ascending pathways are important for conveying sensory information from the periphery to the brain (see Chapters 1–3). Dorsal horn sensory synapses receive descending modulation from supraspinal structures. Most of these descending modulatory influences relay at brainstem nuclei including the RVM (Chapter 1). Activation of these modulatory systems could facilitate, or inhibit, spinal nociceptive transmission and behavioral reflexive responses to noxious stimulation depending on the stimulation parameters used. Thus, the fine regulation of dorsal horn sensory synaptic transmission not only affects the amount of information entering the brain but also influences the behavioral responses to such stimuli.

Glutamate is a major neurotransmitter between primary afferent fibers and dorsal horn neurons [1–3]. Postsynaptic sensory responses are mainly mediated by glutamate AMPA/KA and NMDA receptors (see Chapter 3). Glutamatergic synapses are heterogeneous in the spinal dorsal horn. At least three different types of glutamatergic synapses are found: (1) Silent synapses: in some synapses, only functional NMDA receptors are found.

Neither AMPA nor KA receptors are present, or functional, in these synapses. (2) AMPA receptor-containing synapses: at sensory synapses that receive low-threshold inputs, only AMPA receptors are found and no functional KA receptors exist. (3) KA/AMPA receptor mixed synapses: at synapses receiving high-threshold inputs, both AMPA and KA receptors are found. In both (2) and (3), NMDA receptors are always detected.

Besides being reliable and fast, glutamatergic synaptic transmission in the spinal cord is dynamic and plastic. Recent progress in molecular and cellular aspects of glutamate receptors shows that postsynaptic glutamate receptors are put into the place through a family of proteins containing PDZ domains. Furthermore, these postsynaptic protein–protein interactions are very dynamic and may be involved in the clustering, removal, or insertion of postsynaptic receptors, providing a novel and efficient way to regulate synaptic strength. This type of interaction could switch the phenotype of dorsal horn sensory synapses, i.e., changing silent synapses into functional synapses. In this chapter, silent glutamatergic synapses in spinal dorsal horn neurons will be reviewed, and their possible physiological functions will be discussed.

Spinal LTP

Studies of spinal LTP and its related synaptic mechanisms are limited by the technical difficulty related to spinal cord slices and the complexity of the local spinal neuronal network. Electrophysiological experiments using intracellular, or whole-cell patch-clamp, recordings from the spinal dorsal horn neurons generate some important findings related to spinal LTP [4–6] (see Table 1). Strong tetanic stimulation (100 Hz, 1 sec for three times at 10-sec intervals) of the dorsal root induces long-lasting enhancement of synaptic responses to presynaptic stimulation. The enhancement is relatively long-lasting (ranging from 25 min to 90 min) and input-specific. Postsynaptic depolarization of dorsal horn neurons is critical for the induction of spinal LTP. Pairing postsynaptic depolarization with synaptic activity also induces long-lasting enhancement of synaptic responses (Figure 1). Interestingly, the level of postsynaptic depolarization may be important in

Table 1. Experimental protocols for inducing LTP in the spinal cord.

Stimulation site	Response	Protocol	Refs.
Dorsal root	C response	HFS *(100 Hz for 1 second, three times with 10-s interval)*	[26]
Dorsal root	C response	LFS *(2 Hz for 2 minutes)*	[27]
Entry zone	Unidentified	Pairing protocol *(80 pulses at 2 Hz with postsynaptic depolarization at + 30 mV)*	[4]
Entry zone	C and A_δ responses	Spike timing *(paired three presynaptic stimuli (10 ms ahead) with three postsynaptic APs at 30 Hz, paired 15 times every 5 s)*	Unpublished data

determining whether synaptic transmission will be potentiated or depressed. In some experiments, synaptic depression can also be induced by pairing synaptic activity with modest postsynaptic depolarization.

The induction of spinal LTP requires the activation of NMDA receptors and/or the SP (NK1) receptors. Activation of NMDA receptors in spinal dorsal horn neurons leads to increases in intracellular Ca^{2+}. Pretreatment of spinal cord slices with NMDA receptor antagonist AP5 prevents the induction of LTP. Activation of postsynaptic calcium signaling pathways is important, and the inhibition of postsynaptic calcium signaling pathways blocked the induction of LTP. The contribution of SP to spinal LTP is likely through enhancing NMDA receptor-mediated currents in spinal dorsal projecting neurons. The intracellular signal pathways of spinal LTP have yet to be identified and characterized (see Table 2). Evidence from other studies indirectly indicates that several protein kinases may be important for spinal LTP, such as phospholipid-dependent protein kinase (PKC). Phorbol ester induces long-lasting facilitation of evoked EPSP or EPSC amplitude to the stimulation of presynaptic fibers. One key mechanism for PKC-dependent spinal LTP is through the recruitment of postsynaptic silent synapses or the insertion of AMPA receptors. Possible presynaptic mechanisms of spinal LTP have not been investigated.

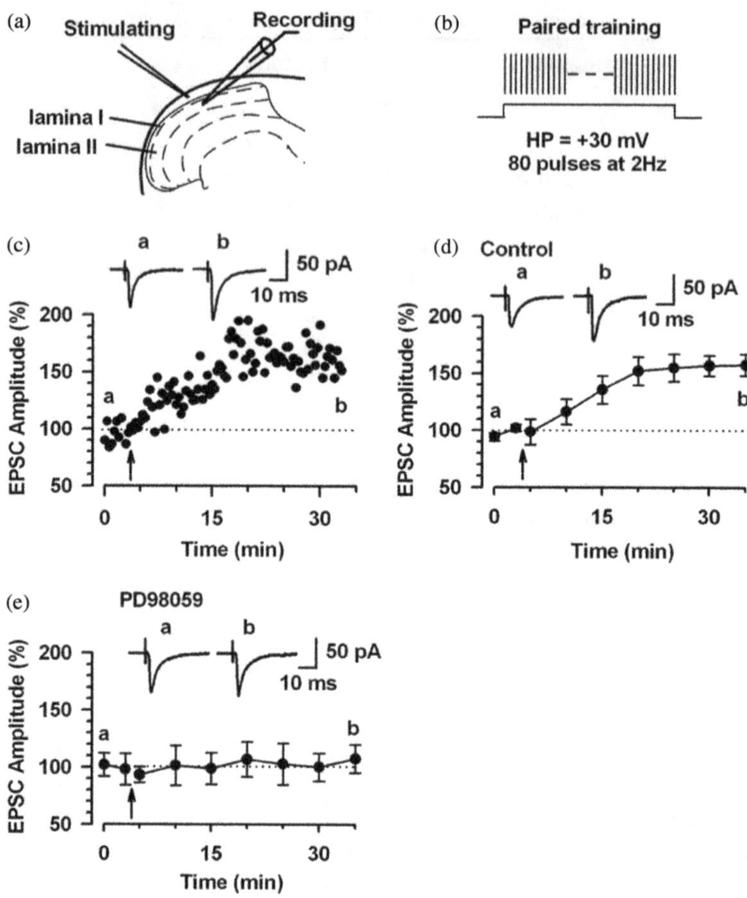

Figure 1.   LTP recorded by whole-cell patch in mice. Activation of Erk for the induction of LTP in the superficial dorsal horn neurons. (A) Diagram of a slice showing the placement of a whole-cell patch-clamp recording and a stimulation electrode in the superficial dorsal horn of the spinal cord. (B) Schematic illustrating the induction protocol consisting of 80 pulses at 2 Hz while holding at +30 mV (paired training). (C) LTP was induced by paired training in the superficial dorsal horn neurons. (D) Summary result of the LTP experiments under control conditions ($n = 9$ neurons). EPSC responses were averaged over 5 min intervals. (E) The MEK inhibitor PD98059 (50 μM) in the intracellular solution completely blocked LTP induction ($n = 7$ neurons). EPSC responses were averaged over 5 min intervals. (c–e) Traces show averages of six EPSCs 3 min before (a) and 25 min after (b) the paired training (arrow). The dashed line indicates the mean basal synaptic responses. Error bars indicate SEM. (adapted from Wei *et al.* [4]).

Table 2. Synaptic mechanisms for spinal LTP.

Stimulation site	Protocol	Postsynaptic component	Refs.
Dorsal root	HFS	NK1 receptor, PLC, IP3 receptor, NMDAR, T-type VGCC	[26]
Dorsal root	LFS	PLC, PKC, CAMKII, IP3 NO	[27]
Entry zone	Pairing protocol	MEK	[4]

Spinal sensitization and synaptic potentiation

Studies of LTP in spinal dorsal horn neurons have drawn much attention because it is believed that the potentiation of sensory responses after injury may explain chronic pain. It has been consistently demonstrated that the spike responses of dorsal horn neurons to peripheral stimulation are enhanced after the injury, however, whether the enhanced spike responses are simply due to enhanced synaptic transmission between the DRG cells and dorsal horn neurons remains to be investigated. Unlike synapses in other areas, such as hippocampus, synaptic potentiation in the spinal dorsal horn neurons is not induced by strong tetanic stimulation. Recent studies further show that LTP only occurs in some of the spinal projecting cells [11]. Spinal cord dorsal horn neurons that did not express SP receptors did not undergo potentiation. Furthermore, activation of NK1 receptors or NMDA receptors is required for LTP. However, in other areas of the brain such as hippocampus and cortex, there is no requirement for SP or any similar neuropeptide for the induction of NMDA receptor-dependent LTP (see Chapter 5). It will be important to investigate why LTP cannot be induced at other neurons that do not express SP receptors.

Spinal LTP caused by peripheral injury

In vivo field recordings, or *in vitro* spinal cord slice recordings, found that noxious stimuli, inflammation, and nerve injury all induced LTP at the spinal dorsal horn synapses (Figure 2). The induction of such spinal LTP is controlled by descending inhibitory systems from supraspinal

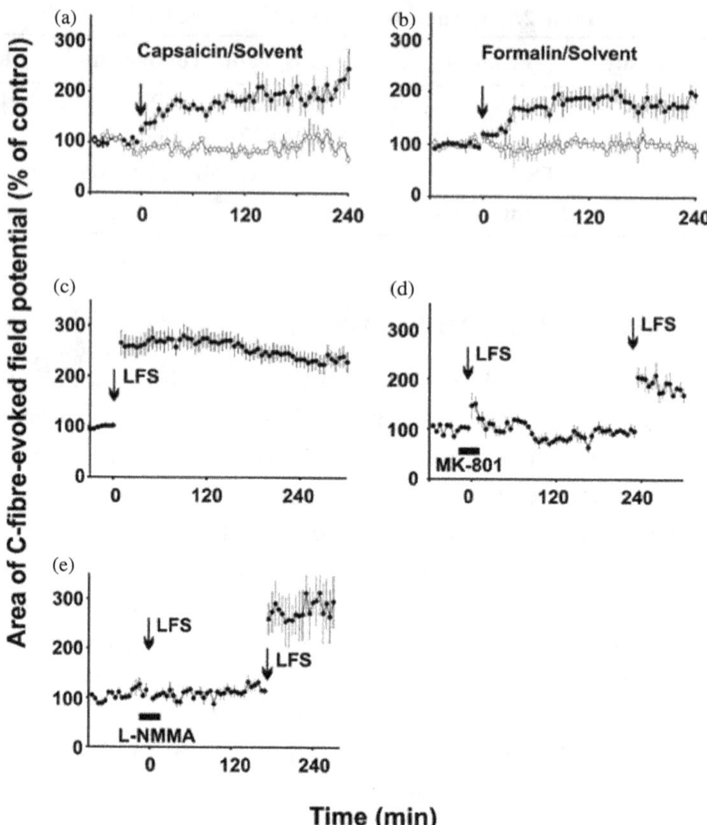

Figure 2.   LTP can be induced by low-frequency afferent barrage evoked by inflammation of peripheral tissue. Mean time courses of C fiber-evoked field potentials recorded extracellularly in superficial spinal dorsal horn in response to electrical stimulation of left sciatic nerve of deeply anesthetized adult rats with spinal cords and afferent nerves intact. Subcutaneous injections of transient receptor potential vanilloid 1 channel agonist capsaicin (1%, 100 μl, $n = 5$) (a) or formalin (5%, 100 μl, $n = 6$) (b) into the glabrous skin at the ipsilateral hind paw, within the innervation territory of the sciatic nerve at time zero (arrows), induced LTP (closed circles), whereas injections of the respective solvents (open circles) had no effect ($n = 3$ in each group). Conditioning electrical LFS (2 Hz, 2 min at C fiber intensity) of sciatic nerve at time zero (arrow) also induced LTP ($n = 28$) (*C*), which was prevented by NMDAR antagonist MK-801 [3 mg kg^{-1}, intravenous (iv) infusion over 30 min: horizontal bar, $n = 5$] (*D*). A second conditioning LFS 4 hours later (arrow) was partially effective in inducing LTP. NOS inhibitor NG-monomethyl-L-arginine (L-NMMA) (100 mg kg^{-1} hour^{-1}, iv infusion: horizontal bar, $n = 5$) (*E*) also blocked LTP induction. This block was fully reversible, as shown by a second LFS 3 hours later (arrow) (adapted from Ikeda *et al.* [27]).

structures. The induction of spinal LTP requires the activation of glutamate NMDA receptors and NK1 receptors. Both receptors raise the postsynaptic calcium level and trigger intracellular calcium-related signaling pathways.

Early studies of ineffective synapses in the spinal cord

The existence of ineffective, or silent, sensory synapses in the spinal cord dorsal horn was first proposed by P.D. Wall in 1977. Due to the limits of the recording method at that time, it was unclear whether ineffective sensory transmission was a result of the failure of a small postsynaptic current to depolarize a neuron to action potential threshold or because of an increased threshold in a neuron with normal synaptic responses. Further indirect evidence for silent synapses comes from intracellular recordings of synaptic responses to stimulation of a single group of Ia afferents in cat spinal motor neurons. To examine the size fluctuations of excitatory postsynaptic potentials (EPSPs), an application of 4-AP or tetanization of the afferent led to the discovery that some previously 'silent' synapses became active following stimulation. Despite the limitations of these recording techniques, these studies provide strong evidence for the existence of silent synapses in the central nervous system. Considering the high-level difficulty of spinal cord preparations, however, only a few studies have followed up on these initial findings.

Silent glutamatergic synapses in the spinal cord dorsal horn

By using whole-cell patch-clamp recording techniques in brain slices, silent glutamatergic synapses are reported in various regions of the CNS including the hippocampus, neocortex, spinal cord dorsal horn, and ventral horn motor neurons [2,7–13]. In silent synapses, no effective AMPA/KA receptors are available to detect the release of glutamate from presynaptic terminals. Consequently, these synapses do not conduct synaptic transmission at the resting membrane potential. The existence of such synapses between sensory fibers and dorsal horn neurons was shown in the lumbar spinal cord [2]. Sensory neurons in the superficial dorsal horn of spinal cord slices were recorded by using whole-cell patch-clamp recording techniques and tested for the possible existence of silent glutamatergic synapses (Figure 3). Fast monosynaptic, excitatory postsynaptic currents (EPSCs) were induced when cells were held at −70 mV. In order to detect silent synapses, the intensity of stimulation was decreased so that no fast EPSC was detected at −70 mV. The holding potential was then changed to +40 mV to detect NMDA receptor-mediated EPSCs. In about

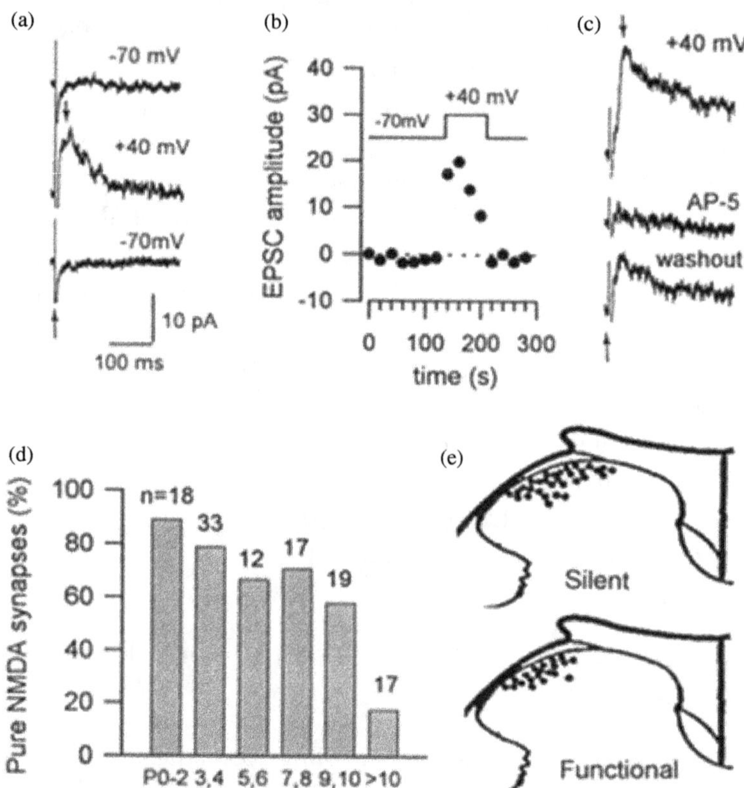

Figure 3. Silent glutamatergic synapses in the lumbar spinal cord. (a) Examples of responses (the average of three continuous traces) at -70 or +40 mV holding potential. (b) Time course of the experiment shown in A. (c) Responses at + 40 mV were reversibly inhibited by 50 μM AP-5. Upward arrow indicates the time of stimulation. Downward arrow indicates the peak currents measured. (d) Developmentally related distribution of silent synapses in the spinal cord. The percentage of silent synapses in the superficial dorsal horn of the lumbar spinal cord at P2-17. The number of cells tested is indicated above the bars. (e) Distribution of labeled spinal neurons (adapted from Li and Zhuo [2]).

59% of the dorsal horn neurons tested, synaptic responses were found at +40 mV but not at −70 mV. Synaptic responses at +40 mV were abolished by the selective NMDA receptor antagonist AP5 (50 μM). The sensitivity to AP5 and the voltage dependence of channel activation indicates that synaptic responses measured at +40 mV were NMDA receptor-mediated. The intensity of stimulation used in these experiments was low, and it is likely that only low-threshold afferent fibers were activated. However, it does not rule out the possibility that silent synapses may exist in other neurons that receive high-threshold afferent inputs in the spinal cord.

Silent synaptic transmission and silent glutamatergic synapses

It is important to point out that silent synapses should not be confused with potential "silent synaptic transmission." The term "silent synapses" refers to a condition where the postsynaptic cell is clamped at −70 mV and NMDA receptors are abundantly located. In an unclamped cell, these NMDA receptors may contribute to sensory synaptic transmission, for example, in the case of high-intensity sensory fiber activity induced by tissue injury. These results consistently suggest that different types of glutamatergic synapses exist in spinal sensory connections between primary afferent fibers and dorsal horn neurons.

Recruitment of silent synapses

Sensory transmission in the spinal cord receives descending modulation from supraspinal structures including the RVM. 5-HT-containing neurons in the RVM send descending projection fibers to targets in the spinal cord, like the superficial dorsal horn. Activation of these descending pathways can facilitate, or inhibit, spinal nociceptive transmission. Previous studies of this descending modulation focused on behavioral, pharmacological, and electrophysiological recordings of spike firing from dorsal horn neurons. It is important to show that these modulatory effects are due to changes in spinal sensory synaptic transmission and not to the modulation of pre-motor spinal interneurons or spinal local inhibitory synapses.

Consistent with the biphasic modulatory effect of 5-HT on spinal nociceptive transmission and behavioral reflexes, 5-HT produced biphasic modulation of excitatory synaptic responses in spinal slices [2,14,15]. 5-HT at high doses inhibited AMPA receptor-mediated EPSCs, while a low dose of 5-HT — or a selective 5-HT$_2$ receptor agonist — induced facilitation of fast EPSCs in the lumbar spinal cord (Figure 4). 5-HT at low doses could facilitate fast EPSCs in the presence of an NMDA receptor antagonist AP-5 (50 μM), indicating that the facilitatory effect is NMDA receptor-independent. Furthermore, the facilitatory effect induced by 5-HT at low doses persisted during the washout of 5-HT. While the activation of 5-HT receptors is important for the induction of facilitation, continuous activation of these receptors is not necessary for the expression of facilitation. Application of methysergide after the serotonergic receptor

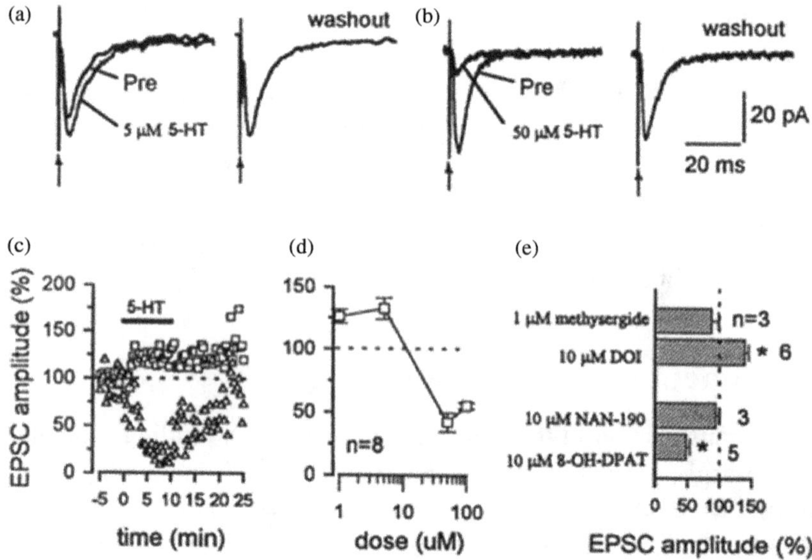

Figure 4. Biphasic modulation induced by 5-HT. (a–b) Examples of 5-HT experiments at two doses. Upward arrows indicate the time of stimulation. (c) The effect of 5-HT on amplitudes of EPSCs in experiments shown in a (squares) and b (triangles). (d) Summary data of 5-HT at four different doses ($n = 8$ for each dose). (e) Different effects of 5-HT$_{1A}$ (8-OH-DPAT) and 5-HT$_2$ (DOI) receptor agonists. The effects were blocked by their receptor antagonists NAN-190 (5-HT$_{1A}$) and methysergide, respectively. *$P < 0.05$ (adapted from Li and Zhuo [2]).

agonist DOI failed to reverse the facilitatory effect. One synaptic mechanism for the 5-HT-produced facilitation is due to the recruitment of silent glutamatergic synapses. Application of 5-HT (5 μM) caused typically fast EPSCs to appear at synapses initially lacking AMPA/KA receptor-mediated responses (Figure 5).

5-HT may affect spinal sensory transmission by acting on presynaptic or postsynaptic receptors. Postsynaptic application of G protein inhibitors, introduced through the recording pipette, abolished the effect of 5-HT on synaptic transmission, suggesting that postsynaptic 5-HT receptors are critical for this effect. In support of this notion, postsynaptic

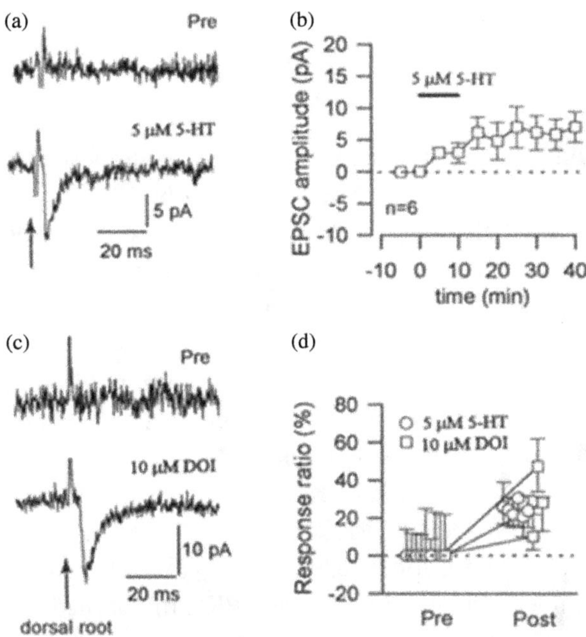

Figure 5. Transformation of silent synapses into functional synapses by 5-HT or DOI. (a) The average of 15 continuous responses (collected at 0.05 Hz) before and 15 min after the end of 5-HT application. (b) Summary data of 5-HT. (c) Responses before and 15 min after DOI application using dorsal root nerve stimulation. (d) Summary of response ratios before and after 5-HT or DOI application. Symbols for a given cell are connected; symbols and errors show response ratio (= the number of stimuli with responses/total number of stimuli $\times$ 100) and the calculated 95% confidence intervals (adapted from Li and Zhuo [2]).

Ca^{2+}-dependent processes were shown to be required for 5-HT-induced facilitation. In experiments with BAPTA (1,2-bis-(o-aminophenoxy) ethane-N,N,N',N'-tetracetic acid) in the pipette solution, the facilitatory effect of 5-HT was abolished, demonstrating the requirement for an increase in postsynaptic Ca^{2+}. Additional evidence arguing against a mechanism of 5-HT-induced synaptic facilitation involving modulation of presynaptic glutamate release comes from the observation that while 5-HT application clearly caused AMPA receptor-mediated EPSCs, NMDA receptor-mediated EPSCs were significantly decreased by 5-HT in the same neurons. This result suggests that postsynaptic enhancement of AMPA receptor-mediated currents by 5-HT is selective.

Intracellular mechanisms for 5-HT-induced facilitation

In central glutamatergic synapses, PKC plays an important role in long-lasting synaptic enhancement. Application of the PKC activator phorbol 12,13-dibutyrate (PDBu) produced long-lasting enhancement of AMPA receptor-mediated responses in hippocampal neurons, and inhibition of PKC prevented the induction of long-term potentiation induced by tetanic stimulation. In dorsal horn neurons of the spinal cord, the excitatory effect of PKC on spinal nociceptive transmission has been reported. Activation of PKC enhances sensory synaptic transmission in the dorsal horn, sensitizing the responses of ascending projection neurons to noxious peripheral stimuli. Mice lacking PKCγ show reduced pain behaviors following nerve injury. Because PKC acts downstream of 5-HT$_2$ receptors and these receptors are critical for the facilitatory effect of low doses of 5-HT, we examined whether postsynaptic PKC might mediate the facilitatory effect of 5-HT in spinal dorsal horn neurons. When the PKC peptide inhibitor PKCI was present in the patch pipette, application of 5-HT failed to cause any synaptic facilitation. Consistently, bath application of the PKC activator PDBu caused an enhancement of postsynaptic responses in spinal dorsal horn neurons. The effect of PDBu persisted during the washout of the drug and was independent of NMDA receptors. In a similar manner as 5-HT, the facilitatory effect induced by PKC involved the recruitment of silent glutamatergic synapses.

AMPA receptor and the expression of facilitation

One possible mechanism for the recruitment of silent synapses is through the interaction of glutamate AMPA receptors and proteins containing postsynaptic density-95/Discs large/zona occludens-1 (PDZ) domains [16,17]. GluR2/3 is widely expressed in sensory neurons in the superficial dorsal horn of the spinal cord. Unlike GluR1 which is mainly expressed in spinal interneurons, GluR2/3 is mainly expressed in non-local inhibitory neurons. Glutamate receptor-interacting protein (GRIP), a protein with 7 PDZ domains that binds specifically to the C-terminus of GluR2/3, is also expressed in spinal dorsal horn neurons. In many dorsal horn neurons, GluR2/3 and GRIP coexist. Long-term overexpression of the C-terminus of GluR2 in hippocampal neurons reduces the number of synaptic AMPA receptor clusters, suggesting that the interaction between GluR2/3 and PDZ proteins is involved in the postsynaptic targeting of AMPA receptors [14]. To examine the functional significance of GluR2/3–PDZ interactions in sensory synaptic transmission, a synthetic peptide was made corresponding to the last 10 amino acids of GluR2 ("GluR2-SVKI": NVYGIESVKI) which disrupts the binding of GluR2 to GRIP. As expected, the GluR2-SVKI peptide blocked the facilitatory effect of 5-HT. The effect of GluR2-SVKI on synaptic facilitation is rather selective, as the baseline level of evoked EPSCs and currents evoked by glutamate application did not change over time in these neurons (Figure 6). Experiments with different control peptides consistently indicate that the interaction between the C-terminus of GluR2/3 and GRIP/ABP (or called GRIP1 and GRIP2) is important for 5-HT-induced facilitation (Figures 4–7). Furthermore, synaptic facilitation induced by PDBu is also blocked by GluR2-SVKI, suggesting that synaptic facilitation mediated by PKC activation is similar to that produced by 5-HT in its dependence on GluR2/3 C-terminal interactions (Figure 7).

Coactivation of cAMP signaling pathways in facilitation of adult sensory synapses

cAMP signaling pathways are implicated in the function of spinal dorsal horn neurons. Activation of several receptors for sensory transmitters such

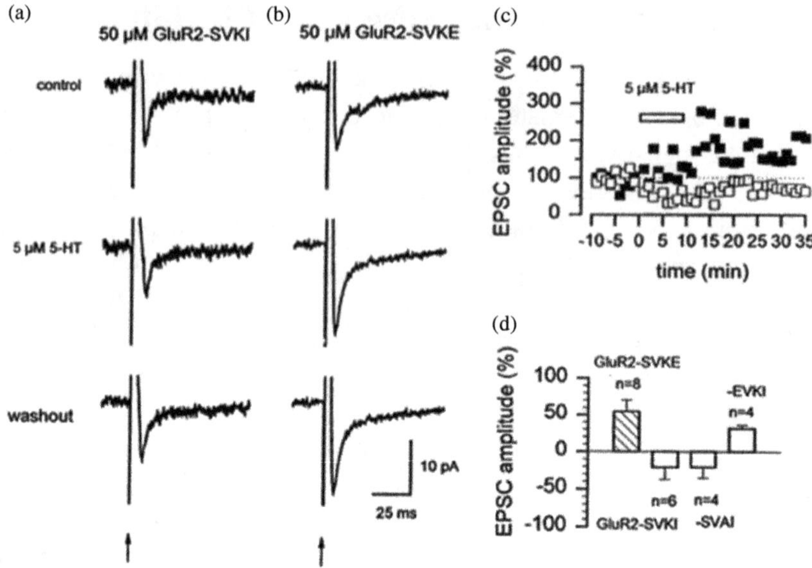

Figure 6.   AMPA receptor–PDZ interactions. (a) 5 μM 5-HT activates silent synapses in dorsal horn neurons. Representative traces show that this activation was prevented when 50 μM GluR2-SVKI was present in the patch pipet. Traces are shown at time points before 5-HT exposure (Pre), at the end of a 10-min bath application of 5 μM 5-HT, and 30 min after return to control bath solution (washout). Representative traces (b) and pooled data (c) show that silent synapses in dorsal horn neurons were activated by 0.5 μM PDBu in the presence of 50 μM GluR2-SVKE ($n = 5$) or no peptide ($n = 5$; data were not different between these two control conditions and were pooled; open squares). However, PDBu failed to activate silent synapses in most neurons loaded with 50 μM GluR2-SVKI ($n = 6$; see text; closed squares). Traces are shown at time points before PDBu exposure (Pre), at the end of a 10-min application of 0.5 μM PDBu, and 30 min after return to control bath solution (washout) (adapted from Li *et al.* [14]).

as glutamate and calcitonin gene-related peptide (CGRP) is reported to raise cAMP levels. In a recent study, application of forskolin did not significantly affect the synaptic responses induced by dorsal root stimulation in slices of adult mice. However, co-application of 5-HT and forskolin produced long-lasting facilitation of synaptic responses [18] (Figure 8). Possible contributors to the increase in cAMP levels are calcium-sensitive ACs. The facilitatory effect induced by 5-HT and forskolin was

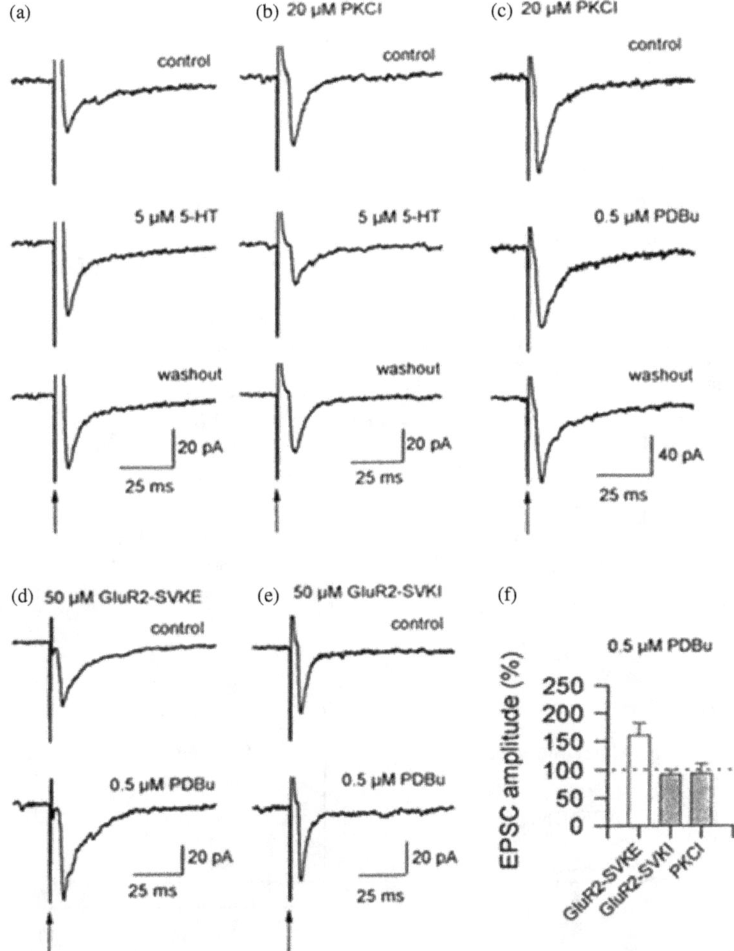

Figure 7. Postsynaptic PKC activation is necessary and sufficient for synaptic facilitation by 5-HT. (a) 5 μM 5-HT induces a persistent potentiation of EPSC amplitude in representative traces from a dorsal horn neuron. The three traces illustrate EPSCs before drug application (control), at the end of a 10-min bath application of 5 μM 5-HT, and 30 min after return to control bath solution (washout). (b, c) Loading postsynaptic neurons with the PKC inhibitor PKCI (20 μM) abolished the facilitatory effect of both 5-HT (5 μM; b) and the PKC activator PDBu (0.5 μM; c) on EPSCs in experiments conducted as in (a). (d–e) Application through the patch pipet of 50 μM GluR2-SVKI (e) but not 50 μM GluR2-SVKE (d) prevented facilitation of dorsal horn synaptic transmission by bath application of 0.5 μM PDBu. (f) Pooled data show that 50 μM GluR2-SVKI ($n = 6$) and 20 μM PKCI ($n = 5$), but not 50 μM GluR2-SVKE ($n = 7$), blocked the potentiation of EPSCs by 0.5 μM PDBu (adapted from Li *et al.* [14]).

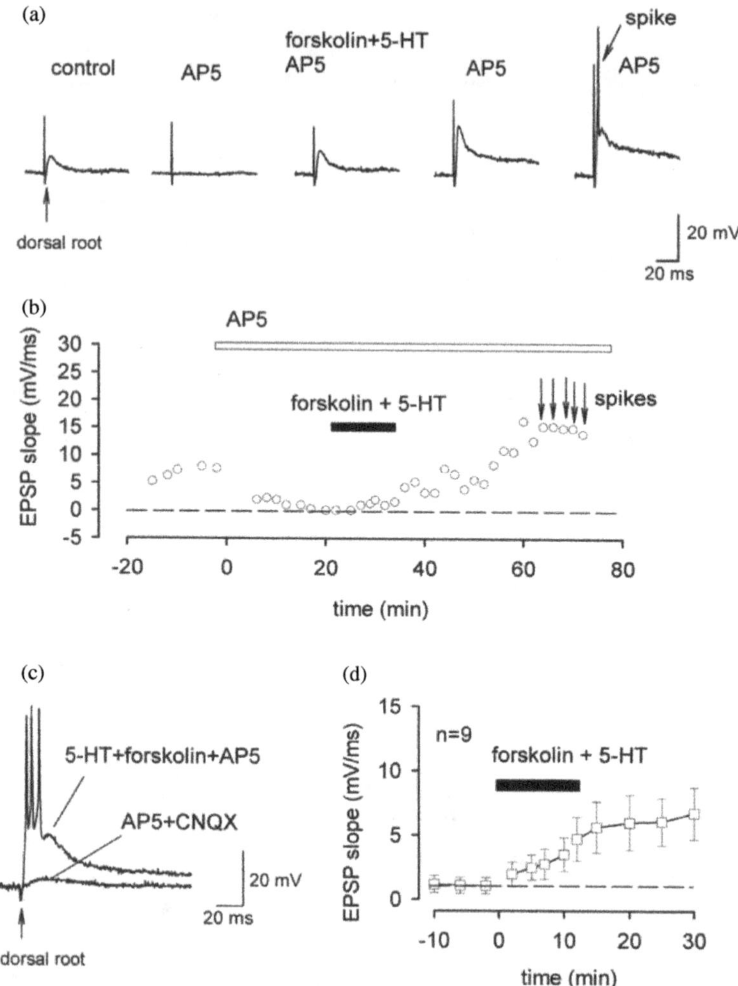

Figure 8.   Synergistic recruitment of AMPA/KA receptor-mediated responses at pure NMDA synapses. (a–c) Examples of EPSPs showing synaptic responses before, during, and after co-application of 5 μM 5-HT and 10 μM forskolin in the presence of 100 μM AP5. The effect of 5-HT and forskolin on the EPSP slopes in the experiment shown in A. Spike responses to stimulation of the dorsal root nerves at the subthreshold intensity were observed during the washout (indicated by arrows). In a separate experiment, 20 μM CNQX completely blocked both EPSPs and action potentials induced by 5 μM 5-HT and 10 μM forskolin in the presence of AP5. (d) Summary data of 5 μM 5-HT and 10 μM forskolin. Data contain two sets of experiments: pure NMDA synapses ($n = 3$) and mixed AMPA/KA and NMDA synapses ($n = 6$). In both cases, significant enhancement was observed and data were pooled together (adapted from Wang and Zhuo [18]).

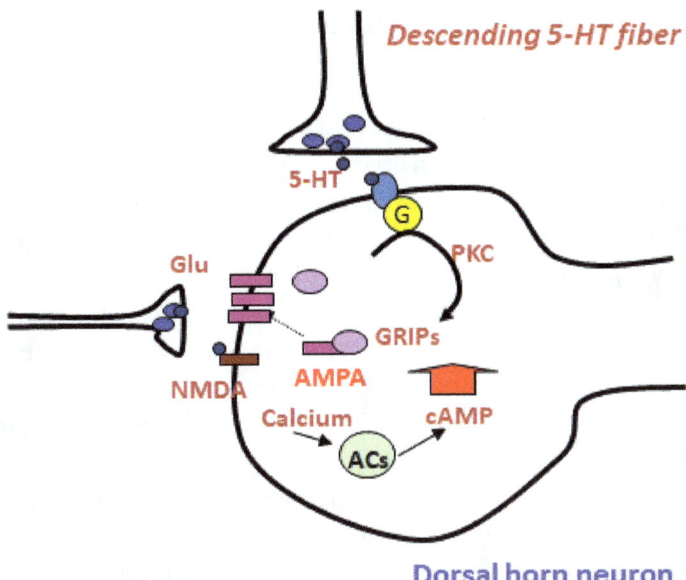

Figure 9.   Model for 5-HT-induced facilitation in the spinal cord. The model pathway explains how activation of 5-HT receptor may recruit silent synapses in spinal dorsal horn neurons. 5-HT released from descending projection fibers activates postsynaptic PKC through G protein receptors. PKC activation and subsequent AMPA receptor and GRIP interactions cause the recruitment of AMPA receptors to the synapse (see Li *et al.* [14] for details).

completely blocked in mice lacking AC1 or AC8, demonstrating the importance of calcium-sensitive ACs. These results show that in adult sensory synapses, cAMP signaling pathways determine whether the activation of 5-HT receptors causes facilitatory, or inhibitory, effects on synaptic responses. This finding provides a possible explanation for the regulation of two different signaling pathways under physiological or pathological conditions. Postsynaptic increases in cAMP levels by sensory transmitters may favor 5-HT-induced facilitation (Figure 9). The interaction between cAMP and 5-HT may provide an associative heterosynaptic form of central plasticity in the spinal dorsal horn which may allow sensory inputs from the periphery to act synergistically with central modulatory influences descending from the brainstem RVM.

Implications for silent synapses and LTP in pathological or chronic pain

One potential function of silent synapses is to contribute to the formation of a descending modulatory network within the spinal cord [1,19,20]. Most 5-HT-containing nerve fibers in the spinal cord originate from the RVM. 5-HT-induced recruitment of silent synapses could strengthen spinal sensory synapses receiving innervation from descending 5-HT projecting fibers. Through an activity-dependent mechanism, the effect of 5-HT could make those receiving innervation more likely to compete, and survive, in adult animals.

Silent synapses may also be involved in pathological conditions like persistent pain. As suggested by Wall [21,22], the recruitment of silent, or ineffective, synapses could significantly enhance spinal sensory transmission, including nociceptive transmission. AMPA receptors are expressed in dorsal horn neurons at synapses receiving high- and low-threshold inputs (see Figure 10). Activation of silent synapses with low-threshold afferents might cause dorsal horn neurons to exhibit an increased firing rate in response to non-noxious stimuli. Increases in postsynaptic AMPA receptor density and activation of silent synapses could contribute to plastic changes in the electrophysiological properties of dorsal horn sensory neurons, including ascending spinothalamic tract projecting neurons. These include increased receptive fields, enhanced responses to noxious and non-noxious stimuli, decreased firing thresholds, and increased background activity.

Two pieces of evidence indirectly support the notion that alterations of descending serotonergic influences and spinal glutamate AMPA receptors may contribute to persistent pain. First, descending serotonergic systems from the RVM to the spinal cord have been implicated in different types of hyperalgesia after tissue injury [20,23–25]. Formalin-induced inflammation causes increased activity in 5-HT-containing neurons in the RVM for up to two hours. Second, changes have been reported in the expression of AMPA receptors on spinal dorsal horn neurons after tissue injury. Therefore, activation of silent synapses could serve as an important cellular mechanism underlying two features of chronic pain: hyperalgesia, where the intensity of responses to noxious stimuli is increased over baseline, and allodynia, where the nociceptive threshold is decreased and a normal non-noxious stimulus (such a gentle touch) can induce pain.

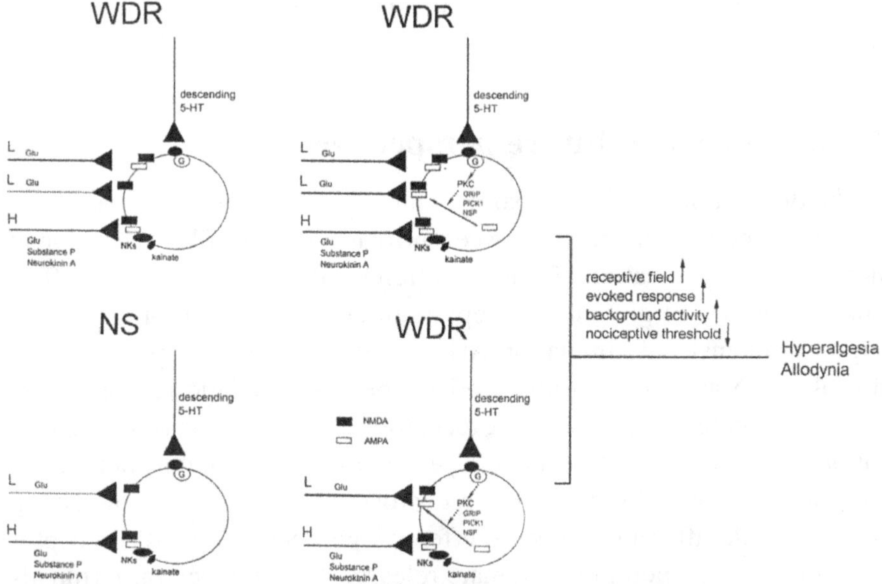

Figure 10.   Contribution of silent synapses and LTP to behavioral allodynia and hyperalgesia. Models showing a hypothetical mechanism of silent synapse activation in two types of nociceptive sensory neurons and a possible role for this activation in hyperalgesia and allodynia. (Top) In a wide dynamic range (WDR) cell, some synapses receiving low-threshold input (indicated by L) are silent (indicated by a dotted line). Release of 5-HT from descending projection fibers converts silent synapses into functional ones. (Bottom) In a nociceptive specific (NS) cell, synapses receiving low-threshold input are all silent. Activation of these synapses renders the cell responsive to low-threshold stimulation, effectively converting it into a WDR cell. The conversion of silent synapses to functional synapses in either WDR or NS cells alters the electrophysiological properties of these neurons, thereby potentially increasing their receptive fields, enhancing their responses to sensory (noxious and non-noxious) stimuli, increasing background firing frequencies, and decreasing their nociceptive threshold. These changes, each of which is indeed observed in recordings from dorsal horn neurons in whole animals experiencing chronic pain, could contribute to both hyperalgesia and allodynia (adapted from Zhuo [1]).

Behavioral studies of GluR2 interfering peptides

Behavioral studies using intrathecal administration of GluR2 interaction inhibiting peptides found that it produced a significant analgesic effect in animal models of neuropathic pain. Consistent with this finding, control peptides did not produce any significant effects. These findings indicate that AMPA receptor interaction is critical for maintaining chronic pain at the level of spinal cord.

Conclusions and future perspectives

In the dorsal horn of the lumbar spinal cord, glutamate serves as a fast, excitatory neurotransmitter between primary afferent fibers and dorsal horn neurons. Activation of primary afferent fibers could activate different populations of glutamate receptors in dorsal horn neurons, depending on the intensity of stimulation. At low intensities of stimulation—which are likely to activate non-nociceptive fibers — synaptic responses are mainly mediated through AMPA receptors. At high intensities of stimulation which activate nociceptive A_δ and C fibers, synaptic responses are mediated by both AMPA and KA receptors. At some synapses receiving low-threshold afferent inputs, no effective postsynaptic AMPA receptor is available to respond to glutamate released from the central terminals of primary afferent fibers. These silent synapses are plastic and can be recruited by 5-HT, a neurotransmitter from descending projection fibers. 5-HT acts through postsynaptic PKC and requires an interaction between PDZ proteins and the C-terminus of GluR2/3 to affect synaptic facilitation. One possibility, which has not been tested directly, is that 5-HT could cause a GRIP-mediated insertion of GluR2/3 into the post-synaptic membrane. The effect of 5-HT is likely not due to the regulation of glutamate release from presynaptic terminals. The facilitatory effect of 5-HT is selective for AMPA receptors, as both postsynaptic NMDA receptors and KA receptors are not significantly affected in the same cells. One obvious direction for future studies is to elucidate the processes regulating GluR2/3 C-terminal interactions, the trafficking and membrane insertion of AMPA receptors, and PKC-mediated intracellular signaling pathways. It is also equally important to identify possible ways to silence already-activated synapses. Understanding these signaling pathways will serve to both prevent sensory sensitization before it is induced (i.e., pre-emptive analgesia), as well as to decrease sensitization after its establishment. Progress in these areas will provide us with opportunities to design better drugs to treat chronic pain in patients.

References

[1] Zhuo, M. (2000) Silent glutamatergic synapses and long-term facilitation in spinal dorsal horn neurons. *Prog Brain Res* 129, 101–113. 10.1016/S0079-6123(00)29008-0.

[2] Li, P. and Zhuo, M. (1998) Silent glutamatergic synapses and nociception in mammalian spinal cord. *Nature* 393, 695–698. 10.1038/31496.

[3] Yoshimura, M. and Jessell, T. (1990) Amino acid-mediated EPSPs at primary afferent synapses with substantia gelatinosa neurones in the rat spinal cord. *J Physiol* 430, 315–335. 10.1113/jphysiol.1990.sp018293.

[4] Wei, F. *et al.* (2006) Calcium calmodulin-stimulated adenylyl cyclases contribute to activation of extracellular signal-regulated kinase in spinal dorsal horn neurons in adult rats and mice. *J Neurosci* 26, 851–861. 10.1523/JNEUROSCI.3292-05.2006.

[5] Li, X.H. *et al.* (2019) NMDA receptor dependent long-term potentiation in chronic pain. *Neurochem Res* 44, 531–538. 10.1007/s11064-018-2614-8.

[6] Ruscheweyh, R. *et al.* (2011) Long-term potentiation in spinal nociceptive pathways as a novel target for pain therapy. *Mol Pain* 7, 20. 10.1186/1744-8069-7-20.

[7] Rumpel, S. *et al.* (1998) Silent synapses in the developing rat visual cortex: evidence for postsynaptic expression of synaptic plasticity. *J Neurosci* 18, 8863–8874.

[8] Liao, D. *et al.* (1995) Activation of postsynaptically silent synapses during pairing-induced LTP in CA1 region of hippocampal slice. *Nature* 375, 400–404. 10.1038/375400a0.

[9] Liao, D. *et al.* (1999) Regulation of morphological postsynaptic silent synapses in developing hippocampal neurons. *Nat Neurosci* 2, 37–43. 10.1038/4540.

[10] Isaac, J.T. *et al.* (1995) Evidence for silent synapses: Implications for the expression of LTP. *Neuron* 15, 427–434. 10.1016/0896-6273(95)90046-2.

[11] Isaac, J.T. *et al.* (1997) Silent synapses during development of thalamocortical inputs. *Neuron* 18, 269–280. 10.1016/s0896-6273(00)80267-6.

[12] Gil, Z. and Amitai, Y. (1996) Adult thalamocortical transmission involves both NMDA and non-NMDA receptors. *J Neurophysiol* 76, 2547–2554. 10.1152/jn.1996.76.4.2547.

[13] Bardoni, R. *et al.* (1998) NMDA EPSCs at glutamatergic synapses in the spinal cord dorsal horn of the postnatal rat. *J Neurosci* 18, 6558–6567.

[14] Li, P. *et al.* (1999) AMPA receptor-PDZ interactions in facilitation of spinal sensory synapses. *Nat Neurosci* 2, 972–977. 10.1038/14771.

[15] Hori, Y. *et al.* (1996) Long-lasting synaptic facilitation induced by serotonin in superficial dorsal horn neurones of the rat spinal cord. *J Physiol* 492 (Pt 3), 867–876. 10.1113/jphysiol.1996.sp021352.

[16] Garner, C.C. *et al.* (2000) PDZ domains in synapse assembly and signalling. *Trends Cell Biol* 10, 274–280. 10.1016/s0962-8924(00)01783-9.

[17] Hayashi, Y. *et al.* (2000) Driving AMPA receptors into synapses by LTP and CaMKII: requirement for GluR1 and PDZ domain interaction. *Science* 287, 2262–2267. 10.1126/science.287.5461.2262.

[18] Wang, G.D. and Zhuo, M. (2002) Synergistic enhancement of glutamate-mediated responses by serotonin and forskolin in adult mouse spinal dorsal horn neurons. *J Neurophysiol* 87, 732–739. 10.1152/jn.00423.2001.

[19] Zhuo, M. (2004) Central plasticity in pathological pain. *Novartis Found Symp* 261, 132–145; discussion 145–154.

[20] Zhuo, M. (2017) Descending facilitation. *Mol Pain* 13, 1744806917699212. 10.1177/1744806917699212.

[21] Wall, P.D. (1988) Recruitment of ineffective synapses after injury. *Adv Neurol* 47, 387–400.

[22] Wall, P.D. (1977) The presence of ineffective synapses and the circumstances which unmask them. *Philos Trans R Soc Lond B Biol Sci* 278, 361–372. 10.1098/rstb.1977.0048.

[23] Zhuo, M. and Gebhart, G.F. (1991) Spinal serotonin receptors mediate descending facilitation of a nociceptive reflex from the nuclei reticularis gigantocellularis and gigantocellularis pars alpha in the rat. *Brain Res* 550, 35–48. 10.1016/0006-8993(91)90402-h.

[24] Zhuo, M. and Gebhart, G.F. (1992) Characterization of descending facilitation and inhibition of spinal nociceptive transmission from the nuclei reticularis gigantocellularis and gigantocellularis pars alpha in the rat. *J Neurophysiol* 67, 1599–1614. 10.1152/jn.1992.67.6.1599.

[25] Zhuo, M. and Gebhart, G.F. (1997) Biphasic modulation of spinal nociceptive transmission from the medullary raphe nuclei in the rat. *J Neurophysiol* 78, 746–758. 10.1152/jn.1997.78.2.746.

[26] Ikeda, H. *et al.* (2003) Synaptic plasticity in spinal lamina I projection neurons that mediate hyperalgesia. *Science* 299(5610), 1237–1240.

[27] Ikeda, H. *et al.* (2006) Synaptic amplifier of inflammatory pain in the spinal dorsal horn. *Science* 312(5780), 1659–1662.

Chapter 5

Cortical LTP

News and Views

Converging on LTP

Anxiety disorders are common in individuals with chronic pain and the development of such pain is more likely in people with anxiety. The anterior cingulate cortex (ACC) is implicated in both conditions, although the underlying neurobiological relationship between pain and anxiety is unclear. Now, Koga *et al.* show in mice that chronic pain and anxiety may be mechanistically linked by a presynaptic form of long-term potentiation (LTP) that occurs at synapses in the ACC.

LTP can be induced in the ACC by the activation of postsynaptic NMDA receptors, with the expression of such LTP being largely mediated by postsynaptic enhancement of AMPA receptor (AMPAR) responses, and evidence suggests that this mechanism is involved in pain perception. However, *ex vivo* studies have revealed that peripheral nerve injury, which can lead to chronic pain, causes enhanced excitatory transmission in the ACC that also involves enhanced glutamate release. This led the authors to examine whether a presynaptic form of LTP in the ACC contributes to chronic pain.

To determine whether presynaptic LTP can be induced at synapses in the ACC, the authors stimulated mouse brain slices that included the ACC using a protocol that is known to induce presynaptic LTP in the amygdala. Stimulation in layer 5/6 increased the amplitudes of evoked excitatory postsynaptic currents (eEPSCs) in layer 2/3 pyramidal neurons in the ACC for at least 1 hour, indicating that presynaptic LTP had been induced.

The authors examined the underlying mechanisms of such LTP and found that genetic deletion of the gene encoding the GLUK1 kainate receptor (KAR) subunit reduced presynaptic LTP in the ACC, and application of a GLUK1 KAR agonist induced presynaptic LTP. These findings show that GLUK1-containing KARs are necessary and sufficient for presynaptic LTP induction in the ACC.

The authors also found that pharmacological blockade of hyperpolarization-activated cyclic nucleotide-gated (HCN) channels before stimulation blocked the induction of presynaptic LTP and that HCN channel inhibition after LTP expression reduced eEPSCs to baseline levels. Thus, HCN channels are necessary for the induction and maintenance of presynaptic LTP at synapses in the ACC.

The authors next examined the effects of chronic pain on presynaptic LTP. In mouse models of chronic inflammatory or neuropathic pain, they were unable to induce presynaptic LTP. However, they reasoned that the apparent absence of presynaptic LTP could have resulted from a saturation of this mechanism and tested this possibility by applying an HCN channel inhibitor to slices taken from mice that underwent nerve injury (a model of chronic pain). This treatment reduced eEPSCs in ACC neurons activated by the injury, suggesting that chronic pain had indeed activated and saturated presynaptic LTP. Interestingly, such LTP could also not be induced in mouse models of anxiety.

Finally, the authors examined whether presynaptic LTP might explain the increase in anxiety that often accompanies chronic pain. To do so, they microinjected an HCN channel inhibitor into the ACC of mice with nerve injury and showed that, in two different anxiety-like behaviour paradigms, these animals had lower levels of anxiety than saline-injected nerve injury controls. They also exhibited a reduction in mechanical allodynia, confirming the importance of presynaptic LTP in the development of chronic pain.

Together, these findings suggest that induction of a form of presynaptic LTP at synapses in the ACC may, at least in part, explain the interaction between chronic pain and anxiety at the clinical level.

Darran Yates

ORIGINAL RESEARCH PAPER Koga, K. *et al.* Coexistence of two forms of LTP in ACC provides a synaptic mechanism for the interactions between anxiety and chronic pain. *Neuron* http://dx.doi.org/10.1016/j.neuron.2014.12.021 (2014)

> These findings show that GLUK1-containing KARs are necessary and sufficient for presynaptic LTP induction in the ACC.

Summary

The traditional view of the cortex in adult animals and humans is that the cortex is relatively non-plastic. Early studies of cortical plasticity suggested that robust LTP could only be found in young animals. The disappearance of LTP in older animals was accepted as being the consequence of cortical synapse maturation. However, recent studies using whole-cell patch-clamp recordings in adult animal brain slices have suggested a different view. In the adult neurons of the ACC, different protocols have induced long-lasting LTP, both *in vivo* and *in vitro*. In this chapter, recent studies on the ACC LTP, as well as the role of glutamate receptors in pain-related cortical plasticity, will be examined.

Keywords: ACC; IC; LTP; NMDA receptor; AMPA receptor; AC1; PKMζ; L-LTP; CREB; pre-LTP; psot-LTP; chronic pain

Introduction

Animal and human studies have consistently demonstrated that forebrain neurons play an important role in nociception and pain perception [1–3]. In animal studies, lesions of the medial frontal cortex including the ACC significantly increased acute nociceptive responses, and formalin injections induced aversive memory behaviors. In patients with frontal lobotomies or cingulotomies, the unpleasantness of pain was abolished [3–5]. Electrophysiological recordings found that cortical neurons within the ACC responded to noxious stimuli, including nociceptive-specific neurons [6]. Neuroimaging studies further confirmed these observations and showed that the ACC, together with other cortical structures, are activated by acute noxious stimuli [1,3,7,8]. Thus, understanding the synaptic mechanisms of cortical areas including the ACC will greatly help us gain insights into the pain-related plastic changes within the brain.

Multiple functions of the ACC neurons

Many cortical areas such as the ACC are multifunctional. For example, human brain imaging studies have demonstrated that the ACC region is activated by different factors including motivational drive, reward, gain or loss, conflict monitoring or error prediction, and attention or anticipation [1]. The neuronal mechanism in the ACC for these different functions largely remains unknown. The ACC is also one of the major brain areas activated by injury and is thought to be important for pain perception and persistent pain. Traditionally, most pain researchers avoid such "non-selective" cortical regions, preferring instead to focus on the spinal dorsal horn. However, studying cortical molecular mechanisms should not be avoided, especially considering the fact that pain interferes with numerous cognitive functions, such as attention, memory, emotion, and mood. The "side-effect" or non-selective roles of the ACC lend further support for the critical importance of the ACC in chronic pain-related mental disorders, as the ACC is involved in not only pain but also pain-related depression, drug addiction, thoughts of suicide, and loss of interests.

Cortical areas related to pain

Recent brain imaging techniques have allowed us to investigate the cortical regions involved in pain, from pain sensation to the severity of unpleasantness (see Table 1). These techniques help bridge the gap created by the limitations in animal research and provide detailed information about the cortical areas that respond to painful stimuli. Furthermore, imaging studies allow for correlation analyses so that brain areas where the activity occurs can be correlated with psychological pain reports. Significant progress has been made in the study of acute pain since testing stimuli are acute and repeatable, and it can be performed in healthy human subjects with similar backgrounds and health history (Table 1).

Among the different cortical regions, there are five major cortical areas that consistently respond to acute pain [9]. They are ACC, IC, primary somatosensory cortex (S1), secondary somatosensory cortex (S2), and PFC (see Figure 1). The ACC has been found to be the most reliable area to be activated by different noxious or painful stimuli. One example of ACC activation from acute pain perception is the thermal grill illusion test [10]. In this experiment, the sensation of warmth or cold alone is not sufficient to activate the ACC, but the warm and cold stimulations together

Table 1.   Stimulation protocols for inducing LTP in the ACC slices.

	Parameters	**Recording method**
Theta-burst stimulation	Five trains of bursts (four pulses at 100 Hz) of stimuli delivered every 200 ms	Field potential recording
	Five trains of bursts (four pulses at 100 Hz) of stimuli delivered every 200 ms, repeated four times at an interval of 10 s	Whole-cell patch-clamp recording
Pairing training	Paired presynaptic 80 pulses at 2 Hz with postsynaptic depolarization at +30 mV	Whole-cell patch-clamp recording
Spike-EPSP pairing (spike-timing) protocol	Paired three presynaptic stimuli that caused three EPSPs (10 ms ahead) with three postsynaptic APs at 30 Hz, paired 15 times every 5 s	Whole-cell patch-clamp recording

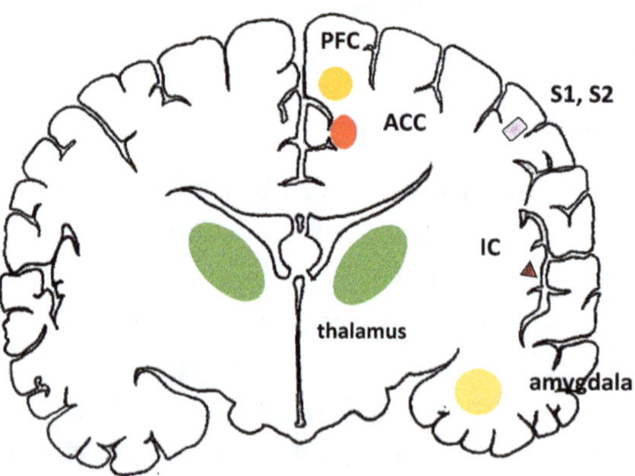

Figure 1.   Major cortical regions that contribute to pain perception and modulation. ACC and IC are thought to be two key cortical areas for pain perception, or the code of unpleasantness and related emotion and fear.

cause a burning sensation and activate the ACC. This report provides direct evidence for the selective involvement of the ACC in pain perception. In addition, the IC is also commonly activated by different painful stimuli. Neuronal activities in the ACC and IC are believed to be important for pain perception and unpleasantness.

Neurons in the ACC and IC are likely to encode multiple forms of pain. The activation of the ACC and IC has been reported to be caused by noxious heat, cold, and chemical stimuli. Results from human brain imaging studies suggest that not all pain is processed by similar brain regions, and different brain areas code each particular form of pain. For example, noxious, visceral, and somatosensory noxious stimuli activated the IC, ACC, and S2; interestingly, the perigenual cingulate cortex was deactivated by visceral pain [11]. Deactivation of cortical areas — presumably due to reduction in neuronal activities — is also very intriguing. It may suggest that excitation (E) and inhibition (I) occur not only at the neuronal level but also in cortical networks (i.e., one cortical area is activated while the adjacent area is inhibited). The ACC and/or IC can also be triggered by psychological pain and social exclusion. Singer *et al.* reported that neurons in the ACC and IC were likely to be activated during pain empathy [12], providing further evidence for their importance in pain.

Imaging chronic pain in the brain

Much less information is available on brain activation during chronic pain compared to acute pain. Unlike acute pain, where stimulation can be easily controlled and healthy subjects are easy to find, chronic pain patients are usually heterogeneous, and many factors affect the evaluation of suffering or pain they report. Despite these difficulties, several reports have investigated some of the aspects of chronic pain in patients. It has been shown that in neuropathic pain, the physiological discriminative function is down-regulated while pain-related activity is enhanced. In patients with irritable bowel syndrome, rectal distension that induces pronounced pain/urge activation (hyperalgesia) is accompanied with increased activities in the IC and the ACC [13]. Sustained high levels of spontaneous pain result in increases in activity within the medical PFC (mPFC), including the ACC [14]. Electrophysiological studies using the electroencephalogram (EEG) recording technique found that in neuropathic pain patients, there was a spontaneous presence of enhanced activation within high theta (6–9 Hz) and low beta frequencies (12–16 Hz) located in the IC, ACC, S1, and S2. These results suggest that frequency-specific ongoing overexcitation may contribute to spontaneous pain reported in patients with chronic pain.

Phantom pain, which is caused by amputation, is another form of chronic neuropathic pain. Early animal studies have suggested that cortical plasticity and reorganization may be the cause of phantom pain

[15,16]. Recently, it has been reported that phantom pain is indeed correlated with cortical activation (or changes in cortical mapping) [17]. It is likely that the sprouting of nerve terminals into inactivated somatosensory areas and cortical overexcitation in pain areas such as the ACC and IC contribute to phantom pain. Recent studies in human patients with other forms of chronic pain found opposite structural changes when compared to phantom pain. Apkarian *et al.* reported that chronic back pain is associated with decreased prefrontal and thalamic gray matter density. Similarly, Schmidt-Wilcke *et al.* reported that chronic back pain is correlated with decreased gray matter in the brainstem and somatosensory cortex [18]. Loss of gray matter in patients with fibromyalgia including in the ACC, IC, mPFC, and parahippocampal gyri is also found to correlate with chronic pain [19]. It is unclear whether the loss of neurons is limited to local inhibitory neurons. If so, chronic pain may be explained as disinhibition or the shift in E/I balance in the brain. Furthermore, it remains to be determined if these structural changes are a consequence of chronic pain or secondary changes due to pain-related mental disorders (such as depression). Draganski *et al.* reported that the loss of gray matter correlates well with the duration of amputation but not the amount of phantom pain [20], suggesting that the loss of neurons may be a secondary response to injury rather than the cause of chronic pain. One possibility is that both synaptic elimination and synaptic outgrowth occur within the cortical network, and consequently, the neuronal circuits related to pain are overexcited despite the loss of neurons. Heightened excitation is consistently reported in cortical areas related to pain in patients with phantom pain or other forms of chronic pain.

Technical limitations of human brain imaging

Human brain imaging studies have a technical limit. For example, it is difficult to identify the source of neuronal activity, i.e., excitatory vs. inhibitory neurons. The activity of the ACC has been reported to be correlated with pain as well as with placebo analgesia/opioid analgesia, two opposite physiological processes [8,12,21,22]. It is impossible to tell if increased metabolic activity in pain vs. analgesic conditions may be due to changes in the activity of inhibitory neurons or excitatory neurons. These results can be explained if excitatory neurons are activated during pain, while inhibitory neurons are activated during placebo/opioid analgesia. More importantly, recent studies have demonstrated that imaged human brain activities do not correlate with neuronal spike activities in the ACC and IC [23–25]. It is clear that the basic mechanisms of chronic pain cannot be revealed by human imaging studies alone, and the use of animal models is still the key approach to studying chronic pain.

Anatomy of the ACC

The ACC is a major forebrain area for pain-related perception [1]. The ACC contains layers of pyramidal cells and local interneurons. The layers II–III contain mainly pyramidal cells, and neurons in layers II–III receive sensory inputs from the medial thalamus, a key relay nucleus for somatosensory information including pain [1,26]. Neurons in layer V are larger than cells in layers II–III and VI and also receive sensory inputs including noxious information. The ACC neurons form interconnections with the ACC neurons on the opposite side of the hemisphere through callosal projecting fibers, as well as the other cortical areas on the same and opposite sides of the brain. Most ACC neurons respond to both noxious and non-noxious stimuli. Many non-pyramidal cells are inhibitory neurons that contain GABA and/or neuropeptides. Non-pyramidal cells typically show reduced spike responses to peripheral noxious stimuli, while pyramidal cells show excitatory or increased spike responses. It is important to note that while the soma of the pyramidal cells may be located only at layer V, its distal branches can extend to other layers of the ACC, such as layer I. Thus, the evoked responses recorded from layer V cells maybe contain synaptic responses occurring at the outer layers of the ACC. Within the local circuits, inhibitory neurons often receive innervations from pyramidal cells (glutamatergic) and then release GABA onto the perisomatic region of the pyramidal cells. Based on differences in their morphology and the pattern of spike activities, pyramidal and inhibitory neurons can be identified electrophysiologically in brain slice preparations.

Excitatory transmission in the ACC

Glutamate is the major fast excitatory transmitter in the ACC. Different types of glutamate receptors, including AMPA, KA, NMDA receptors, and mGluRs, are found in the ACC. Fast synaptic responses induced by local stimulation or stimulation of thalamocortical projection pathways are mediated by AMPA/KA receptors since bath application of CNQX completely blocks fast synaptic responses [27]. In adult mouse ACC slices at physiological temperatures, NMDA receptor-mediated slow synaptic responses were also recorded from the ACC *in vitro* and *in vivo*, suggesting that NMDA receptors are tonically active in this region [3]. Recent studies using *in vivo* recordings from freely moving mice consistently demonstrate that NMDA receptors contribute to slow EPSP responses in the ACC [28].

Genetic studies of glutamate KA receptors in the ACC

Recent studies using whole-cell patch-clamp recordings from genetically modified mice show that postsynaptic KA receptors contribute to fast synaptic transmission in the ACC pyramidal neurons [27,29]. Single-shock stimulation could induce small KA EPSCs in the presence of picrotoxin, AP-5, and a selective AMPA receptor antagonist, GYKI 53655. KA EPSCs had a significantly slower rise time course and decay time constant compared with AMPA receptor-mediated EPSCs. High-frequency repetitive stimulation significantly facilitated the KA EPSCs. Genetic deletion of the GluK2 (or called GluR6) or GluK1 (or called GluR5) subunit significantly reduced, and GluK1 and 2 DKO completely abolished, KA EPSCs and KA-activated currents in the ACC pyramidal neurons. Our results show that KA receptors contribute to synaptic transmission in adult ACC pyramidal neurons (Figures 2 and 3).

In addition to contributing to synaptic transmission, integrative neurobiological evidence shows that presynaptic GluK1 containing KA receptors plays an important role in the modulation of inhibitory synaptic transmission in the ACC [30].

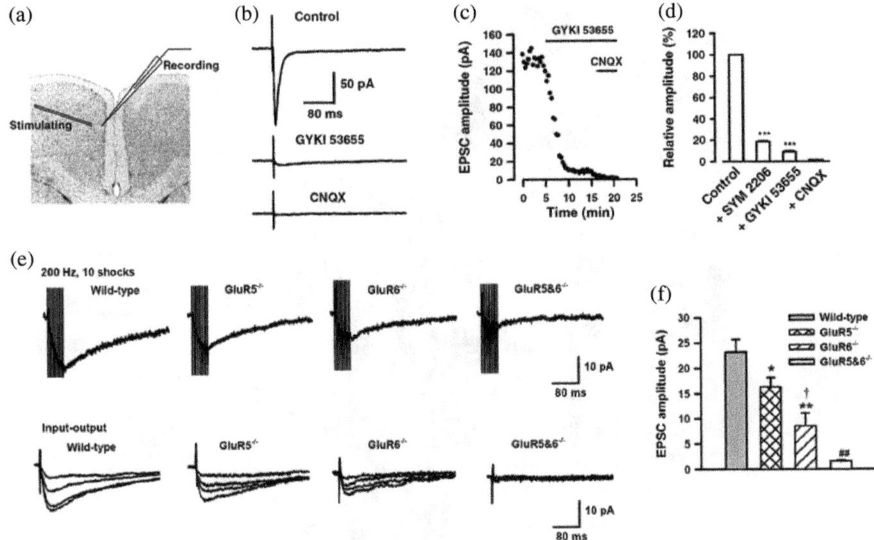

Figure 2.   KA receptor-mediated EPSCs in adult ACC pyramidal neurons. (a) Diagram showing the placement of stimulating and recording electrodes in the ACC. (b) Control EPSCs were recorded in the presence of picrotoxin (PTX, 100 μM) and AP-5 (50 μM). After the perfusion of GYKI 53655 (100 μM), a small residual current remained, which could be totally blocked by CNQX (20 μM). In the following example of EPSCs, each trace represents an average of 5–10 consecutive recordings. (c) Sample points showing the time course of GYKI 53655 and CNQX effects on the neuron shown in B. (d) Statistical results showing the percentage of EPSCs in the presence of SYM 2206, GYKI53655, or CNQX. (e,f) Repetitive stimulation (200 Hz, 10 shocks) induced KA receptor EPSCs in wild-type, GluK1 (GluK1), GluK2 (GluK2), and GluK1&2 (GluK1&6 KO) mice. Amplitude of KA receptor-mediated EPSCs was significantly decreased in GluK2, GluK1, and GluK1&6 KO mice. In this figure, * denotes significant difference between KO and wild-type mice; † denotes significant difference between GluK1 and GluK2 KO mice; # denotes significant difference between GluK1&6 KO and GluK2 KO mice (adapted from Wu *et al.* [27]).

ACC LTP

Synaptic and cellular mechanisms of LTP in the ACC are best studied using brain slice preparations. Different stimulation protocols have been found for inducing long-lasting potentiation of synaptic responses in the ACC cells. LTP can be induced using three different protocols, including

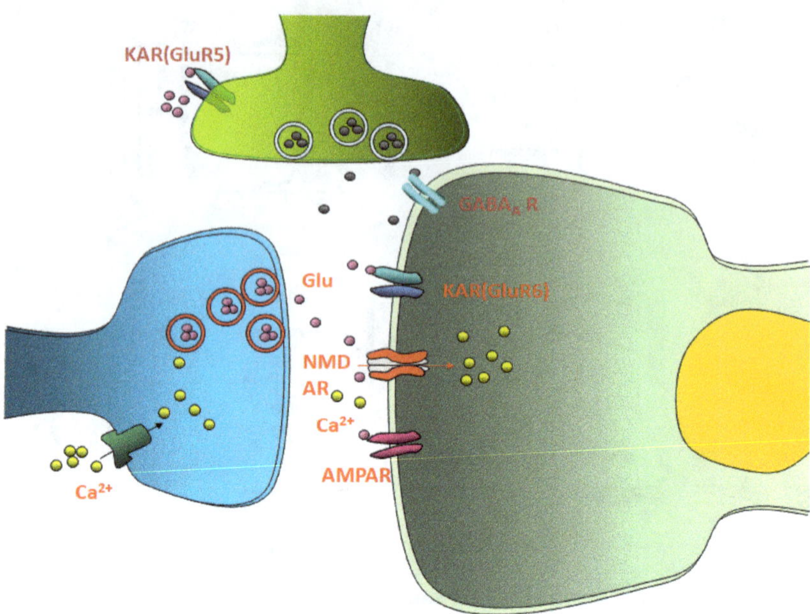

Figure 3. Excitatory and inhibitory synaptic transmission. Action potentials from the thalamic projection fibers trigger release of vesicles containing glutamate. Fast excitatory transmission is mediated by glutamate in the ACC. EPSCs are mediated by AMPA receptors, while small percentages of EPSCs are mediated by GluK2-containing KA receptors. In freely moving animals, synaptic responses are also mediated by postsynaptic NMDA receptors. Inhibitory transmission is mediated by postsynaptic GABA$_A$ receptors. GluK1 (GluK1) containing KA receptors located at presynaptic terminals of inhibitory neurons regulate the release of GABA.

the pairing training protocol, the spike-EPSP pairing protocol, and the TBS protocol [31,32]. LTP of excitatory synaptic responses can be recorded in the ACC using field recording and whole-cell patch-clamp recording techniques. For field recordings from adult rat or mouse ACC slices, glutamatergic synapses in the ACC can undergo long-lasting potentiation (at least 40–120 min) in response to TBS, a paradigm that more closely resembles the activity of the ACC neurons. Unlike in the hippocampus, where LTP contains two different phases, strong tetanic stimulation does not cause reliable long-term potentiation in the ACC. However, whole-cell patch-clamp recordings allow for better investigation of LTP synaptic mechanisms in the ACC. Activation of NMDA receptors has

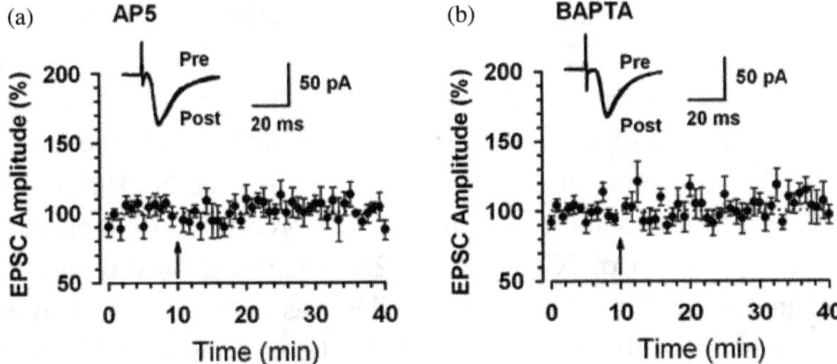

Figure 4. ACC LTP is NMDA receptor-dependent. (a,b) LTP was completely blocked by bath application of AP-5 ($n = 7$) or addition of BAPTA ($n = 7$) in the intracellular solution. The insets show averages of 6 EPSCs 5 min before and 25 min after the pairing training (arrow). The dashed line indicates the mean basal synaptic responses. Error bars represent SEM (adapted from Zhao *et al.* [32]).

been found to be important for ACC LTP (Figure 4). In addition to NMDA receptors, L-type voltage-gated calcium channels (L-VGCCs) are also required for inducing LTP. Unlike the field recordings, LTP induced by the pairing protocol is mainly triggered by the activation of NMDA receptors but not through L-VGCCs. Submerged slice preparations offer good access for pharmacological applications; however, it is poor for studying LTP (see Figure 4 for decaying LTP). These studies provide clear evidence that LTP induced by different stimulation protocols may share some common signaling pathways. It is important to keep in mind that LTP recorded through field recording or whole-cell patch-clamp recording may employ different synaptic signaling pathways for induction and expression.

NMDA receptors and the induction of LTP

NMDA receptors are well known for the induction of LTP in many central synapses [33,34]. In the ACC, the induction of LTP by different protocols is mostly NMDA receptor-dependent, and NMDA receptors containing NR2A or NR2B subunits contribute to most of the NMDA receptor currents. Bath application of NR2A antagonist NVP-AAM077 and NR2B antagonist ifenprodil/RO compounds has been shown to produce an almost complete blockade of NMDA receptor-mediated EPSCs [32].

Several recent studies have reported the non-selective effects of NVP-AAM077 on NMDA NR2A vs. NR2B receptors [35]. However, most of these recordings are performed from *in vitro* cell lines that express cloned receptors. It is hard to determine if such artificially made receptors can mimic NMDA receptors in adult ACC neurons. In fact, NVP-AAM077 at a higher dose failed to abolish all RO- or ifenprodil-sensitive currents. In the ACC, application of NR2A or NR2B antagonist only reduced LTP, rather than completely abolishing it. LTP was only abolished after co-application of both inhibitors. It should also be noted that LTP induced by spike-timing protocol appeared to be more sensitive to NMDA NR2B blockade than the LTP induced by pairing training protocol.

Calcium/CaM-related signaling pathways

Activation of glutamate NMDA receptors leads to an increase in postsynaptic Ca^{2+} in dendritic spines. Ca^{2+} serves as an important intracellular signal for triggering a series of biochemical events that contribute to the expression of LTP. Ca^{2+} binds to CaM and leads to activation of calcium-stimulated signaling pathways. Furthermore, postsynaptic injection of BAPTA completely blocks the induction of LTP, indicating the importance of elevated postsynaptic Ca^{2+} concentrations [32]. A recent work using electroporation of mutant CaM in the ACC suggests that calcium-binding sites of CaM are critical for the induction of ACC LTP [36].

cAMP signaling pathways are the key signaling pathways in biological systems. Among more than 10 subunits, AC1 and AC8 are two AC subtypes that respond positively to calcium-CaM. When compared with AC8, AC1 is more sensitive to calcium increase. In the ACC, AC1 is highly expressed in cingulate neurons located in most layers [37]. AC1 is selective for plastic changes; gene deletion of AC1 did not affect basal glutamate transmission in the ACC. By contrast, LTP induced by TBS or pairing stimulation was abolished in cingulate pyramidal cells [38]. AC1 also contributes to synaptic potentiation induced by forskolin, an AC activator that is non-selective for AC isoform. Gene deletion of AC8 subunit partially contributes to forskolin-induced potentiation (Figure 5). Whole-cell patch-clamp recording also revealed that AC1 activity is required for

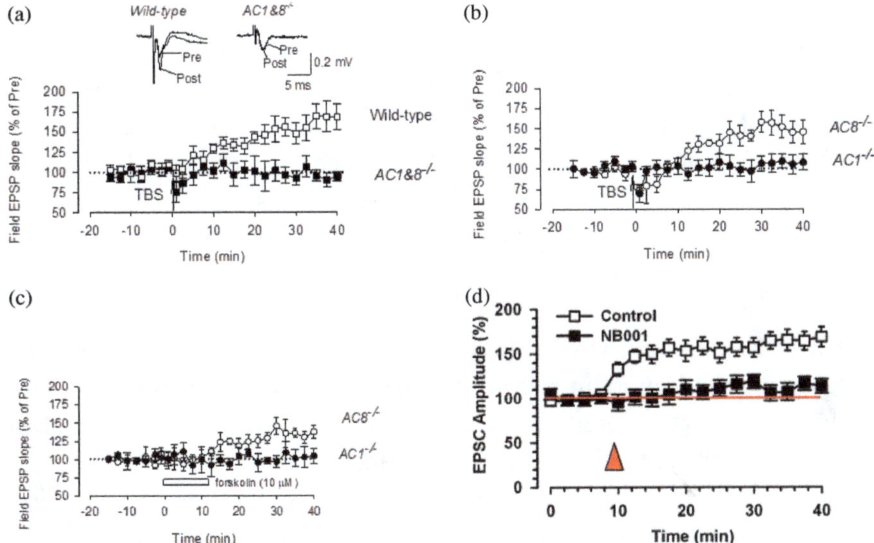

Figure 5. AC1 is required for synaptic potentiation in the ACC of adult mice. (a) LTP was abolished in mice lacking both AC1 and AC8. Insets show example traces of field EPSPs before and after TBS in wild-type and AC1&AC8 KO mice. (b) LTP was completely blocked in AC1 KO mice but not in AC8 KO mice. (c) Bath application of forskolin-induced LTP in wild-type mice but not in the AC1 and AC8 double KO mice. Forskolin-induced potentiation was blocked in AC1 KO mice but not in AC8 KO mice. (d) NB001, a selective inhibitor for AC1, blocked LTP in the ACC (adapted from Liauw *et al.* [38]).

the induction of LTP in ACC pyramidal cells. By using chemical design and biochemical screening, several selective inhibitors of AC1 have been identified. Consistently, pharmacological inhibitions of AC1 in ACC neurons abolished LTP induced by pairing protocol [39].

A major neuronal signaling pathway by which Ca^{2+} activates CREB involves the Ca^{2+}/calmodulin-dependent protein kinase type IV (CaMKIV). CaMKIV is a member of a group of multifunctional CaMK, so-called because of their broad substrate specificity, which also includes CaMKI and CaMKII. CaMKIV is distinguished among the CaM kinases in its capacity to activate CREB-dependent transcription both by virtue of its nuclear localization and catalysis of CREB phosphorylation on Serine 133. CaMKIV also promotes CREB function by activating the transcriptional co-activator CREB binding protein (CBP). Present in the ACC, CaMKIV is required for CaM translocation triggered by neural activity into the nuclei of ACC neurons. Previous studies have shown that CaM translocation reflects the trapping of Ca^{2+}/CaM complexes by nuclear CaM binding proteins. The abolishment of CaM translocation in CaMKIV KO mice identifies CaMKIV as the critical link that traps Ca^{2+}/CaM complexes in neuronal nuclei. This trapping leads to CaMKIV activation and subsequent CREB phosphorylation and activation. Consistently, we have found in both *in vitro* and *in vivo* conditions that activation of CREB was significantly reduced or abolished in CaMKIV KO mice [37]. Considering the important roles of CREB in LTP, it is not surprising that synaptic potentiation induced by TBS was also reduced or abolished in these mice. Further to support the important roles of CaMKIV in cortical potentiation, in CaMKIV transgenic mice with forebrain overexpression of CaMKIV, it was found that cingulate LTP and behavioral trace fear learning were significantly enhanced [40].

GluA1 receptor and expression of ACC LTP

At least four possible mechanisms may contribute to the expression of LTP: (1) presynaptic enhancement of glutamate release, (2) postsynaptic enhancement of glutamate receptor-mediated responses, (3) recruitment of previously "silent" synapses or synaptic trafficking or insertion of AMPA receptors, and (4) structural changes. In certain cases, under *in vitro* brain slice conditions, the LTP mechanism may have appeared to depend on the induction protocol, as paired-pulse facilitation (PPF) was not altered after the induction of cingulate LTP. However, it cannot be ruled out that possible presynaptic changes may occur in the ACC at other physiological/pathological conditions. To address these possibilities, the roles of GluA1 and GluA2/3 were investigated using genetic and pharmacological approaches. GluA1 subunit C-terminal peptide analog Pep1-TGL blocked the induction of cingulate LTP [41]. Thus, in the ACC, the interaction between the C-terminus of GluA1 and PDZ domain proteins is required for the induction of LTP. Synaptic delivery of the GluA1 subunit from extrasynaptic sites is the key mechanism underlying synaptic plasticity, and GluA1-PDZ interactions play a critical role in this plasticity.

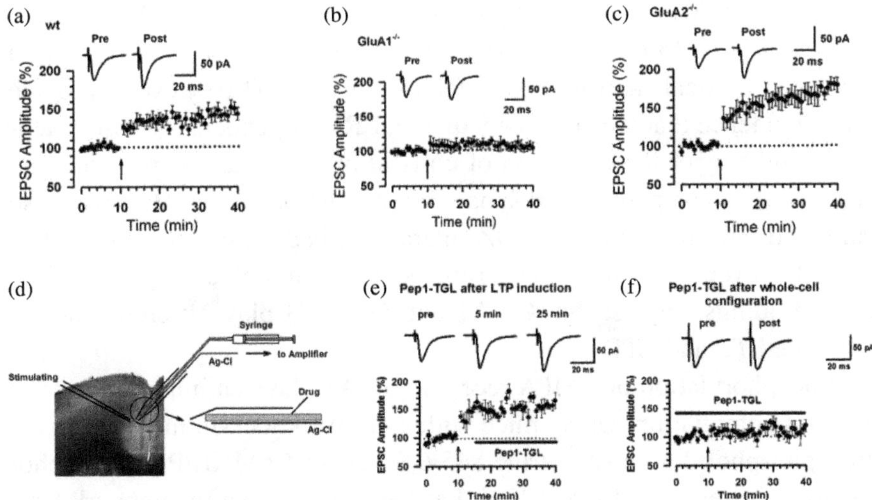

Figure 6. Abolishment of ACC LTP in GluA1$^{-/-}$ mice. (a) LTP was induced in ACC pyramidal neurons in WT mice (n = 13 slices/6 mice). (b) LTP was lost in ACC pyramidal neurons in GluA1$^{-/-}$ mice (n = 8 slices/6 mice). The insets show averages of five EPSCs at baseline responses and 30 min after the pairing procedure (arrow). The dashed line indicates the mean basal synaptic responses. (c) LTP was enhanced in ACC pyramidal neurons in GluA2$^{-/-}$ mice (n = 8 slices/ 5 mice) compared with WTCD1 mice. (d) Diagram for showing the intracellular injection of peptide inhibitor. (e) LTP was not affected by postsynaptic injection of Pep1-TGL. (f) Pre-application of the same peptide prevented LTP (adapted from Toyoda *et al.* [41]).

Pharmacological experiments show that the application of Pep1-TGL 5 min after paired training reduced synaptic potentiation while Pep1-TGL had no effect on basal responses (Figure 6). Therefore, Ca^{2+}-permeable GluA2-lacking receptors may contribute to the maintenance of LTP and are necessary for subsequent LTP stabilization. Although these data do not provide direct evidence for the synaptic trafficking or insertion of GluA1 receptors at the postsynaptic membrane, the present findings suggest selective contribution of AMPA subtype receptors to cingulate LTP.

Whole-cell patch-clamp recordings were performed from the ACC and somatosensory cortex neurons to determine the role of AMPA receptor subunits in long-term synaptic plasticity by using mice lacking the genes for GluA1 or GluA2 [42]. Furthermore, ERK1/2 phosphorylation was analyzed in these cortical regions by using mouse models of inflammation. The AMPA receptor subunits GluA1 and GluA2 differentially contribute to LTP in the ACC. Moreover, GluA1 KO mice showed decreased cortical activation of ERK1/2 *in vivo*. These results provide strong evidence that the GluA1-mediated, ERK-dependent signaling pathway could trigger the induction of cortical plasticity and persistent pain. The role of these peptides in synaptic potentiation in the ACC was also examined, and the GluA2/3-PDZ interaction had no effect on cingulate LTP. However, the same interfering peptides inhibited cingulate LTD. These findings suggest that GluA1 and GluA2/3 play different roles in cingulate LTP and LTD.

Phosphorylation of AMPA receptor GluA1 plays an important role in synaptic potentiation. Using mice with a GluA1 knock-in mutation at the PKA phosphorylation site serine 845 (s845A) or CaMKII/PKC phosphorylation site serine 831 (s831A), it has been shown that the network LTP, which is constructed by multiple recordings of LTP at different locations within the ACC, was investigated. The expression of LTP and network LTP was significantly impaired in the s845A mice but not in the s831A mice (Figure 7). By contrast, basal synaptic transmission and NMDA receptor-mediated responses were not affected [43]. These results strongly indicate that PKA phosphorylation of the GluA1 is important for the network LTP expression in the ACC.

Transcription and translocation mechanisms

While different protein kinases and second messengers have long been hypothesized to be involved in ACC LTP, recent data on the requirement of several immediate early genes and gene-related proteins in ACC LTP are surprising. Potentiation, induced in the ACC synapses within 10–30 min, is affected by the gene deletion of Egr1 and FMRP. These effects are unlikely to be due to indirect inhibition of NMDA receptors since NMDA receptor-mediated currents are not affected. The CREB is a transcription factor that plays an important role in the formation of long-term memory. In the spinal cord and forebrain, neuronal CREB is activated after tissue injury. Although the exact link between activation of CREB and persistent pain remains unclear, phosphorylated CREB

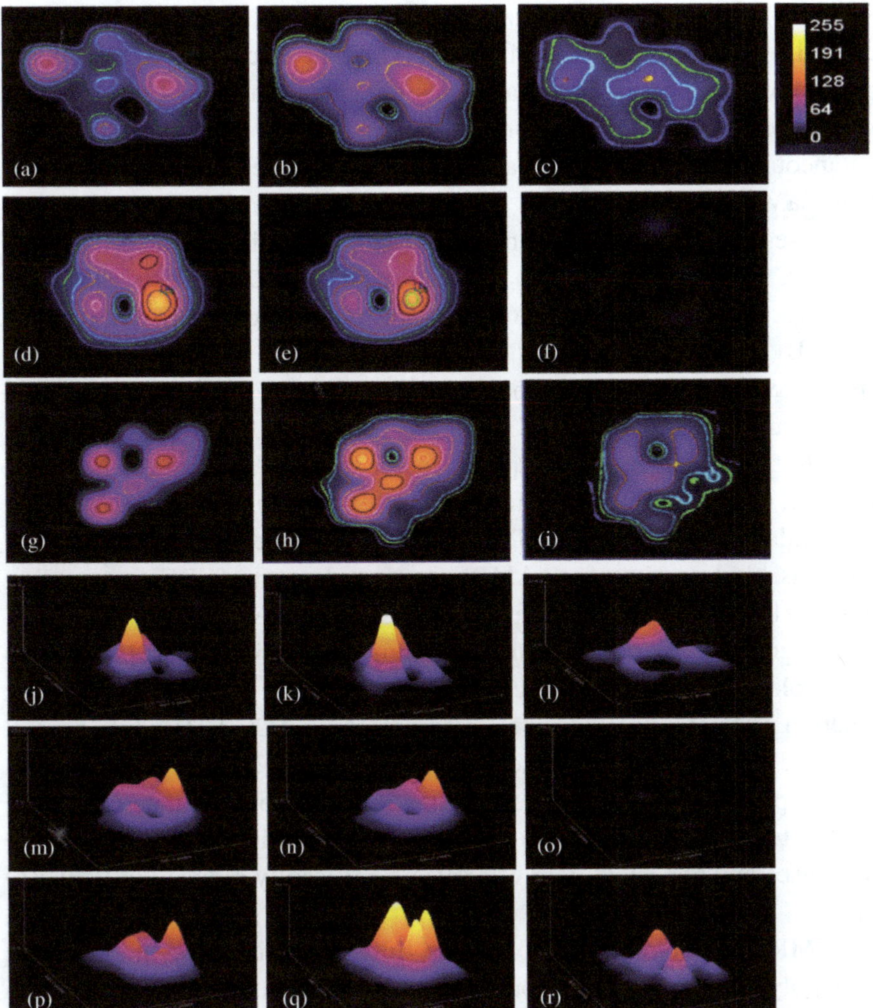

Figure 7.   The spatial distribution pattern of LTP in three types of mice. (a–c) The high-density images showed the value of fEPSP slopes before (a) and 4 h after TBS (b) in one sample slice of WT mice. Their difference is shown in (c). (d–f) The value of fEPSP slope before and after TBS in one sample slice of s845A mice. Their difference is shown in (f). (g–i) The value of fEPSP slope before and after TBS in one sample slice of s831A mice. Their difference is shown in (i). (j–l) The 3D version of (a–c), respectively. (m–o) The 3D version of (d–f), respectively. (p–r) The 3D version of (g–i), respectively (adapted from Song *et al.* [43]).

(pCREB) can be used as a marker for activation of AC1 and AC8 in the central nervous system. In addition to cAMP, the calcium/CaM-dependent protein kinase pathway also activates CREB. Data from CaMKIV KO mice revealed that the behavioral responses to acute noxious stimuli, subcutaneous formalin injection, and CFA injection were normal, indicating that CaMKIV signaling does not significantly contribute to behavioral responses to injury. In addition to CaMKIV, formalin-activated CREB in the ACC and IC was similarly reduced in AC1&8 DKO mice, as well as in AC1&8 DKO mice (Figure 6) [37]. The presence of significant residual pCREB in formalin-injected AC1&8 DKO mice indicates that Ca^{2+}-stimulated AC1 and AC8 are not the only pathways linked to CREB activation during inflammation.

The zinc finger transcription factor Egr1 (also called NGFI-A, Krox24, or zif/268) is critical for coupling extracellular signals to changes in cellular gene expression. The upstream promoter region of the Egr1 contains binding sites for cAMP-response elements (CRE), suggesting that Egr1 may act downstream from the CREB pathway. In the ACC neurons, Egr1 is activated by injury, including amputation [44,45]. One possible role of Egr1 may be to contribute to synaptic potentiation. Genetic deletion of Egr1 mice indeed showed defects in ACC LTP, while normal synaptic transmission was observed. Furthermore, NMDA receptor-mediated responses, a key component for the induction of cingulate LTP, were unaffected. Consistent with synaptic potentiation, behavioral fear responses and persistent pain were also significantly reduced in Egr1 KO mice.

FMRP is a ubiquitously expressed mRNA-binding protein associated with polyribosomes and is thought to be involved in the translational efficiency and trafficking of certain mRNAs. FMRP is predominantly a cytoplasmatic protein, but it does shuttle between the nucleus and cytoplasm, perhaps transporting selective mRNA molecules to their final destination within the cell. In neurons, FMRP is found in the head of the dendritic spine, thereby playing a role in local protein synthesis, which is considered a necessary and important component of synaptic morphology and plasticity. FMRP has been shown to function as a translational repressor for some synaptic proteins, such as Arc, α-CaMKII, and the dendritic microtubule-associated protein 1b. The pairing training has produced a significant, long-lasting potentiation of synaptic responses in WT mice. However, synaptic potentiation in slices of FMR1 KO mice was completely blocked, showing that FMRP may contribute to synaptic potentiation in the ACC neurons [31,46,47].

L-LTP in the ACC

Recently, a long-lasting form of LTP (L-LTP) has been reported from adult mouse ACC slices [48]. TBS stimulation protocol induced long-lasting potentiation of synaptic responses in the ACC slices for at least 3–6 h. This form of L-LTP is thought to at least share some of the common mechanisms with the nerve injury-induced cortical potentiation. In an occlusion experiment, L-LTP induced in ACC slices from animals with nerve injury is significantly smaller than that of normal or sham-treated mice [48]. These findings indicate that many of the excitatory synapses have already undergone potentiation after nerve injury. Genetic and pharmacological studies found that AC1 or AC1 activity was required for L-LTP in the ACC [46,47]. Furthermore, a role of tonic activity of PKMζ has been established in ACC L-LTP (Figure 8). One key function of PKMζ is to maintain cortical potentiation at individual synapses [49].

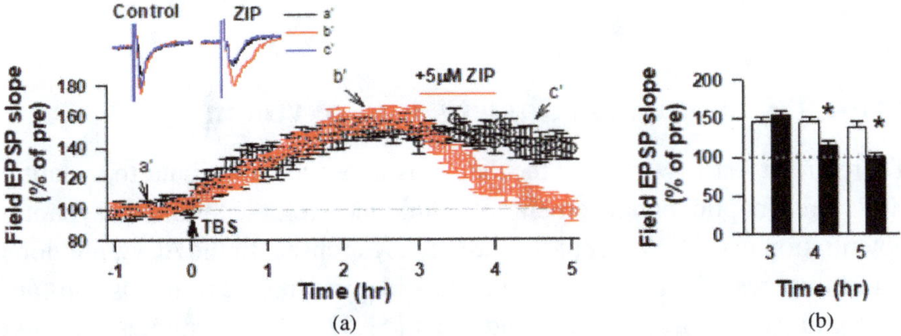

Figure 8. L-LTP in the ACC. (a) L-LTP from naïve ($n = 6$) and sham-operated mice ($n = 5$) were similar, and data were pooled together (black). Blocking of the activity of PKMζ by bath application of ZIP (5 μM) for 1 h can reverse the synaptic potentiation (blue). Sample traces showing the field EPSP recorded at 0.5 h before TBS conditioning (a'), 2.5 h (b'), and 4.5 h (c', 0.5 h after ZIP incubation) after TBS conditioning. (b) Summarized amplitudes of L-LTP at different times (adapted from Li *et al.* [48]).

To test if PKMζ was required for maintaining L-LTP in the ACC, LTP was recorded in ACC slices and bath application of a selective PKMζ inhibitor ZIP (5 μM) 3 h after induction significantly reduced the ACC L-LTP, suggesting that PKMζ is required for the maintenance of L-LTP in the ACC.

Synaptic tagging and injury

Synaptic tagging, which was first reported in the hippocampus, is also found in the ACC [50]. E-LTP and L-LTP can interact with each other in a "synaptic tagging-like" manner, where a weak tetanus inducing E-LTP sets a "tag", which can capture the plasticity-related proteins (PRPs) synthesized following the strong tetanus inducing L-LTP [50]. Consequently, a weak stimulus can induce L-LTP if it is preceded or followed by a strong tetanus given to a separate, independent pathway that converges into the same neuronal population. The synaptic tagging hypothesis is believed to play important roles in long-term memory storage/allocation. Synaptic tagging has been reported in excitatory synapses within the ACC (Figure 9) [50]. Tail amputation-induced peripheral injury caused a loss of this heterosynaptic L-LTP and occluded strong TBS-evoked L-LTP. This loss synaptic tagging-like phenomenon in the ACC of adult mice may contribute to injury-related cognitive changes and phantom limb sensation and pain.

Pre-LTP, behavioral anxiety, and oxytocin

There are at least two major mechanisms that may contribute to synaptic LTP: presynaptic enhancement of glutamate release and postsynaptic potentiation of AMPA receptor-meditated responses. In the ACC, previous studies suggested that both presynaptic and postsynaptic transmissions are up-regulated in chronic pain conditions [51,52]. Recent studies reported the existence of pre-LTP in ACC brain slices from adult mice [53–58]. The induction of pre-LTP is NMDA receptor-independent, a key feature differentiating pre- from post-LTP in the ACC. Furthermore, pre- and

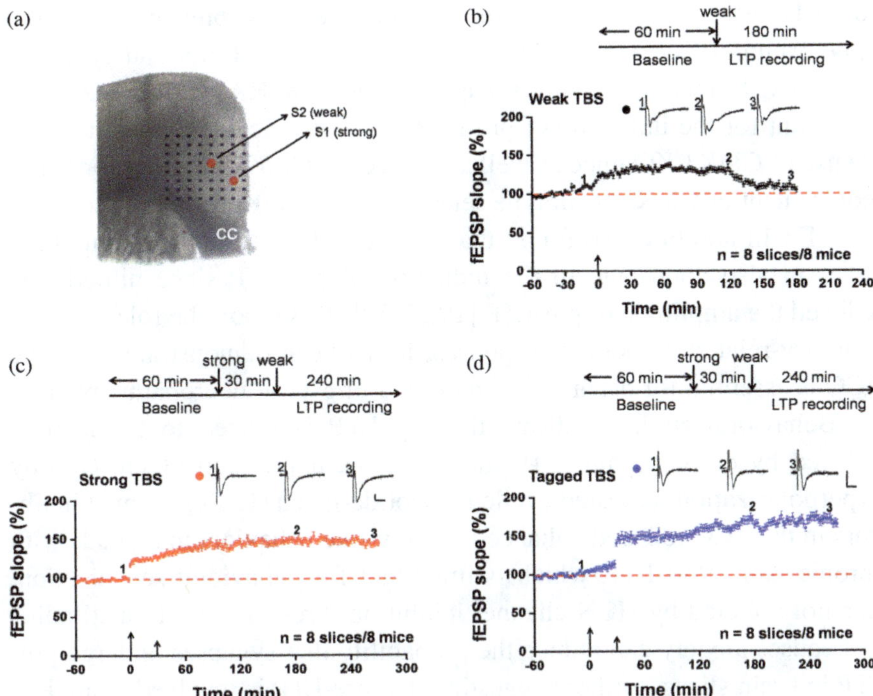

Figure 9.   Synaptic tagging occurs in the adult mouse ACC. (a) Light microscopy photograph showing the location of the MED64 probe relative to one ACC slice. We applied electrical stimulation separately to two distinct inputs as shown by the two red dots in the picture. The S1 stimulation site received strong TBS in the superficial layer, whereas the S2 site was given a weak TBS in the deep layer of ACC. (b) Summarized plot of the fEPSP slope demonstrates the induction of E-LTP by weak TBS alone that gradually decayed to the baseline within 3 h. (c) Summarized plot of the fEPSP slope demonstrates the induction of L-LTP by weak TBS if preceded, 30 min earlier, by a strong TBS (i.e., tagged TBS). Tagged TBS-evoked L-LTP lasts for at least 4 h. (d) Summarized plot of the fEPSP slope shows the induction of persistent L-LTP by strong TBS alone that lasts for 4.5 h. For b–d, schematic diagram of the recording procedure for each experiment is shown at the top of the plot. Insets show representative fEPSP traces taken at the time points indicated by numbers in the graph. Calibration: 100 μV, 10 ms. Large arrows indicate starting point of strong TBS application; small arrows mark the time point of weak TBS delivery (adapted from Liu *et al.* [50]).

post-LTP have been observed in the same neurons, presumably at the same excitatory synapses [54]. These findings suggest pre- and post-LTP can co-exist in the same adult synapses. Glutamate KA GluK1 receptor is important for the induction of pre-LTP. This was nicely demonstrated by the use of GluK1 KO mice as well as a selective GluK1 inhibitor [54]. The requirement of GluK1 is subtype selective since GluK2 is not required for pre-LTP. In addition, we found that L-type voltage-gated calcium channels can also contribute to the induction of pre-LTP since nimodipine reduced the amplitude of pre-LTP [29,53,59]. To support the roles of ACC in injury-related emotional responses, it has been demonstrated that the ACC is essential for driving the emotional impact of neuropathic pain.

Behavioral studies indicate that pre-LTP is related to the anxiety induced by chronic pain [54]. Inhibiting the expression of pre-LTP by hyperpolarization-activated cyclic nucleotide-gated (HCN) channel inhibitors in the ACC reduced enhanced anxiety-like behaviors in animals with chronic pain. Post-LTP, which is important for behavioral sensitization, was not affected by HCN channel inhibition. Pre-exposure to a stimulus that causes anxiety can reduce the probability that synapses undergo pre-LTP in brain slices, further suggesting that pre-LTP is involved in anxiety [54]. These results also suggest that pre-LTP may serve as one of the general mechanisms for heightened anxiety, in addition to anxiety specifically caused by chronic pain. In further support for the role of pre-LTP in injury-induced anxiety, it has been reported that oxytocin reduced nerve injury-induced behavioral anxiety by selectively inhibiting the maintenance of pre-LTP in the ACC [60]. A summary of both pre-LTP and post-LTP is given in Figure 10.

LTP in the IC

The interest in synaptic plasticity in another pain-related cortical area, insular cortex, has gained strength among investigators [58]. LTP, including L-LTP, can be recorded *in vivo* and *in vitro* brain slice preparation [61,62]. IC LTP induced by the pairing protocol was NMDA receptor-dependent and expressed postsynaptically since paired-pulse ratio (PPR) was not affected. Postsynaptic calcium is important for the

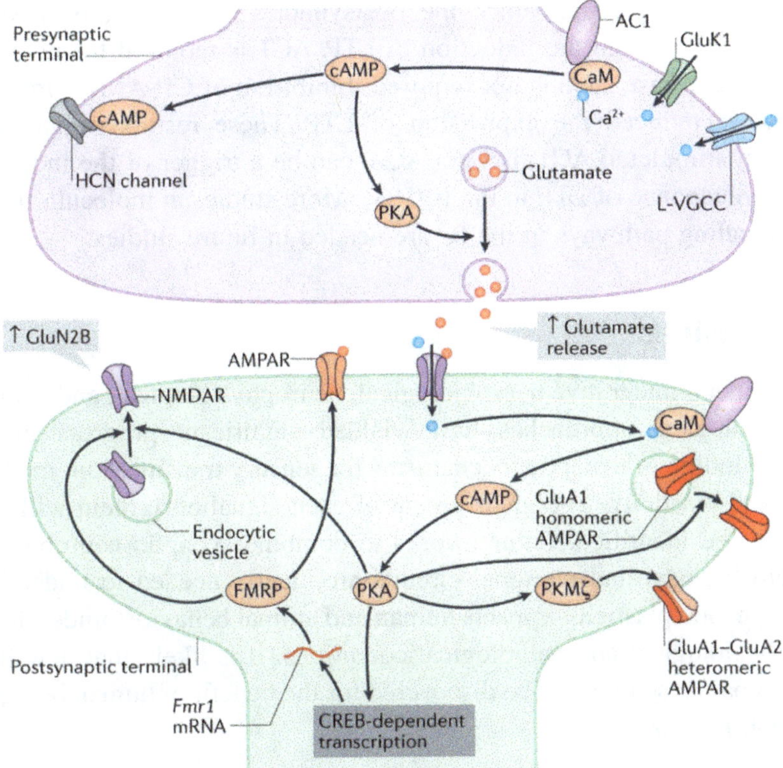

Figure 10. Two major forms of LTP in the ACC. Two forms of LTP can be recorded in the ACC: presynaptic LTP and postsynaptic LTP. The two forms may occur at the same synapses. Induction of postsynaptic LTP requires the activation of NMDA receptors, and the maintenance of postsynaptic LTP expression requires the postsynaptic activity of an atypical PKC isoform (most probably PKMζ). AC1 is also critical for the induction of postsynaptic LTP in the ACC. Activation of PKA drives the insertion of CP-AMPARs (GluA1 homomers). PKA also translocates to the nucleus where it phosphorylates the transcription factor CREB, leading to the synthesis of several downstream plasticity proteins, including FMRP. PKMζ may maintain LTP by upregulating GluA1/2 heteromers. In addition, increased FMRP may enhance the postsynaptic function of AMPA receptors as well as NMDA receptors. Pre-synaptic LTP in the ACC is not dependent on NMDA receptors; instead, presynaptic KA receptors and AC1 activation are necessary for its induction, and the expression of presynaptic LTP may require the activity of HCN channels. cAMP binds to the HCN channel to increase its sensitivity, and PKA enhances vesicle fusion (adapted from Bliss *et al.* [64]).

induction of post-LTP since the postsynaptic application of BAPTA completely blocked the induction of LTP. AC1 is required for potentiation. By contrast, AC8 is not required. Inhibition of CP-AMPA receptor or PKMζ reduced the expression of LTP. These results suggest that calcium-stimulated AC1, but not AC8, can be a trigger of the induction and maintenance of LTP in the IC [63]. More studies at molecular levels for signaling pathways in the IC are needed in future studies.

Conclusions

In summary, integrative neurobiological, neurophysiological, and neuropharmacological approaches have yielded significant progress in our understanding of synaptic mechanisms for sensory transmission, modulation, and plasticity. The involvement of each signaling protein will continue to be identified. Future works to combine synaptic neurobiology, system biology, and neuronal circuits are clearly needed to understand how synaptic plasticity impacts human and animal behaviors under different physiological and pathological conditions. It is likely that new findings in pain research will be discovered for the benefit of human beings in the near future.

References

[1] Vogt, B.A. (2005) Pain and emotion interactions in subregions of the cingulate gyrus. *Nat Rev Neurosci* 6, 533–544. 10.1038/nrn1704.

[2] Neugebauer, V. *et al.* (2009) Forebrain pain mechanisms. *Brain Res Rev* 60, 226–242. 10.1016/j.brainresrev.2008.12.014.

[3] Zhuo, M. (2008) Cortical excitation and chronic pain. *Trends Neurosci* 31, 199–207. 10.1016/j.tins.2008.01.003.

[4] Zhuo, M. (2007) A synaptic model for pain: Long-term potentiation in the anterior cingulate cortex. *Mol Cells* 23, 259–271.

[5] Zhuo, M. (2007) Neuronal mechanism for neuropathic pain. *Mol Pain* 3, 14. 10.1186/1744-8069-3-14.

[6] Hutchison, W.D. *et al.* (1999) Pain-related neurons in the human cingulate cortex. *Nat Neurosci* 2, 403–405. 10.1038/8065.

[7] Price, D.D. (2000) Psychological and neural mechanisms of the affective dimension of pain. *Science* 288, 1769–1772. 10.1126/science.288.5472.1769.

[8]   Rainville, P. (2002) Brain mechanisms of pain affect and pain modulation. *Curr Opin Neurobiol* 12, 195–204. 10.1016/s0959-4388(02)00313-6.

[9]   Apkarian, A.V. *et al.* (2005) Human brain mechanisms of pain perception and regulation in health and disease. *Eur J Pain* 9, 463–484. 10.1016/j.ejpain.2004.11.001.

[10]   Craig, A.D. *et al.* (1996) Functional imaging of an illusion of pain. *Nature* 384, 258–260. 10.1038/384258a0.

[11]   Dunckley, P. *et al.* (2005) Cortical processing of visceral and somatic stimulation: Differentiating pain intensity from unpleasantness. *Neuroscience* 133, 533–542. 10.1016/j.neuroscience.2005.02.041.

[12]   Singer, T. *et al.* (2004) Empathy for pain involves the affective but not sensory components of pain. *Science* 303, 1157–1162. 10.1126/science.1093535.

[13]   Morgan, V. *et al.* (2005) Amitriptyline reduces rectal pain related activation of the anterior cingulate cortex in patients with irritable bowel syndrome. *Gut* 54, 601–607. 10.1136/gut.2004.047423.

[14]   Stern, J. *et al.* (2006) Persistent EEG overactivation in the cortical pain matrix of neurogenic pain patients. *Neuroimage* 31, 721–731. 10.1016/j.neuroimage.2005.12.042.

[15]   Wei, F. and Zhuo, M. (2001) Potentiation of sensory responses in the anterior cingulate cortex following digit amputation in the anaesthetised rat. *J Physiol* 532, 823–833. 10.1111/j.1469-7793.2001.0823e.x.

[16]   Kaas, J.H. *et al.* (1999) Subcortical contributions to massive cortical reorganizations. *Neuron* 22, 657–660. 10.1016/s0896-6273(00)80725-4.

[17]   Flor, H. *et al.* (2006) Phantom limb pain: A case of maladaptive CNS plasticity? *Nat Rev Neurosci* 7, 873–881. 10.1038/nrn1991.

[18]   Schmidt-Wilcke, T. *et al.* (2006) Affective components and intensity of pain correlate with structural differences in gray matter in chronic back pain patients. *Pain* 125, 89–97. 10.1016/j.pain.2006.05.004.

[19]   Kuchinad, A. *et al.* (2007) Accelerated brain gray matter loss in fibromyalgia patients: Premature aging of the brain? *J Neurosci* 27, 4004–4007. 10.1523/JNEUROSCI.0098-07.2007.

[20]   Draganski, B. *et al.* (2006) Decrease of thalamic gray matter following limb amputation. *Neuroimage* 31, 951–957. 10.1016/j.neuroimage.2006.01.018.

[21]   Derbyshire, S.W. *et al.* (2004) Cerebral activation during hypnotically induced and imagined pain. *Neuroimage* 23, 392–401. 10.1016/j.neuroimage.2004.04.033.

[22]   Zhuo, M. (2005) Canadian association of neuroscience review: Cellular and synaptic insights into physiological and pathological pain. EJLB-CIHR

Michael Smith Chair in Neurosciences and Mental Health lecture. *Can J Neurol Sci* 32, 27–36. 10.1017/s031716710001684x.

[23] Logothetis, N.K. *et al.* (2001) Neurophysiological investigation of the basis of the fMRI signal. *Nature* 412, 150–157. 10.1038/35084005.

[24] Logothetis, N.K. (2007) The ins and outs of fMRI signals. *Nat Neurosci* 10, 1230–1232. 10.1038/nn1007-1230.

[25] Viswanathan, A. and Freeman, R.D. (2007) Neurometabolic coupling in cerebral cortex reflects synaptic more than spiking activity. *Nat Neurosci* 10, 1308–1312. 10.1038/nn1977.

[26] Wang, C.C. and Shyu, B.C. (2004) Differential projections from the mediodorsal and centrolateral thalamic nuclei to the frontal cortex in rats. *Brain Res* 995, 226–235. 10.1016/j.brainres.2003.10.006.

[27] Wu, L.J. *et al.* (2005) Kainate receptor-mediated synaptic transmission in the adult anterior cingulate cortex. *J Neurophysiol* 94, 1805–1813. 10.1152/jn.00091.2005.

[28] Wu, L.J. *et al.* (2005) Upregulation of forebrain NMDA NR2B receptors contributes to behavioral sensitization after inflammation. *J Neurosci* 25, 11107–11116. 10.1523/JNEUROSCI.1678-05.2005.

[29] Zhuo, M. (2017) Cortical kainate receptors and behavioral anxiety. *Mol Brain* 10, 16. 10.1186/s13041-017-0297-8.

[30] Wu, L.J. *et al.* (2007) Genetic and pharmacological studies of GluR5 modulation of inhibitory synaptic transmission in the anterior cingulate cortex of adult mice. *Dev Neurobiol* 67, 146–157. 10.1002/dneu.20331.

[31] Zhao, M.G. *et al.* (2005) Deficits in trace fear memory and long-term potentiation in a mouse model for fragile X syndrome. *J Neurosci* 25, 7385–7392. 10.1523/JNEUROSCI.1520-05.2005.

[32] Zhao, M.G. *et al.* (2005) Roles of NMDA NR2B subtype receptor in prefrontal long-term potentiation and contextual fear memory. *Neuron* 47, 859–872. 10.1016/j.neuron.2005.08.014.

[33] Tang, Y.P. *et al.* (1999) Genetic enhancement of learning and memory in mice. *Nature* 401, 63–69. 10.1038/43432.

[34] Wei, F. *et al.* (2001) Genetic enhancement of inflammatory pain by forebrain NR2B overexpression. *Nat Neurosci* 4, 164–169. 10.1038/83993.

[35] Zhang, X.H. *et al.* (2008) Induction- and conditioning-protocol dependent involvement of NR2B-containing NMDA receptors in synaptic potentiation and contextual fear memory in the hippocampal CA1 region of rats. *Mol Brain* 1, 9. 10.1186/1756-6606-1-9.

[36] Wei, F. *et al.* (2003) Calmodulin regulates synaptic plasticity in the anterior cingulate cortex and behavioral responses: A microelectroporation study in adult rodents. *J Neurosci* 23, 8402–8409.

[37] Wei, F. *et al.* (2002) Genetic elimination of behavioral sensitization in mice lacking calmodulin-stimulated adenylyl cyclases. *Neuron* 36, 713–726. 10.1016/s0896-6273(02)01019-x.

[38] Liauw, J. *et al.* (2005) Calcium-stimulated adenylyl cyclases required for long-term potentiation in the anterior cingulate cortex. *J Neurophysiol* 94, 878–882. 10.1152/jn.01205.2004.

[39] Wang, H. *et al.* (2011) Identification of an adenylyl cyclase inhibitor for treating neuropathic and inflammatory pain. *Sci Transl Med* 3, 65ra63. 10.1126/scitranslmed.3001269.

[40] Wu, L.J. *et al.* (2008) Genetic enhancement of trace fear memory and cingulate potentiation in mice overexpressing Ca2+/calmodulin-dependent protein kinase IV. *Eur J Neurosci* 27, 1923–1932. 10.1111/j.1460-9568.2008.06183.x.

[41] Toyoda, H. *et al.* (2007) Time-dependent postsynaptic AMPA GluR1 receptor recruitment in the cingulate synaptic potentiation. *Dev Neurobiol* 67, 498–509. 10.1002/dneu.20380.

[42] Toyoda, H. *et al.* (2009) Roles of the AMPA receptor subunit GluA1 but not GluA2 in synaptic potentiation and activation of ERK in the anterior cingulate cortex. *Mol Pain* 5, 46. 10.1186/1744-8069-5-46.

[43] Song, Q. *et al.* (2017) Selective phosphorylation of AMPA receptor contributes to the network of long-term potentiation in the anterior cingulate cortex. *J Neurosci* 37, 8534–8548. 10.1523/JNEUROSCI.0925-17.2017.

[44] Wei, F. *et al.* (1999) Loss of synaptic depression in mammalian anterior cingulate cortex after amputation. *J Neurosci* 19, 9346–9354.

[45] Ko, S.W. *et al.* (2005) Selective contribution of Egr1 (zif/268) to persistent inflammatory pain. *J Pain* 6, 12–20. 10.1016/j.jpain.2004.10.001.

[46] Chen, T. *et al.* (2014) Pharmacological rescue of cortical synaptic and network potentiation in a mouse model for fragile X syndrome. *Neuropsychopharmacology* 39, 1955–1967. 10.1038/npp.2014.44.

[47] Chen, T. *et al.* (2014) Adenylyl cyclase subtype 1 is essential for late-phase long term potentiation and spatial propagation of synaptic responses in the anterior cingulate cortex of adult mice. *Mol Pain* 10, 65. 10.1186/1744-8069-10-65.

[48] Li, X.Y. *et al.* (2010) Alleviating neuropathic pain hypersensitivity by inhibiting PKMzeta in the anterior cingulate cortex. *Science* 330, 1400–1404. 10.1126/science.1191792.

[49] King, T. *et al.* (2009) Unmasking the tonic-aversive state in neuropathic pain. *Nat Neurosci* 12, 1364–1366. 10.1038/nn.2407.

[50] Liu, M.G. *et al.* (2018) Loss of synaptic tagging in the anterior cingulate cortex after tail amputation in adult mice. *J Neurosci* 38, 8060–8070. 10.1523/JNEUROSCI.0444-18.2018.

[51] Zhao, M.G. *et al.* (2006) Enhanced presynaptic neurotransmitter release in the anterior cingulate cortex of mice with chronic pain. *J Neurosci* 26, 8923–8930. 10.1523/JNEUROSCI.2103-06.2006.

[52] Xu, H. *et al.* (2008) Presynaptic and postsynaptic amplifications of neuropathic pain in the anterior cingulate cortex. *J Neurosci* 28, 7445–7453. 10.1523/JNEUROSCI.1812-08.2008.

[53] Koga, K. *et al.* (2015) Coexistence of two forms of LTP in ACC provides a synaptic mechanism for the interactions between anxiety and chronic pain. *Neuron* 86, 1109. 10.1016/j.neuron.2015.05.016.

[54] Koga, K. *et al.* (2015) Impaired presynaptic long-term potentiation in the anterior cingulate cortex of Fmr1 knock-out mice. *J Neurosci* 35, 2033–2043. 10.1523/JNEUROSCI.2644-14.2015.

[55] Yamanaka, M. *et al.* (2016) Pre-LTP requires extracellular signal-regulated kinase in the ACC. *Mol Pain* 12. 10.1177/1744806916647373.

[56] Koga, K. *et al.* (2017) SCRAPPER selectively contributes to spontaneous release and presynaptic long-term potentiation in the anterior cingulate cortex. *J Neurosci* 37, 3887–3895. 10.1523/JNEUROSCI.0023-16.2017.

[57] Zhuo, M. (2016) Neural mechanisms underlying anxiety-chronic pain interactions. *Trends Neurosci* 39, 136–145. 10.1016/j.tins.2016.01.006.

[58] Zhuo, M. (2016) Contribution of synaptic plasticity in the insular cortex to chronic pain. *Neuroscience* 338, 220–229. 10.1016/j.neuroscience.2016.08.014.

[59] Zhuo, M. (2017) Ionotropic glutamate receptors contribute to pain transmission and chronic pain. *Neuropharmacology* 112, 228–234. 10.1016/j.neuropharm.2016.08.014.

[60] Li, X.H. *et al.* (2021) Oxytocin in the anterior cingulate cortex attenuates neuropathic pain and emotional anxiety by inhibiting presynaptic long-term potentiation. *Cell Rep* 36, 109411. 10.1016/j.celrep.2021.109411.

[61] Jones, M.W. *et al.* (1999) Molecular mechanisms of long-term potentiation in the insular cortex in vivo. *J Neurosci* 19, RC36.

[62] Liu, M.G. *et al.* (2013) Long-term potentiation of synaptic transmission in the adult mouse insular cortex: Multielectrode array recordings. *J Neurophysiol* 110, 505–521. 10.1152/jn.01104.2012.

[63] Yamanaka, M. *et al.* (2017) Calcium-stimulated adenylyl cyclase subtype 1 (AC1) contributes to LTP in the insular cortex of adult mice. *Heliyon* 3, e00338. 10.1016/j.heliyon.2017.e00338.

[64] Bliss, T.V. *et al.* (2016) Synaptic plasticity in the anterior cingulate cortex in acute and chronic pain. *Nat Rev Neurosci* 17, 485–496. 10.1038/nrn.2016.68.

Chapter 6

Cortical Depression

News and Views

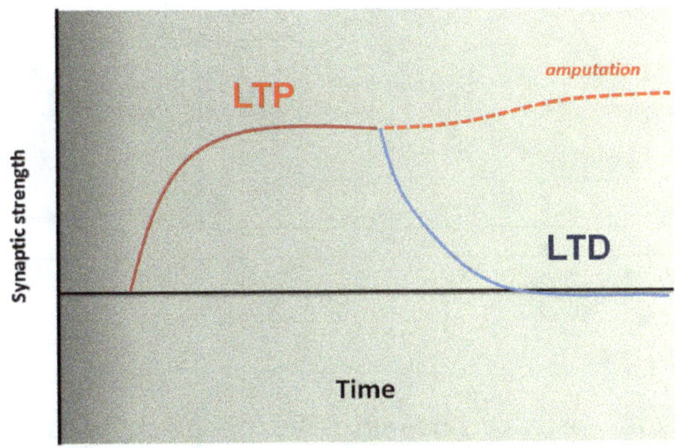

LTD is a form of synaptic plasticity that resets LTD in the ACC. In physi-ological conditions, it serves to bring abnormal synaptic responses back to normal levels. In pathological conditions, the loss of LTD may contrib-ute to the enhancement of cortical responses after the injury. A recent study reported by Wang *et al.* found that resetting ACC LTD reduced behavioral allodynia induced by nerve injury (see Wang *et al.*, 2020).

Graphical Abstract

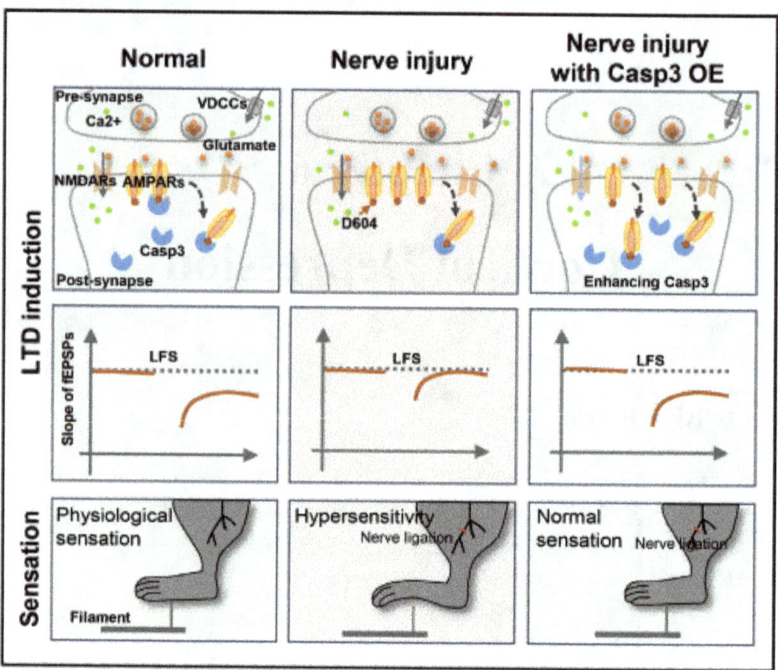

Summary

Synaptic LTD is mostly carried out in the hippocampus, a region critical for certain forms of learning and memory. ACC LTD has been reported in adult animals. However, few reports of LTD are available for pain-related cortical areas. We know that cortical LTD can be modified by peripheral injury. Synaptic LTD in the ACC may contribute to enhanced neuronal responses to subsequent somatosensory stimuli after peripheral injuries, and the investigation of LTD mechanisms may provide a new way to treat chronic pain.

Keywords: ACC; IC; LTD; NMDA receptor; L-VGCCs; mGluRs; amputation; chronic pain

Introduction

As described in Chapter 5, synaptic transmission in pain-related synapses can undergo LTP. If memory and the loss of memory depend on synaptic plasticity mechanisms, synaptic strength should be bidirectional. That is to say, both LTP and LTD should correlate with each other and LTD should play a complementary role to LTP. LTD is an activity-dependent, long-lasting decrease in synaptic strength [1]. Homosynaptic LTD has been observed in many brain regions, such as the hippocampus, visual cortex, striatum, perirhinal cortex, and prefrontal cortex [1–3]. In the cerebellum, it has been clearly shown that a decrease in synaptic transmission during LTD may form the basis for physiological functions. It has also been demonstrated that LTD can be induced *in vivo*. Thus, research has focused mainly on the relationship between LTD and learning and memory.

The ACC is a key cortical area for pain perception and emotion [4–7]. Modification of ACC synaptic transmission affects behavioral and/or emotional responses to painful stimuli. Thus, it is clear that ACC LTD plays an important role in pain and emotion.

LTD in mammalian central nervous system

LTP and LTD are artificially induced phenomena used to demonstrate the capability of the synapses to undergo long-lasting modifications [4]. Throughout the brain, there are various forms of LTD at excitatory synapses, and the brain seems to take the advantage of this kind of neuronal capability as one of the cellular mechanisms for learning and memory. The first evidence of LTD was found in the cerebellum at parallel fiber to Purkinje cell. This LTD has been implicated in several forms of motor learning, including associative eyelid conditioning and adaptation of the vestibulo-ocular reflex.

The existence of LTD has been fully established in mammalian central nervous systems in various preparations including the intact brain, slices, and cell cultures [3,8,9]. Several mechanistically distinct forms of homosynaptic LTD are found to exist, or coexist, in the hippocampus and cerebellum. One form is dependent on the activation of NMDA receptors and postsynaptic protein phosphatases or presynaptic NMDA receptors. The second form of LTD requires activation of the metabotropic gluta-mate receptors (mGluR1 or mGluR5). Stable expression of this kind of LTD requires rapid and dendritic mRNA expression, and protein synthesis inhibitors to block the synaptic changes. The third form of LTD depends on postsynaptic release of endocannabinoids, which then act retrogradely on presynaptic CB1 receptors and presynaptic NMDA receptors, or coin-cident activation of postsynaptic mGluR1.

Frequency-dependent relationship for inducing LTD

One key feature for inducing LTP vs LTD is the frequency of stimulation used. In case of LTP, higher frequency of stimulation is typically required. Tetanic stimulation is sufficient to maximally activate postsynaptic NMDA receptors, thus trigger postsynaptic signaling pathways that are required for LTP. Unlike LTP, low-frequency stimulation for a longer period of time is required for inducing LTD. It is often reported that LTD is age-dependent as the frequency of LTD is reduced in adult hippocampal synapses. Unlike LTP, it is difficult to obtain long-lasting LTD in adult hippocampus (greater than 3–6 h). In ACC slices, LFS induced LTD in slices of adult mice (Figure 1) [5]. Recent studies using multiple elec-trodes recording system further confirm LTD in adult ACC slices [12].

LTD in ACC pyramidal neurons

Conventional whole-cell patch-clamp recordings from visually identified pyramidal neurons in layer III of ACC slices were used to study ACC LTD. Fast EPSCs were obtained by delivering focal electrical stimulation to layer V. In addition to visual identification, the recordings were performed from cortical pyramidal cells by injecting depolarizing currents into the neuron. Injection of depolarizing currents into neurons induced repetitive action potentials with frequency adaptation that is typical of the

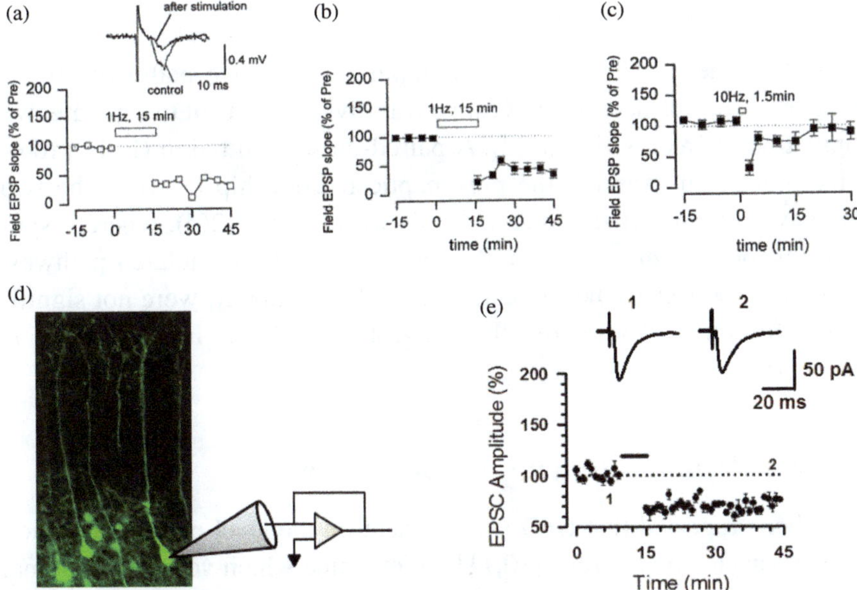

Figure 1.   ACC LTD recorded by field recording or whole-cell patch-clamp. (a) Low-frequency stimulation produced LTD in the ACC. An example that low-frequency stimulation (1 Hz) produced a long-lasting depression of synaptic responses. Inset is representative records of the EPSP recorded before and 30 min after 1 Hz stimulation. (b) Summary results for LTD ($n = 10$; $36.3 \pm 6.3\%$ of control 30 min after the stimulation; $P < 0.01$ compared with EPSPs before stimulation). (c) Repetitive stimulation at 10 Hz stimulation for 1.5 min did not produce any depression (squares; $n = 5$; $80.8 \pm 11.9\%$ of control). (d) Diagram of a cingulate cortical slice showing the placement of a whole-cell patch recording and a stimulation electrode. LTD of whole-cell EPSCs is recorded in layer II/III of the ACC in response to stimulation of layer V of the ACC. (e) LTD is induced by 300 pulses at 1 Hz while holding at −45 mV. The insets show averaged EPSCs during baseline responses, 5 and 25 min after the pairing procedure (bar) (adapted from Wei *et al.* [24]).

firing pattern of pyramidal neurons. The pairing of synaptic activity with postsynaptic depolarization (pairing training) (300 pulses of presynaptic stimulation at 1 Hz in layer V with postsynaptic depolarization at −45 mV) induced LTD of synaptic responses in the ACC (Figure 1). In the control group, in which neurons were not subjected to pairing training, no change in EPSC amplitude was observed for the duration of recordings.

ACC LTD is input specific

To test whether LTD in the ACC is input-specific, two-pathway experiments were performed. An ACC slice was divided by a cut to separate two populations of axons (Figure 2). A paired-pulse facilitation (PPF) protocol was used to examine the presynaptic relationship between the two pathways (Figures 2(b) and (c)). As shown in Figure 2(d), whereas synaptic responses were significantly depressed in the stimulated pathway, synaptic responses in the second, independent pathway, were not significantly affected. These results suggest that LTD in the ACC is input-specific.

Voltage-dependent synaptic plasticity

It has been reported that LTD is dependent on the level of depolarization of the postsynaptic neurons [10,11]. To examine which voltage is optimal for induction of LTD, we used different holding potentials (−70, −30, −15, and +30 mV) during presynaptic stimulation. Holding potentials at −70, −30, and −15 mV could not induce LTD (Figure 3). By contrast, a holding potential at +30 mV induced long-lasting potentiation of responses (Figure 3). These results suggest that neurons that were only slightly depolarized (−45 mV) during presynaptic stimulation were in a suitable condition for the induction of LTD.

Requirement of NMDA receptor activation and postsynaptic calcium

In many brain regions — including the hippocampus, visual cortex, somatosensory cortex, and peripheral cortex — the induction of LTD was dependent on the synaptic activation of NMDA receptors [2,12,13]. First, NMDA receptor EPSCs were isolated pharmacologically, and these NMDA receptor EPSCs were completely blocked by bath application of a selective NMDA receptor antagonist, 50 μM AP5 (Figure 4). To determine if NMDA receptor activation is required for LTD induction within

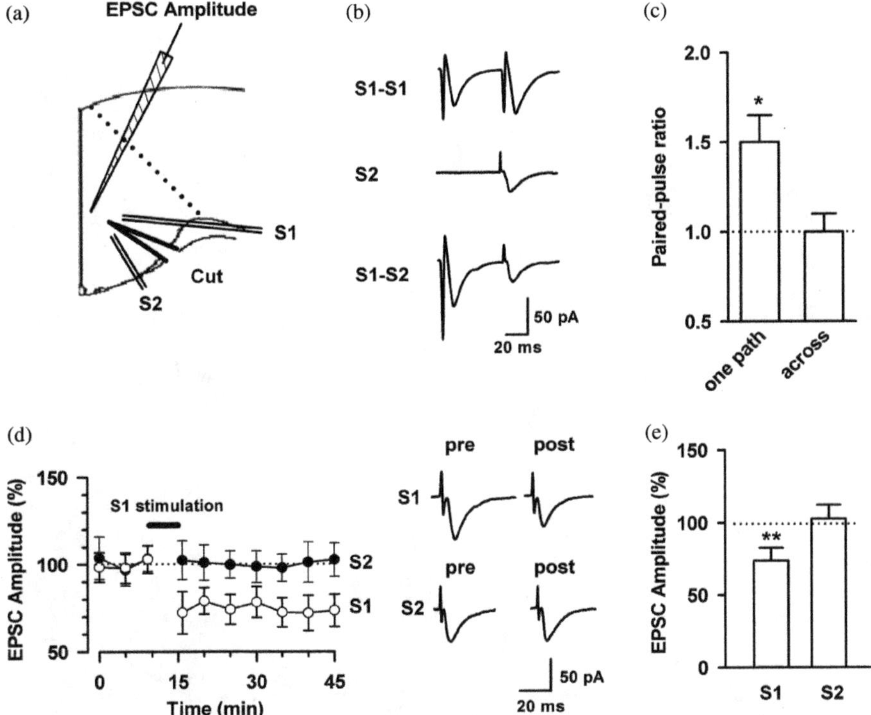

Figure 2. Two-pathway experiments show that LTD is input-specific. (a) Diagram of an ACC slice showing the placement of stimulating electrodes in two divided pathways (S1, S2). EPSCs are recorded in ACC neurons in whole-cell patch-clamp. (b) Paired pulse to one pathway shows PPF, while cross-facilitation is not observed between pathways. (c) Statistical summary of paired-pulse ratio between two pathways ($n = 6$). PPF is calculated as the ratio of the second EPSC amplitude to the first EPSC amplitude. $*P < 0.05$ compared with across. (d) Summary of results for two pathway experiments ($n = 6$). Stimulated pathway (S1) only shows LTD. Sample traces show averaged EPSCs during baseline responses and 25 min after the pairing procedure (bar). (e) Statistical summary of EPSCs between two pathways ($n = 6$). $**P < 0.01$ compared with baseline responses (adapted from Toyoda *et al.* [15]).

the ACC, AP5 (50 μM) was applied and LTD induced by pairing training was completely blocked (Figure 4).

It has been clearly shown that influx of Ca^{2+} into the postsynaptic cell and the resulting rise in intracellular Ca^{2+} concentration are

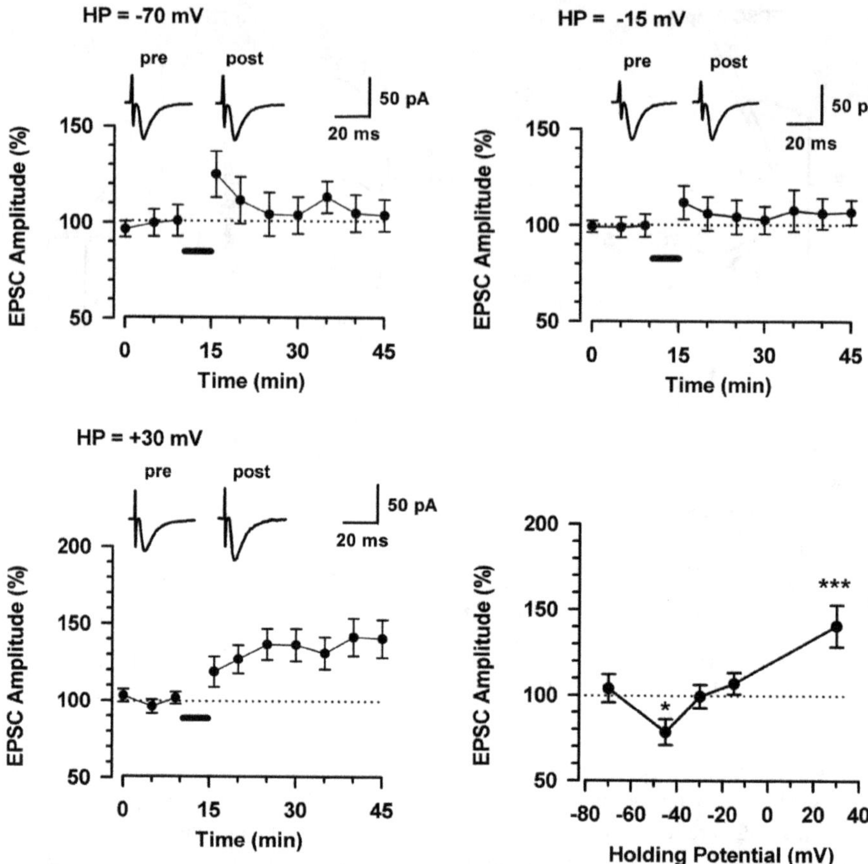

Figure 3.   Effects of voltage-dependent thresholds on induction of LTD. (a,b) Different holding potentials (HP) at −70 and −15 mV fail to induce LTD (−70 mV, 103.9 ± 8.2% of baseline responses, $n = 6$; −15 mV, 106.9 ± 6.3% of baseline responses, $n = 7$). The insets show averaged EPSCs during baseline responses and 25 min after the pairing procedure (bar). (c) Synaptic potentiation is induced by holding potentials at +30 mV in pyramidal ACC neurons (140.6 ± 12.2% of baseline responses, $n = 10$). The insets show averaged EPSCs during baseline responses and 25 min after the pairing procedure (bar). (d) Summary results for the effects of different holding potentials (−70, −45, −30, −15, and +30 mV). Only a slightly depolarized (−45 mV) holding potential can induce LTD. *$P < 0.05$ and ***$P < 0.001$ compared with baseline responses (adapted from Toyoda *et al.* [15]).

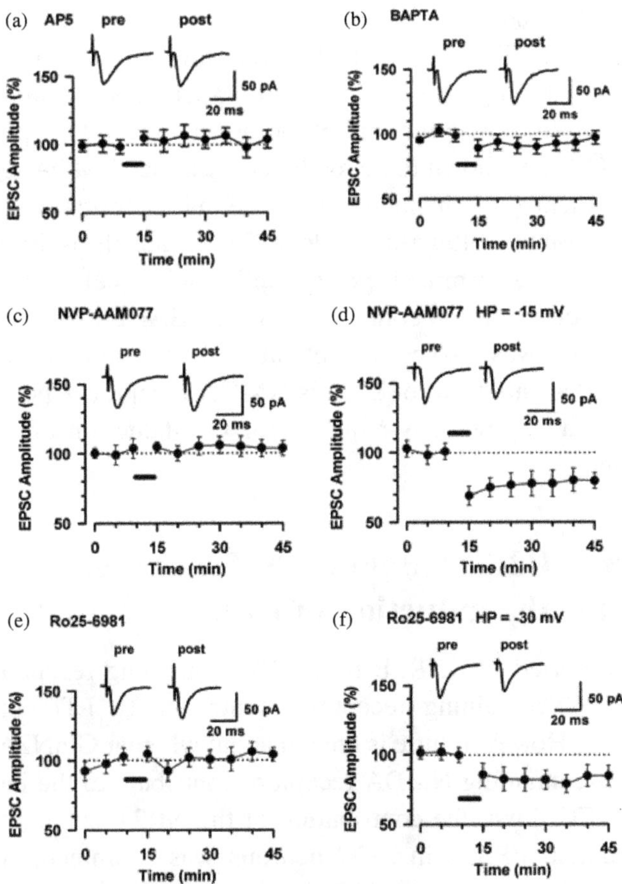

Figure 4.   LTD induced by postsynaptic NMDA receptor activation. (a,b) LTD is completely blocked by bath applied AP5 (50 μm, $n = 8$) and BAPTA (11 mm, $n = 9$) in the intracellular solution. Insets show averaged EPSCs recorded during baseline responses and 25 min after the pairing procedure (bar). (c) LTD is completely blocked by NVP-AAM077 (0.4 μm, $n = 6$, 103.8 $\pm$ 5.4% of baseline responses). (d) In the presence of NVP-AAM077, a holding potential at -15 mV during presynaptic stimulation induces LTD (79.3 $\pm$ 4.8%, $n = 6$, $P < 0.05$ compared with baseline responses). Insets show averaged EPSCs recorded during baseline responses and 25 min after the pairing procedure (bar). (e) LTD is completely blocked by Ro25-6981 (0.3 μM, $n = 7$, 104.9 $\pm$ 4.9% of baseline responses) or ifenprodil (3 μM, $n = 9$, 97.0 $\pm$ 6.7% of baseline responses). (e,f) In the presence of Ro25-6981, a holding potential at -15 mV during presynaptic stimulation does not induce LTD (96.7 $\pm$ 4.6%, $n = 7$, $P > 0.05$ compared with baseline responses), but a holding potential at -30 mV in the presence of Ro25-6981 induces LTD (84.8 $\pm$ 7.3%, $n = 6$, $P < 0.05$ compared with baseline responses). Insets show averaged EPSCs recorded during baseline responses and 25 min after the pairing procedure (bar) (adapted from Toyoda *et al.* [15]).

necessary for the induction of both LTP and LTD. Whether the synapse undergoes LTP or LTD is dependent on the intracellular Ca^{2+} concentration. For example, large Ca^{2+} influx leads to LTP and slight Ca^{2+} influx leads to LTD. In the induction of homosynaptic LTD, postsynaptic increases in Ca^{2+} are shown to be mediated by either NMDA receptors or voltage-dependent Ca^{2+} channels. There are many reports which say that Ca^{2+} influx through voltage-dependent Ca^{2+} channels is important for LTD induction. To determine if postsynaptic Ca^{2+} signaling pathways are involved, a pipette solution containing 11 mM BAPTA was used and the induction of LTD was also completely abolished (Figure 4). These findings indicate that this form of LTD is NMDA receptor-dependent and a postsynaptic Ca^{2+} increase is required for the induction of LTD in the ACC neurons.

NR2A- vs. NR2B-containing NMDA receptors are required for the induction of LTD

It has been reported that NR2B (GluN2B)-containing receptors, but not NR2A(GluN2A)-containing receptors, contribute to LTD in the hippocampus [14]. However, little is known about whether GluN2A (NR2A)- and/or NR2B-containing NMDA receptors contribute to the induction of LTD in the ACC. First, the contribution of the NR2A subunit to NMDA receptor-mediated EPSCs in ACC neurons was examined. A selective pharmacological antagonist for NR2A was used to examine synaptically induced NMDA receptor-mediated EPSCs. The averaged decay time constant and rising time (10–90%) of NR2A-mediated currents were 95 ± 8 and 13 ± 1 ms, respectively (Figure 4). The effect of NVP-AAM077 on PPF was also examined, which is thought to be a presynaptic process. In the presence of NVP-AAM077, PPF was unchanged, suggesting that this drug did not affect neurotransmitter release [15].

Blocking the NR2A subunit with NVP-AAM077 (0.4 μM) abolished LTD (Figure 4). To test whether LTD could be induced in the presence of NR2A blockade, a different form of stimulus (holding potential at -15 mV during presynaptic stimulation) was used. Interestingly, this form of stimulation in the presence of NVP-AAM077 was able to induce LTD, indicating that NR2A is not absolutely required for ACC LTD.

The contribution of the NR2B subunit to NMDA receptor-mediated EPSCs was also investigated in ACC neurons. A selective antagonist for the NR2B subunit was used to examine synaptically induced NMDA receptor-mediated EPSCs. Bath application of a selective NR2B subunit antagonist, Ro25-6981 (0.3 μM), depressed the total NMDA receptor-mediated currents by 19% of control, indicating that the NMDA receptor EPSCs in ACC neurons are mostly mediated by NR2A-containing receptors. As with NVP-AAM077, application of Ro25-6981 had no effect on AMPA receptor-mediated EPSCs. The average decay time constant and rising time (10–90%) of NR2B-mediated currents were 129 ± 10 and 19 ± 2 ms, respectively. The rising time (10–90%) and decay time constant of NR2B-mediated EPSCs were significantly longer than those of NR2A-mediated currents in ACC pyramidal neurons. Furthermore, Ro25-6981 did not affect PPF. By contrast, a blockade of the NR2B subunit with Ro25-6981 or ifenprodil abolished the induction of LTD. These results show that the NR2B subunit also contributes to the induction of LTD in the ACC. To test whether LTD could be induced with a different form of stimulus (holding potential at −30 and −15 mV during presynaptic stimulation) in the presence of Ro25-6981, LTD was repeated in the presence of NR2B blockade. Unlike NVP-AAM077, this induction protocol at a holding potential of −15 mV failed to induce LTD in the presence of Ro25-6981 (Figure 4). However, a holding potential at −30 mV in the presence of Ro25-6981 induced LTD (Figure 4). Taken together, these results indicate that both NR2A and NR2B subunits are required for the formation of LTD in the ACC and the abolishment of LTD by these compounds could be due to the reduction in calcium influx via NMDA R receptors.

NMDA receptor-independent form of LTD

In addition to NMDA receptor-dependent LTD, there is NMDA receptor-independent LTD in the ACC as well. Using field recording of EPSP in adult ACC slices, repetitive stimulation at a low frequency induced LTD.

Both voltage-gated calcium channels and mGluRs contribute to ACC LTD [16]. Nimodipine (10 μM) completely blocked the induction of LTD (Figure 5), although basal synaptic responses were not significantly affected. By contrast, 50 μM AP-5 did not affect field LTD (Figure 5). MCPG (500 μM), a metabotropic glutamatergic receptor antagonist, also blocked LTD. These results suggest that both L-type calcium channels and

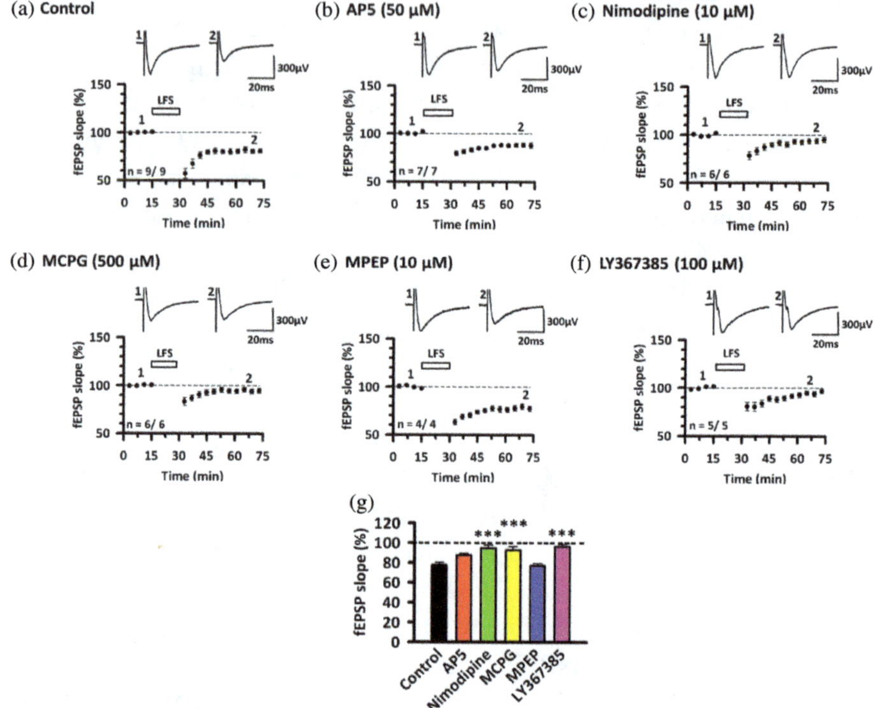

Figure 5. Pharmacological aspects of ACC LTD. (a) Controls show LTD after 1 Hz, 15 min low-frequency stimulation (78 $\pm$ 2%; n = 9/9). (b) NMDA receptor antagonist, AP5 (50 μM) partially blocked LTD (88 $\pm$ 2%; n = 7/ 7). (c) L-VGCC blocker, nimodipine (10 μM) blocked LTD (95 $\pm$ 3%; n = 6/6). (d) Groups I and II mGluR antagonist, MCPG (500 μM) also blocked LTD (93 $\pm$ 3%; n = 6/ 6). (e) mGluR5 antagonist, MPEP (10 μM) had no effect on ACC LTD (77 $\pm$ 2%; n = 4/4). (f) mGluR1 antagonist, LY367385 (100 μM) blocked LTD (96 $\pm$ 2%; n = 5/5). (g) Summarized results of the averaged fEPSP slope of the last 10 min of each experiment ($F_{(5,31)}$ = 10.64, p < 0.01; one-way ANOVA with Bonferroni *post hoc*; p < 0.001 for control versus nimodipine, MCPG and LY367385) (adapted from Kang *et al.* [16]).

mGluRs are critical for the induction of LTD in the ACC. LTD was not affected in the presence of picrotoxin (100 μM), a GABA$_A$ receptor antagonist, indicating that inhibitory influences are not required for LTD.

Postsynaptic expression of LTD

Two major possible mechanisms may contribute to LTD: the reduction of the release of glutamate from presynaptic terminals and decreases in post-synaptic AMPA receptor-mediated responses [11,17–20]. To test if the interaction between the GluR2/3 C-terminal and its specific PDZ binding partners is important for the induction of LTD, several synthetic peptides that disrupt the interaction between AMPA subunits and PDZ-containing proteins were used. Peptides were applied through the patch recording electrode into postsynaptic neurons. First, a control peptide (Pep2-SVKE), in which the PDZ interaction motif is inhibited by substituting the last amino acid (isoleucine) with glutamate was used. Postsynaptic application of Pep2-SVKE (100 μM) did not affect basal synaptic transmission in ACC slices (Figure 6). In the presence of Pep2-SVKE, LTD was not affected (Figure 6). Next, a Pep2-SVKI peptide, which interferes with interactions between GluR2 and GRIP (glutamate receptor interaction protein), ABP (AMPA receptor binding protein), and PICK1, was used. In the presence of Pep2-SVKI (100 μM), basal synaptic responses were also unchanged. However, LTD was completely blocked by the presence of Pep2-SVKI. These results provide direct evidence that postsynaptic inter-action between AMPA GluR2 receptor and PDZ protein is critical for the induction of cingulate LTD.

Another Pep2-AVKI peptide, which disrupts binding of GluR2 to PICK1 (protein interaction with C kinase), was also used. Postsynaptic application of Pep2-AVKI (100 μM) had no effect on basal synaptic trans-mission, but LTD was blocked. These results strongly suggest that both GluR2/3-PICK1 and GluR2/3-GRIP1/ABP interactions are likely contrib-uting to cingulate LTD. Cingulate LTD requires activation of postsynaptic NMDA receptors. To exclude the possible inhibition of NMDA receptors by these peptides, we measured the effects of postsynaptic injection of the peptides on NMDA receptor-mediated EPSCs. We found that these pep-tides had no effect on NMDA receptor-mediated EPSCs, indicating that

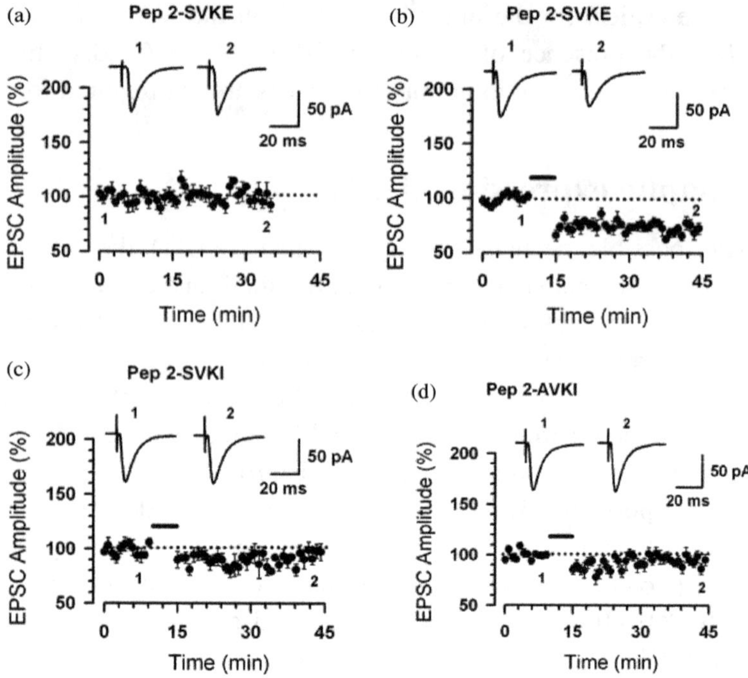

Figure 6.   Pep2-SVKI blocks the induction of LTD. (a, c) Pep2-SVKE (100 μM) and Pep2-SVKI (100 μM) do not affect baseline response. The insets show averages of six EPSCs at the time points of 5 (1) and 35 min (2) during the recording. The dashed line indicates the mean basal synaptic response. (b) Pep2-SVKE (100 μM, $n=8$) has no effect on the induction of LTD. (d) In the presence of Pep2-SVKI (100 μM) in the intracellular solution, LTD is blocked ($n = 10$). (b and d) Traces show averages of six EPSCs at baseline responses (1) and 30 min (2) after the paired training (bar). The dashed line indicates the mean basal synaptic response (adapted from Toyoda *et al* [27]).

the blocking effects are not simply due to the inhibition of NMDA receptor functions in the cingulate neurons.

To examine if GluR2 subunit contributes to the expression of LTD, Pep2-SVKI was injected into neurons 5 min after the induction of LTD. No significant effect was found on the expression of LTD during the 25 min treatment with Pep2-SVKI. This result suggests that GluR2 receptor-mediated possible trafficking events are completed within 5–10 min after LTD induction.

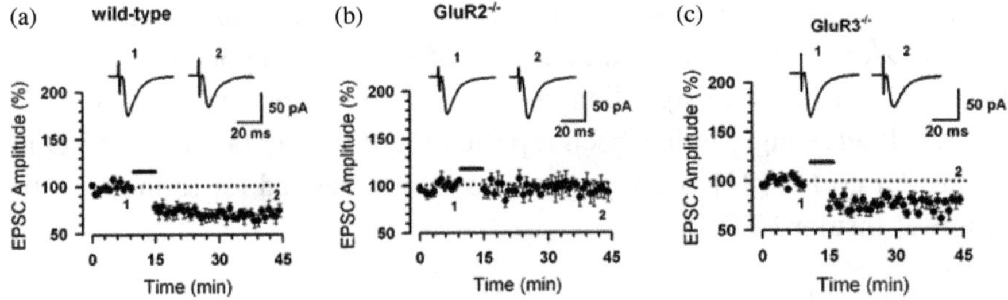

Figure 7. Synaptic depression in GluR2 KO mice. (a) LTD is induced in ACC neurons in wild-type mice ($n = 7$ slices/6 mice). (b) LTD is absent in ACC neurons in GluR2 KO mice (GluR2$^{-/-}$) ($n = 9$ slices/6 mice). (c) LTD is absent in ACC neurons in GluR3 KO mice (GluR3$^{-/-}$) ($n = 9$ slices/6 mice) (adapted from Toyoda *et al.* [27]).

Genetic studies of LTD

Although the use of peptide inhibitors offers possible insights for the involvement of AMPA receptor subtypes in LTD, it is difficult to rule out possible non-selective interactions. To study the subunit-specific function of GluR2 and GluR3 in cingulate LTD, LTD was recorded from ACC neurons of GluR2 or GluR3 KO mice (GluR2$^{-/-}$ or GluR3$^{-/-}$). As expected, LTD in GluR2$^{-/-}$ mice was completely abolished (Figure 7), while normal LTD was induced in wild-type controls (Figure 7). These results suggest that the GluR2 contributes to the induction of cingulate LTD. Additional experiments found that GluR2 deletion did not affect basal excitatory synaptic transmission, PPF, and NMDA receptor-mediated responses.

The possible role of GluR3 in LTD was also examined. LTD in ACC slices from GluR3$^{-/-}$ and wild-type mice was comparable in both groups, suggesting that synaptic depression was not affected by the deletion of GluR3 subunit.

LTD in other pain-related cortical areas

The IC is another cortical area that is critical for pain perception and emotion. LTD has been reported in the IC using multiple electrode field recording system or whole-cell patch-clamp recording [21–23]. In field recording studies, the induction of LTD was found to require activation of the NMDA receptors, mGluR5, and L-VGCCs. Protein phosphatase 1/2A and endocannabinoid signaling are also critical for the insular LTD [21,22]. Interestingly, it has been reported that nicotine facilitated LTD in the insular cortex [23]. More studies are clearly needed for future investigation of insular LTD.

Functional implication for cortical LTD

Considering LTP is important for cortical excitation after peripheral injury, it is conceivable that ACC and IC LTD may contribute to cortical inhibition in behavioral animals. LTD may also help to reset enhanced synaptic transmission. For pain regulation, such resetting effects may be analgesic in behavioral status, and the loss of LTD may lead to further potentiation of cortical responses. Studies with different types of animal injury models found that ACC LTD was diminished after injury (Figure 8). Tail amputation, a potential animal model for the study of phantom pain, led to loss of LFS-induced LTD in the ACC and IC [16,21,24]. Similar results were found in animal models of bone cancer pain and nerve injury [25,26]. In supporting this hypothesis, it has been reported that restoration of ACC LTD alleviated peripheral pain hypersensitivity [25].

Conclusion

ACC and IC play important roles in pain perception and emotion. Both LTP and LTD give excitatory synapses in cortical neurons the ability to store and erase the information. The existence of different forms of LTD may indicate that they may contribute to various physiological and pathological functions (see Table 1). In case of injuries, the loss of LTD plays a crucial role in cortical excitation caused by the injury. Postsynaptic changes are likely the key mechanisms for LTD, although presynaptic LTD is still possible in these synapses. Restoration of LTD in the cortex may serve as a new way to reduce chronic pain and related emotional responses.

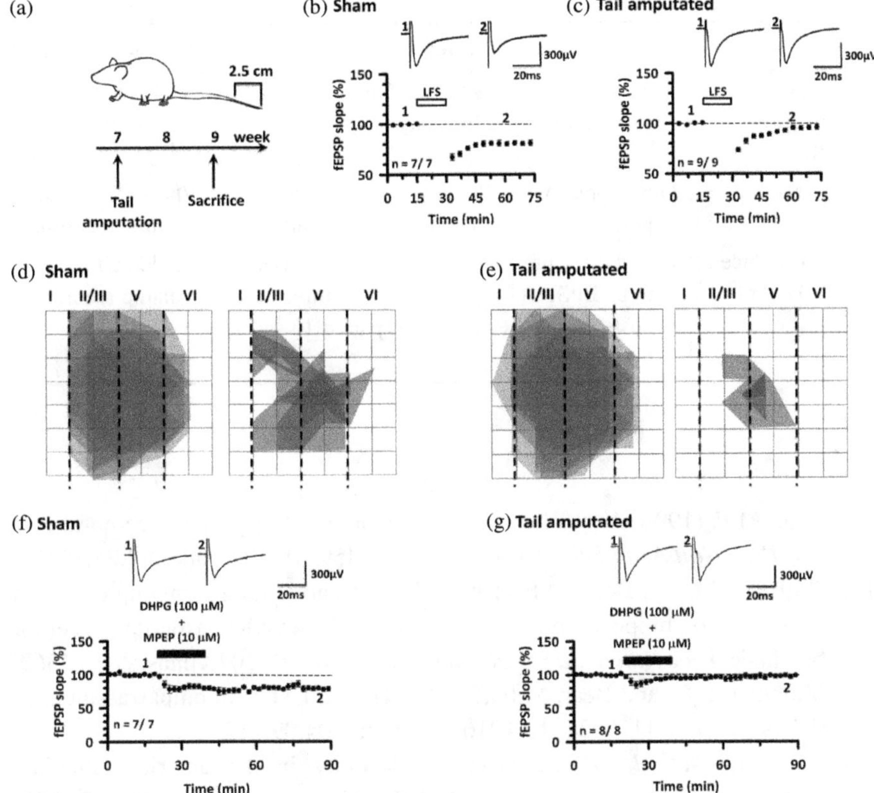

Figure 8.   Tail amputation impairs ACC LTD. (a) The schematic view of tail amputation experiment process. (b) The sham groups showed similar LTD as the control group ($81 \pm 2\%$; $n = 7/7$). (c) The 2-week tail-amputated group showed impaired ACC LTD ($96 \pm 2\%$; $n = 9/9$). (d, e) The spatial distribution of activated channels during baseline (blue) and the channels that underwent LTD (red). The spatial distribution of activated channels during baseline in sham and amputated models is similar. However, sham group shows a wider distribution of LTD-occurring channels compared with the tail amputated group, which rarely shows some LTD near the stimulation site. (f) Sham showed normal chemical LTD by applying DHPG ($100 \mu M$) and MPEP ($10 \mu M$) for 20 min ($86 \pm 4\%$; n = 7/7). (g) Tail-amputated group showed no chemical LTD by applying DHPG ($100 \mu M$) and MPEP ($10 \mu M$) for 20 min ($99 \pm 3\%$; $n = 8/8$) (adapted from Kang *et al* [16]).

Table 1. Stimulation protocols for inducing LTD in the ACC slices.

	Parameters	**Recording method**
Low-frequency stimulation (LFS)	1 Hz for 15 min or 3 Hz for 5 min at the same stimulation intensity	Field potential recording
Pairing training	Paired presynaptic 80 pulses at 2 Hz with postsynaptic depolarization at +30 mV	Whole-cell patch-clamp recording
Chemical-induced LTD: DHPG	Paired three presynaptic stimuli that caused three EPSPs (10 ms ahead) with three postsynaptic APs at 30 Hz, paired 15 times every 5 s	Whole-cell patch-clamp recording

References

[1] Bear, M.F. (1996) A synaptic basis for memory storage in the cerebral cortex. *Proc Natl Acad Sci USA* 93, 13453–13459. 10.1073/pnas.93.24.13453.

[2] Dudek, S.M. and Bear, M.F. (1992) Homosynaptic long-term depression in area CA1 of hippocampus and effects of N-methyl-D-aspartate receptor blockade. *Proc Natl Acad Sci USA* 89, 4363–4367. 10.1073/pnas.89.10.4363.

[3] Malenka, R.C. and Bear, M.F. (2004) LTP and LTD: An embarrassment of riches. *Neuron* 44, 5–21. 10.1016/j.neuron.2004.09.012.

[4] Bliss, T.V. *et al.* (2016) Synaptic plasticity in the anterior cingulate cortex in acute and chronic pain. *Nat Rev Neurosci* 17, 485–496. 10.1038/nrn.2016.68.

[5] Zhuo, M. (2008) Cortical excitation and chronic pain. *Trends Neurosci* 31, 199–207. 10.1016/j.tins.2008.01.003.

[6] Zhuo, M. (2016) Neural mechanisms underlying anxiety-chronic pain interactions. *Trends Neurosci* 39, 136–145. 10.1016/j.tins.2016.01.006.

[7] Vogt, B.A. (2005) Pain and emotion interactions in subregions of the cingulate gyrus. *Nat Rev Neurosci* 6, 533–544. 10.1038/nrn1704.

[8] Bear, M.F. and Abraham, W.C. (1996) Long-term depression in hippocampus. *Annu Rev Neurosci* 19, 437–462. 10.1146/annurev.ne.19.030196.002253.

[9] Toyoda, H. *et al.* (2006) NMDA receptor-dependent long-term depression in the anterior cingulate cortex. *Rev Neurosci* 17, 403–413. 10.1515/revneuro.2006.17.4.403.

[10] Artola, A. *et al.* (1990) Different voltage-dependent thresholds for inducing long-term depression and long-term potentiation in slices of rat visual cortex. *Nature* 347, 69–72. 10.1038/347069a0.

[11] Bolshakov, V.Y. and Siegelbaum, S.A. (1994) Postsynaptic induction and presynaptic expression of hippocampal long-term depression. *Science* 264, 1148–1152. 10.1126/science.7909958.

[12] Chen, Q.Y. *et al.* (2021) NMDA receptors and synaptic plasticity in the anterior cingulate cortex. *Neuropharmacology* 197, 108749. 10.1016/j.neuropharm.2021.108749.

[13] Lee, H.K. *et al.* (1998) NMDA induces long-term synaptic depression and dephosphorylation of the GluR1 subunit of AMPA receptors in hippocampus. *Neuron* 21, 1151–1162. 10.1016/s0896-6273(00)80632-7.

[14] Liu, L. *et al.* (2004) Role of NMDA receptor subtypes in governing the direction of hippocampal synaptic plasticity. *Science* 304, 1021–1024. 10.1126/science.1096615.

[15] Toyoda, H. *et al.* (2005) Roles of NMDA receptor NR2A and NR2B subtypes for long-term depression in the anterior cingulate cortex. *Eur J Neurosci* 22, 485–494. 10.1111/j.1460-9568.2005.04236.x.

[16] Kang, S.J. *et al.* (2012) Plasticity of metabotropic glutamate receptor-dependent long-term depression in the anterior cingulate cortex after amputation. *J Neurosci* 32, 11318–11329. 10.1523/JNEUROSCI.0146-12.2012.

[17] Zakharenko, S.S. *et al.* (2002) Altered presynaptic vesicle release and cycling during mGluR-dependent LTD. *Neuron* 35, 1099–1110. 10.1016/s0896-6273(02)00898-x.

[18] Luthi, A. *et al.* (1999) Hippocampal LTD expression involves a pool of AMPARs regulated by the NSF-GluR2 interaction. *Neuron* 24, 389–399. 10.1016/s0896-6273(00)80852-1.

[19] Mulkey, R.M. and Malenka, R.C. (1992) Mechanisms underlying induction of homosynaptic long-term depression in area CA1 of the hippocampus. *Neuron* 9, 967–975. 10.1016/0896-6273(92)90248-c.

[20] Mulkey, R.M. *et al.* (1993) An essential role for protein phosphatases in hippocampal long-term depression. *Science* 261, 1051–1055. 10.1126/science.8394601.

[21] Liu, M.G. *et al.* (2013) Long-term depression of synaptic transmission in the adult mouse insular cortex in vitro. *Eur J Neurosci* 38, 3128–3145. 10.1111/ejn.12330.

[22] Liu, M.G. and Zhuo, M. (2014) Loss of long-term depression in the insular cortex after tail amputation in adult mice. *Mol Pain* 10, 1. 10.1186/1744-8069-10-1.

[23] Toyoda, H. (2018) Nicotine facilitates synaptic depression in layer V pyramidal neurons of the mouse insular cortex. *Neurosci Lett* 672, 78–83. 10.1016/j.neulet.2018.02.046.

[24] Wei, F. *et al.* (1999) Loss of synaptic depression in mammalian anterior cingulate cortex after amputation. *J Neurosci* 19, 9346–9354.

[25] Wang, Y.J. *et al.* (2020) Restoration of cingulate long-term depression by enhancing non-apoptotic caspase 3 alleviates peripheral pain hypersensitivity. *Cell Rep* 33, 108369. 10.1016/j.celrep.2020.108369.

[26] Chiou, C.S. *et al.* (2012) Impairment of long-term depression in the anterior cingulate cortex of mice with bone cancer pain. *Pain* 153, 2097–2108. 10.1016/j.pain.2012.06.031.

[27] Toyoda, H. *et al.* (2007) Long-term depression requires postsynaptic AMPA GluR2 receptor in adult mouse cingulate cortex. *J Cell Physiol* 211, 336–343. 10.1002/jcp.20940.

Chapter 7

Cortical Sensitization and Reorganization

News and Views

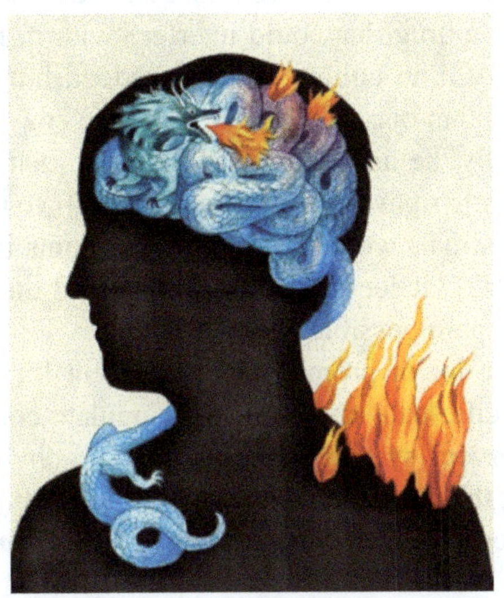

Scientists have traced chronic pain to a defect in one enzyme in a single region of the brain. Could this be a decisive turn in the battle against chronic pain?

Min Zhuo is a neuroscientist at the University of Toronto, who has been testing potential painkillers on mice with the aim of discovering better treatments for chronic pain. In order to test this, he crimped a nerve in the legs of mice to induce a chronic pain condition. In a matter of days, the mice developed many of the symptoms — and even some of the brain alterations — often seen in people suffering from chronic pain.

Zhuo and his colleagues then sought out different compounds that could affect the learning that goes on during chronic pain. They focused on the behavior of neurons in a region of the brain called the anterior cingulate cortex, which shows intense activity in scans of people with chronic pain. The cingulate cortex contains an abundant amount of an enzyme called AC1. Zhuo wondered if the neural learning that leads to chronic pain was accelerated when levels of the enzyme were high.

As a test, Zhuo's team genetically engineered mice so that they could not make enzyme AC1. The animals turned out almost entirely normal. They could even sense regular types of pain, however when the mice were subject to the leg crimp procedure, they did not develop chronic pain.

Once Zhuo recognized that AC1 is essential for chronic pain, he started the hunt for a drug that could interfere with it. He grew cells that produced enzyme AC1 in culture and then added hundreds of different compounds, hoping that one would latch on to the enzyme and thus block its action. Eventually, he and his team found one that did and named it NB001. When scientists gave an oral dose of NB001 to rats suffering from chronic pain, the animals were rid of their symptoms in just 45 minutes. By latching onto AC1, it seems the drug prevented the neuronal activity that makes chronic pain possible.

NB001 shows a lot of promise as a treatment for chronic pain as it focuses on the specific pain engine in the cingulate cortex instead of the entire brain and nervous system. There are also no obvious side effects in laboratory animals; the rats suffered no harm to their memory or their ability to learn. Zhuo hopes to launch clinical trials in humans in the near future.

Summary

Investigation of the cortical synapses in *in vitro* brain slices found that these sensory synapses in the brain are highly plastic, and such plastic changes may be long-lasting. In this chapter, we will review the evidence that supports the long-term plastic changes within the ACC synapses after periphery injury, from inflammation and nerve injury to amputation. It is consistently demonstrated that excitatory synaptic transmission in the layer

II-II synapses is significantly enhanced after peripheral injury, and such enhancement is long-lasting. Long-term potentiation of excitatory synaptic transmission within the ACC likely contributes to cortical sensitization and subsequent behavioral responses. The use of *in vitro* brain slice models and transgenic mice provides useful tools for identifying leading protein targets involved in the induction and expression of chronic pain.

Keywords: ACC; LTP; AC1; intrinsic plasticity; nerve injury; PKMζ; cortical reorganization

Introduction

The adult human somatosensory cortex is often illustrated with a distorted human figure in order to outline the somatotopic map. Historically, it was thought that this figure remained relatively stable in adults. However, studies over the past 20 years have dramatically changed this perception. Cortical representations in the mammalian brain are dynamic and can be modified by experience [1,2]. Not only do plastic changes occur in adults but they can also happen on a rapid time scale (from a few minutes to several hours). It has been proposed that use-dependent changes in synaptic strength, such as LTP and LTD, may serve as key synaptic mechanisms of cortical plasticity.

Cortical reorganization

Cortical reorganization has been reported in the somatosensory cortex of adult animals after peripheral deafferentation or amputation. Cortical reorganization acts as an adaptive mechanism during development and learning, and it could also play a detrimental role in traumatic events, such as the loss of a limb. It has been demonstrated that cortical reorganization occurs after limb or digit amputation. After digit amputation, neuronal terminals invade adjacent cortical areas representing deafferented fingers. Similarly, human amputees experience phantom limb sensations — otherwise known as phantom pain — and the amount of cortical reorganization directly correlates with the severity of the phantom pain.

Molecular and cellular mechanisms for cortical reorganization are still largely unknown. However, recent progress made in animal models of injury may provide more insight into the molecular and cellular mechanism for such cortical reorganization [3–5]. In an animal model of nerve injury, it was reported that activation of the ACC induced by peripheral nerve injury increases the turnover of specific synaptic proteins in a persistent manner [6]. Neural cell adhesion molecule 1 (NCAM1) is one of the molecules involved and it mediates spine reorganization and contributes to behavioral sensitization. This study demonstrates a synaptic mechanism for cortical reorganization and points to potential avenues for neuropathic pain treatment.

LTP in the ACC after injury

Expression of activity-dependent immediate early genes, such as c-fos and Egr1, is linked to the induction of late LTP in the hippocampus, and their expression is commonly used as a marker for LTP in studies of learning and memory [7]. Consistent with the idea that chronic pain is linked to persistent LTP in the ACC, rodent models of chronic inflammatory pain, neuropathic pain, bone cancer pain, and chronic visceral pain are all associated with the activation of different immediate early genes including c-fos and Egr1 in ACC neurons [8–12] (see Figure 1 for an example after

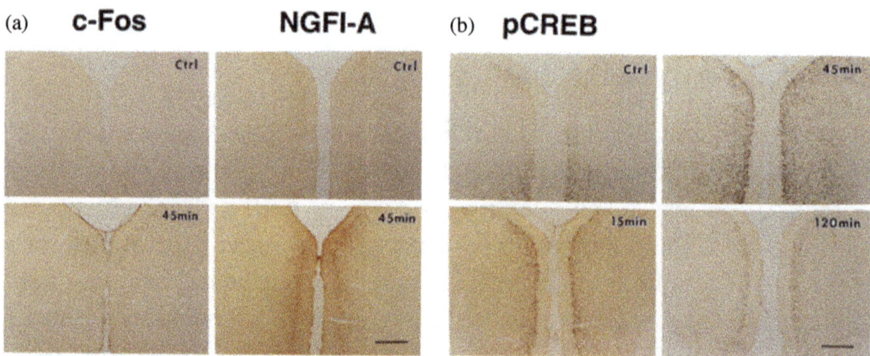

Figure 1. Potentiation of IEG expression in the ACC after the amputation. Photomicrographs showing expression of c-Fos and NGFI-A (a) and phosphorylation of CREB (b) in the coronal ACC sections from sham animals (Ctrl) and animals at different times after the amputation of the unilateral hind paw third digit. Scale bars: 500 μm. The numbers of c-Fos-, NGFI-A-, and pCREB-immunoreactive cells increased bilaterally after the amputation (see a summary). AC1 and AC8 contributed to CREB activation following formalin injection (adapted from Wei *et al.* [10]).

amputation). In addition, there is evidence of increased neuronal excitability and synaptic potentiation within the ACC. For example, peripheral digit amputation (an animal model for phantom pain) produces a long-lasting enhancement of synaptic responses within the ACC to either local or peripheral stimulation *in vivo* [13] (Figure 2). Peripheral nerve block experiments have demonstrated that peripheral nerve activity is not required for the maintenance of amputation-induced potentiation of evoked synaptic responses in the ACC, indicating that synaptic plasticity in the ACC is maintained by central mechanisms.

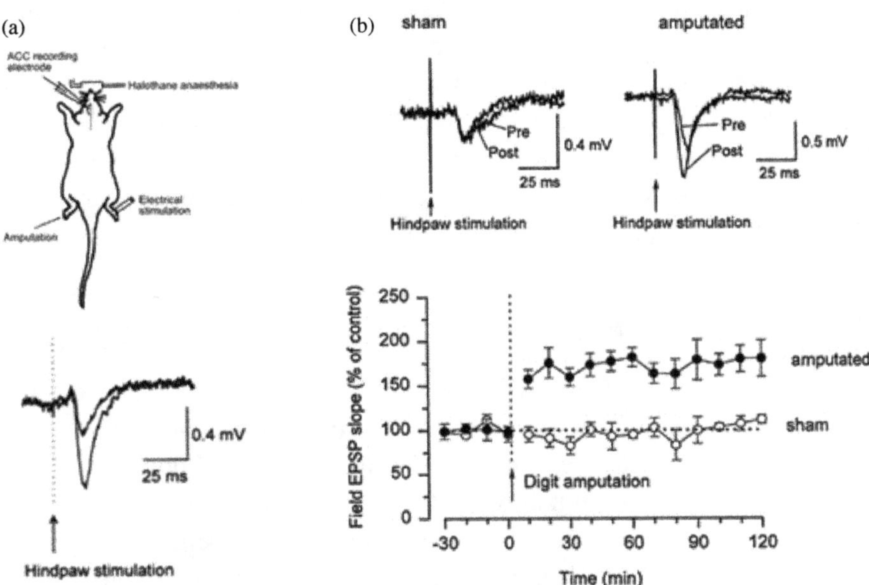

Figure 2. LTP recorded *in vivo* recording from the ACC. (a) Diagram of *in vivo* recording from the ACC in an anaesthetized rat; animals were maintained in a lightly anaesthetized state by halothane. The recording electrode was placed into the ACC contralateral to the peripheral stimulation electrode. Amputation (the removal of the third digit of the hind paw) was performed on the non-stimulated hind paw. During amputation, a higher concentration of halothane was used. (b) Representative traces of EPSPs 5 min before amputation (Pre) and 115–120 min after (Post) sham treatment or amputation. The latency of sensory responses was not changed after the amputation, while the EPSP slope was increased. Amputation of a single digit of the contralateral hind paw (indicated by an arrow) caused long-lasting enhancement of sensory responses. Sensory responses were not significantly changed in sham-treated animals. The test stimulation frequency was 0.01 Hz (adapted from Wei and Zhuo [13]).

Experimentally induced LTP and LTD

In a rodent model of neuropathic pain, synaptic responses in the ACC are potentiated at the time that allodynia develops, assessed 1–2 weeks after nerve injury (see Chapter 6). At this time, the late component of LTP can no longer be induced by theta-burst stimulation, indicating that maximal potentiation has already occurred. These findings suggest that neuropathic pain is linked to the mechanisms that underlie the expression of LTP in the ACC. Pain-related long-term changes in synaptic transmission are not limited to potentiation. LTD induced by repetitive stimulation of ACC neurons is also strikingly affected after peripheral tissue injury. For example, in rodents with digit or tail tip amputation, the induction of LTD in the ACC, measured 45 min and 2 weeks after amputation, is suppressed [10,14]. Furthermore, in a mouse model of bone cancer pain, LTD in the ACC is also impaired (see Chapter 6). Based on these observations, we propose that chronic pain is associated with a saturated late component of LTP, and with a suppression of LTD, in the ACC.

Presynaptic glutamate release

In addition to postsynaptic changes, the release of glutamate is also enhanced in the ACC in animal models of chronic pain [15–17] (Figure 3). One to two weeks after peripheral nerve ligation (or 3–5 days after CFA injection to induce inflammation of the hind paw in adult mice), PPF is reduced in the ACC, indicating presynaptic enhancement of excitatory synaptic transmission. Furthermore, an increase in the frequency of AMPA receptor-mediated mEPSCs occurs in ACC neurons after peripheral nerve injury or inflammation in mice [18].

Postsynaptic AMPA receptor changes

AMPA receptor-dependent EPSCs, recorded from pyramidal neurons of layers II/III and V, and evoked by focal electrical stimulation within the ACC, are increased following peripheral nerve ligation in mice [18–21] (Figure 3). Similar changes are found in ACC neurons in a murine model

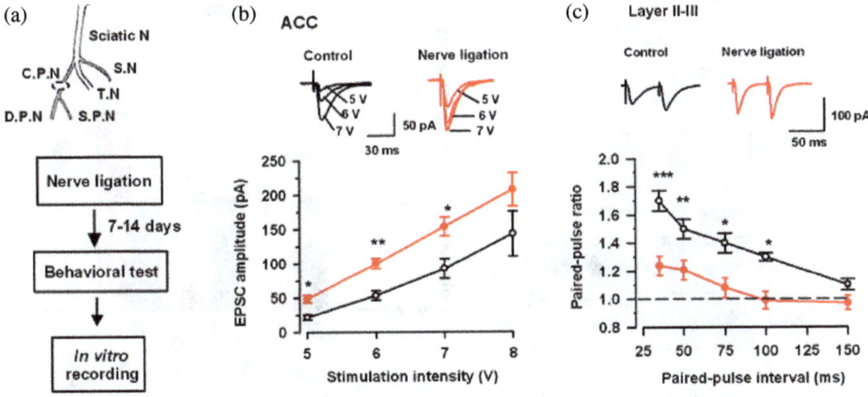

Figure 3. Increased synaptic transmission in the ACC following peripheral nerve ligation. (a) Diagram illustrates the nerve injury (CPN) and experimental design. (b) Synaptic
input–output curves in slices from control ($n = 6$ neurons) and nerve-ligated mice ($n = 7$
neurons). *$P < 0.05$ and **$P < 0.01$ compared with those of control group. Open circles:
neurons from control mice; solid circles: neurons from mice with nerve ligation. (c)
Representative traces with an interval of 50 ms recorded in the layer II/III of the ACC.
Paired pulse facilitation (PPF: the ratio of EPSC2/EPSC1) was recorded at intervals of 35,
50, 75, 100, and 150 ms from control and nerve-ligated mice. Open circles: neurons from
control mice ($n = 17$ neurons); solid circles: neurons from mice with nerve ligation ($n =
19$ neurons). *$P < 0.05$, **$P < 0.01$, and ***$P < 0.001$ (adapted from Xu *et al.* [18]).

of inflammation, induced by CFA. These changes are associated with an
alteration in the rectification of AMPA receptor-mediated synaptic transmission and the development of sensitivity to an inhibitor of CP-AMPA
receptors. Consistent with this, the population of membrane-bound
GluA1-containing AMPA receptors is increased, whereas GluA2/3-
containing AMPA receptors are not significantly affected. The selective
contribution of GluA1 to the increase in EPSCs is further supported by
genetic studies. Deletion of GluA1, but GluA2, significantly reduces
peripheral injury-triggered c-fos activation in the ACC, as well as in the
spinal cord dorsal horn. Behavioral responses to the peripheral injury are
also reduced [22]. Recent electron microscopy data further supports the
conclusion that postsynaptic GluA1-containing receptors are increased
after peripheral injury (Figure 5).

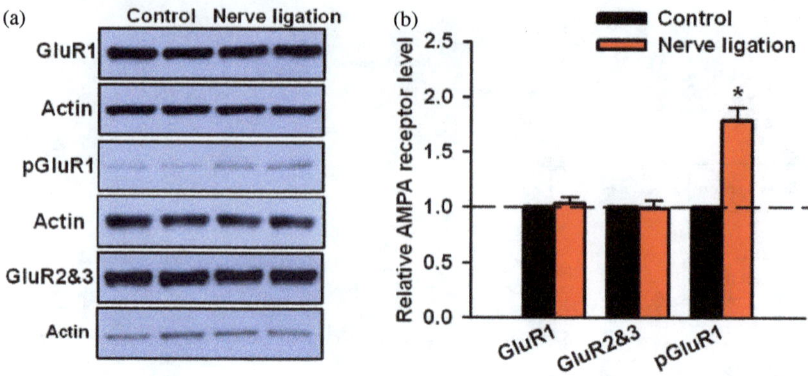

Figure 4.   Altered phosphorylation of GluR1 and rectification index of AMPA receptor-mediated current in the ACC after nerve injury. (a) Representative expression of GluR1 and phosphorylation of GluR1 and GluR2/3 by western blot in the ACC from control and nerve-ligated mice. (b) Pooled data showing that phosphorylation of GluR1 was upregulated in mice with nerve ligation (adapted from Xu *et al.* [18]).

Postsynaptic NMDA receptors

In addition to the enhancement of AMPA receptor EPSCs, peripheral nerve injury leads to increased GluN2B-containing NMDA receptor-mediated responses [23] (see Chapter 8). Furthermore, in models of persistent inflammation, the expression of GluN2B subunits in the ACC is upregulated, thereby increasing the GluN2B component in NMDAR-mediated responses. Administration of GluN2B receptor-selective antagonists — either systemically or directly to the ACC — inhibits pain hypersensitivity associated with peripheral inflammation. Interestingly, recent studies of the IC and PFC have shown that upregulation of GluN2B-containing NMDA receptors after nerve injury also occurs in these structures. This may therefore constitute a generalized cortical response to peripheral injury that facilitates the induction of NMDA receptor-dependent LTP. Such metaplasticity could contribute to the sustained enhancement of synaptic transmission that is associated with the chronic pain state.

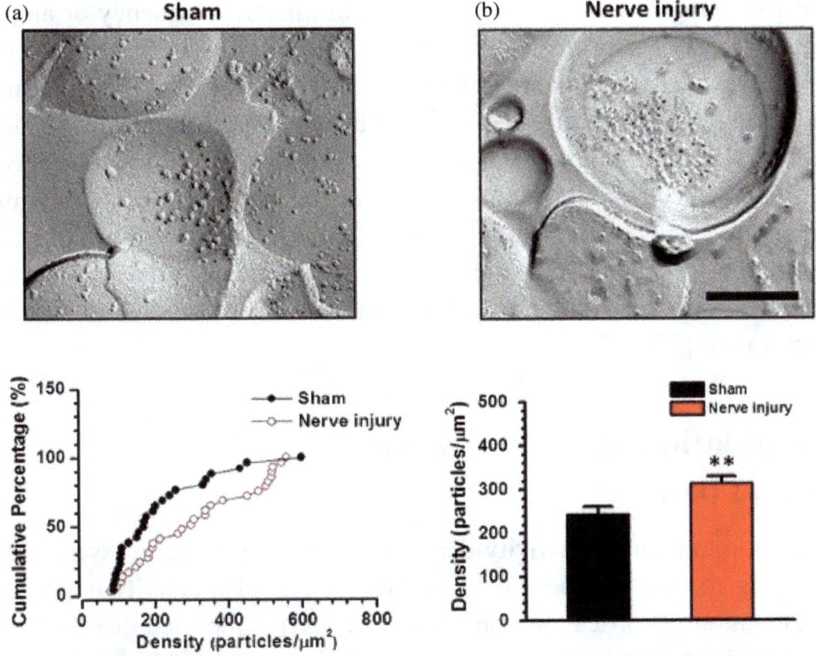

Figure 5.   Postsynaptic accumulation of GluA1 in the ACC after nerve injury. (a, b) SDS-digested freeze-fracture replica labeling EM samples showing increased GluA1 particles in the synaptic region on ACC layer V neurons in mice with nerve injury compared with mice with sham surgery. Bar equals 200 nm. (C) Cumulative histograms showing the distribution of postsynaptic GluA1 density. (d) Averaged density of the postsynaptic GluA1. **$P < 0.01$ (adapted from Chen *et al.* [19]).

Calcium-stimulated AC1

Calcium-stimulated AC1 is critical for cortical excitation triggered by injury [7,24–26]. For example, in $AC1^{-/-}$ mice, reduced chronic pain sensitization was found in inflammatory pain and neuropathic pain models. Moreover, following nerve injury, the $AC1^{-/-}$ mice did not demonstrate a reduction in PPF of AMPA receptor-mediated EPSCs as shown by the wild-type mice after the nerve injury [18]. While frequency and amplitude of mEPSCs in ACC neurons were consistently increased in wild-type mice following nerve injury, no increase in mEPSC frequency or amplitude was observed in $AC1^{-/-}$ mice [18]. These results indicate that both presynaptic and postsynaptic enhancement of excitatory synaptic transmission in the ACC is dependent on AC1 in neuropathic pain. Biochemical studies found that injury-induced increases of the phosphorylation levels of GluR1 induced by nerve injury were blocked in the ACC of $AC1^{-/-}$ mice compared to wild-type mice, indicating that AC1 is involved in the phosphorylation of GluR1 receptors in the ACC during neuropathic pain (Figure 4). Similar results are also found in animal models of chronic visceral pain [27].

Upregulation of AC1 in chronic visceral pain

In addition to being an activity-dependent signaling protein, essential for producing the second messenger cAMP, AC1 is also activity-dependent for regulation in cortical neurons. Liu *et al.* (2019) reported that AC1 significantly increased in the ACC in an animal model of IBS and persisted for at least a few weeks. Furthermore, inhibiting AC1 activity by NB001 significantly reduced the upregulation of AC1 proteins in the ACC, suggesting that AC1 activity itself is critical for AC1 protein upregulation [28]. This finding indicates that AC1 may form a positive regulation in the cortex during chronic visceral pain. Figure 9 is a proposed model for AC1 positive feedback control in disease condition.

Upregulation of PKMζ

An atypical PKC isoform (either PKMζ or PKCι/λ) is critical for maintaining enhanced synaptic responses in the ACC in rodent models of neuropathic pain [29] (Figure 6). Peripheral nerve injury causes activation of PKMζ in the ACC, and locally applied ZIP erases synaptic potentiation caused by nerve injury (Figure 7). In addition, microinjection of ZIP into the ACC blocks mechanical allodynia in mice [29]. These observations suggest that chronic pain activates signaling cascades that are similar to

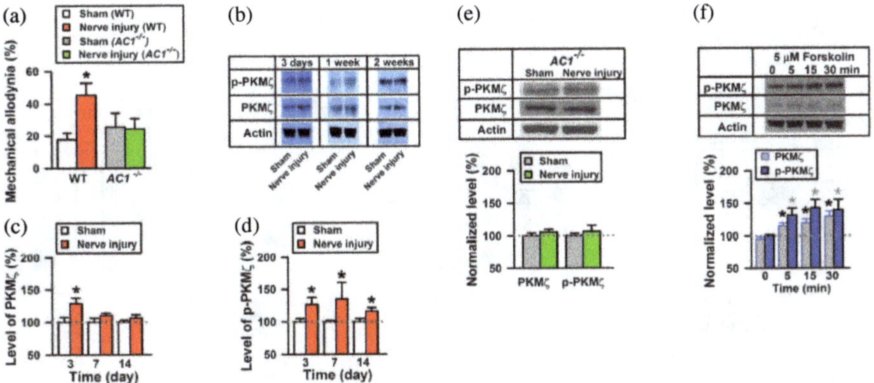

Figure 6. Expression and phosphorylation of PKMζ in the ACC during neuropathic pain. (a) Mechanical allodynia was tested in the sham and nerve injury groups from wild-type (WT) and AC1$^{-/-}$ animals 3 days after nerve injury. $*P < 0.05$. (b) Western blots for PKMζ and p-PKMζ in the ACC obtained between 3 days and 2 weeks after nerve injury (the ACC of mice from the sham and nerve injury groups was used for western blot analysis 90 min after the allodynia test). (c) Level of PKMζ in the ACC of mice from the sham and nerve injury groups. Level of PKMζ increased significantly 3 days after nerve injury compared with PKMζ in the sham group. (d) Level of p-PKMζ increased significantly 3 days after nerve injury. This effect lasted over 2 weeks after nerve injury compared with p-PKMζ in the sham group. (e) Peripheral nerve injury did not increase the levels of PKMζ and p-PKMζ in the ACC of AC1$^{-/-}$ mice. (f) Forskolin incubation increased PKMζ (gray) and p-PKMζ (blue) levels in the ACC of WT mice (adapted from Li *et al.* [29]).

those activated by experimental stimuli used to induce NMDAR-dependent LTP at ACC synapses.

Altered intrinsic properties after injury

In addition to changes in synaptic transmission, long-term changes in firing patterns and intrinsic electrical properties have been noted in ACC neurons in chronic pain states [30,31]. There are at least three major types of pyramidal cells in the ACC, classified according to their action potential firing pattern: regular spiking, intrinsic bursting, and intermediate cells. The population distribution and the single action potential properties of these three groups are not affected 1–2 weeks after nerve injury in

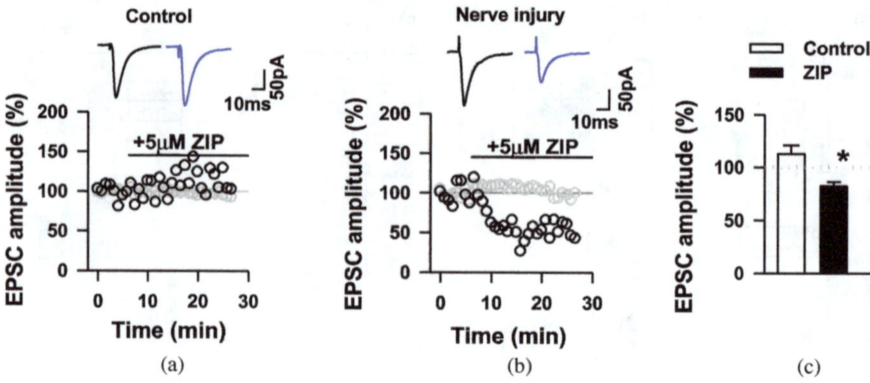

Figure 7.   Inhibition of PKMζ selectively decreased the amplitude of eEPSCs in ACC neurons of mice with neuropathic pain by reducing the number of active AMPA receptors. (a, b) Samples showed the effect of ZIP (5 mM) on the amplitude of eEPSCs in the ACC neurons of animals from the nerve injury [black circles in (a)] and sham group [black circles in (b)]. The gray open circles represent the change of membrane resistance during recording. Black traces in the upper part of (a) and (b) indicate the averaged response at baseline, and blue traces indicate the average of 2 min responses collected 10 min after ZIP application. (c) Pooled data of effects of ZIP on the eEPSCs recorded from the ACC of mice in the sham (open) and nerve injury (solid) groups 10 min after ZIP application. *$P < 0.05$ (adapted from Li *et al.* [29]).

mice. However, intermediate cells from animals with neuropathic pain showed higher initial firing frequencies. Furthermore, it has been shown that the temporal precision of action potential firing in the ACC is reduced 1–3 days after injection of CFA, or 1–2 weeks after nerve injury, as reflected in increased jitter [31] (Figure 8). These findings suggest that chronic pain does not cause dramatic changes in the frequency of the ACC neuronal firing pattern but that the temporal precision of information coding in the ACC is reduced for long periods of time.

Activation by allodynic stimulation

Recent studies have consistently indicated that the Erk signaling cascade plays an important role in activity-dependent plasticity and may contribute to the molecular mechanisms underlying learning, memory, and

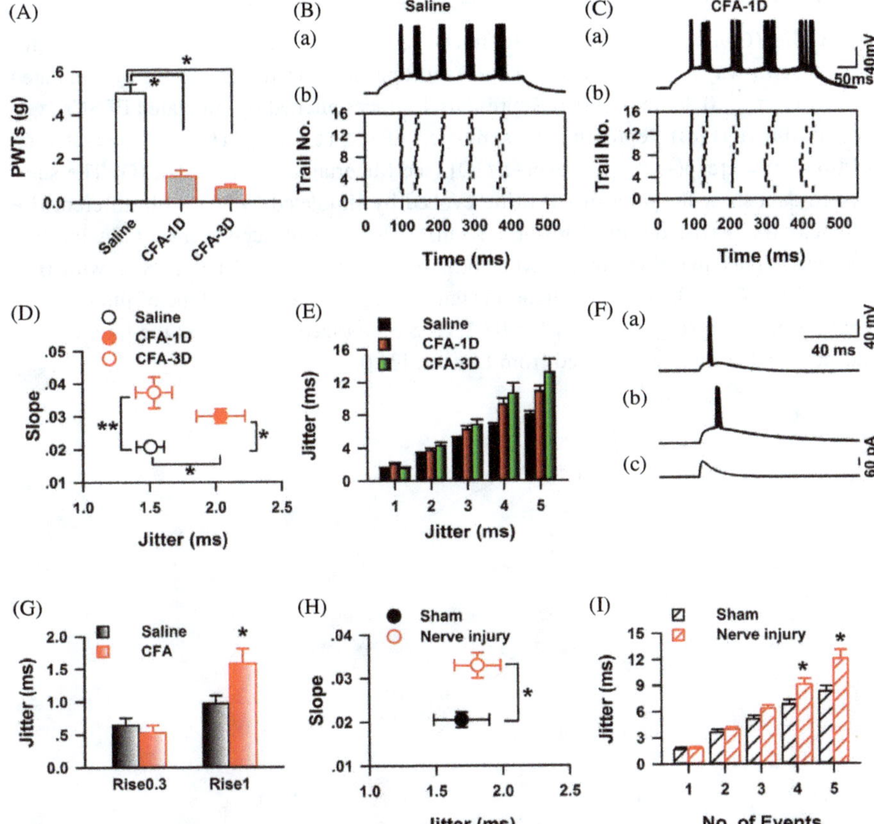

Figure 8. Inflammation of mice decreased the temporal precision of AP firing of neurons in the ACC of mice *in vitro*. (A) CFA significantly decreased the paw withdrawal threshold of mice on 1 and 3 d after injection. (B(a)) The traces showed 15 responses under the steady-state current stimulations recorded in the ACC from control mice. (B(b)) The raster plot shows the APs recorded in (A(a)). (C(a)) The traces showed 15 responses under the steady-state current stimulations recorded in the ACC from CFA-injected mice (CFA-1D). (C(b)) The raster plot shows the APs recorded in (B(a)). (D) Significant differences were detected in the jitter of the first AP and the slope of jitter between saline (open black circle) and CFA-1D (filled red circle), CFA-3D group (open red circle); *$P < 0.05$. (E) The jitter of AP firing from saline-injected mice (black bar) was significantly different from

Figure 8. (*Continued*) CFA-injected mice (red bar for CFA-1D and green bar for CFA-3D; two-way ANOVA; *P < 0.05). (F(a)) Represents 15 traces elicited by simulated EPSCs (τ_{rise}/τ_{decay}:0.3/3 ms). (F(b)) Represents 15 traces elicited by simulated EPSCs (τ_{rise}/τ_{decay}: 1/10 ms). (F(c)) Represents the simulated EPSCs (τ_{rise}/τ_{decay}: 1/10 ms) generated by function $I(t) = \lambda[\exp(-t/\tau_{decay}) - \exp(-t/\tau_{rise})]$ (see Materials and Methods). (G) The summarized data show the jitter of AP firing evoked by simulated EPSCs with different rise and decay times (rise/decay: 0.3/3 or 1/10 ms). The jitter of latency from CFA-injected mice was higher than the control with the stimulations of simulated EPSCs with rise/decay: 1/10. *P < 0.05. (H) CPN ligation significantly increased the slope of jitter change tested 1 week after surgery; *t*-test, P < 0.05. (I) Summarized data present the jitter change induced by CPN ligation (adapted from Li *et al.* [31]).

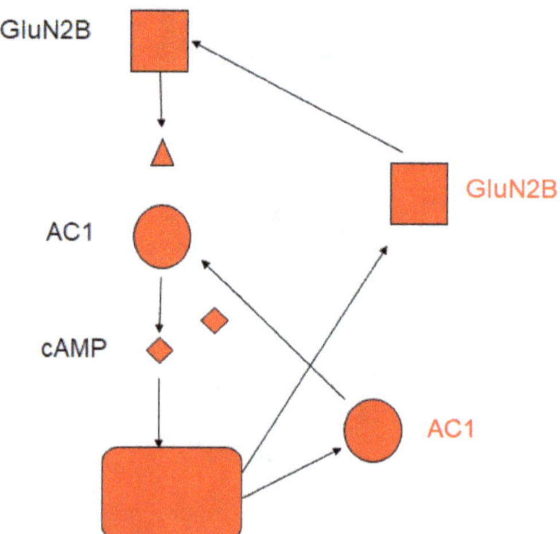

Figure 9. Positive feedback control of intracellular AC1 in case of chronic visceral pain. Peripheral injury activates significant increases in AC1 protein synthesis through NMDA receptor-dependent mechanism. NMDA NR2B containing receptor is required for the induction. In addition to the increase in AC1 protein, NMDA NR2B subtype is also increased. In this positive feedback loop, the activity of AC1 is required.

persistent pain. Electrophysiological experiments have demonstrated that the activity of Erk contributes to ACC LTP and that such inhibitors are relatively selective and do not affect basic synaptic transmission. Interestingly, rapid increases in P-Erk expression were found in some of the layer II neurons in the bilateral ACC at 15 min after tissue injury.

The activated Erk level in the ACC was reduced at 45 min and declined further at 90 min after formalin injection. There was no obvious Erk activation in the deep-layer neurons in the ACC. Thus, the P-Erk expression pattern in the region is different from that of immediate early genes, such as c-Fos, which is widely expressed in ACC neurons located in all layers after formalin injection. It has previously been demonstrated that neurons in the layer II/III receive ascending noxious inputs from the thalamus and communicate with other cortical areas. Thus, the Erk activation in ACC neurons may play a role in Erk-dependent neuronal plasticity in the ACC during the induction and/or development of inflammatory pain. More importantly, 2 weeks after the injury, peripheral gentle non-noxious stimuli can cause wide-spread activation of Erk in ACC neurons, indicating that injury can cause long-term changes in neuronal excitation [12].

Loss of inhibition

Besides changes in excitatory synapses, it has previously been reported that there is a striking loss of connections between excitatory and inhibitory neurons in the ACC after nerve injury [32]. No significant changes in synaptic efficacy in the remaining connected pairs were found. These changes were reflected at the network level by a decrease in the mEPSC and mIPSC frequencies. Additionally, nerve injury resulted in a potentiation of the intrinsic excitability of pyramidal neurons, whereas the cellular properties of interneurons were unchanged. These changes at the cortical network level might therefore contribute to the neuropathic pain condition, via disinhibition of the anterior cingulate cortex.

Possible roles of microglia

While the role of microglia in chronic pain-related spinal plasticity has been researched, microglia seem to be inactive in responding to peripheral injury in cortical areas, such as the ACC and IC [33,34]. Direct whole-cell recordings from microglia in the ACC found that microglia are not responsive to excitatory and inhibitory transmitters [35] while demonstrating ATP P2Y receptor-mediated outward potassium currents. Furthermore, local synaptic activation that triggers excitatory synaptic responses showed no response at all in microglia cells [36]. LTP also did not cause activation of microglia. Thus, it is likely that microglia do not significantly contribute to fast synaptic transmission and synaptic plasticity. Similar results were found in the spinal cord dorsal horn [37]. Consistent with this finding, minocycline, an inhibitor of microglial activation, was not found to have any effect on two different forms of LTP in the ACC of adult animals [38]. It is more likely that microglia or microglia-related chemical factors may be involved in chronic brain injury.

Enhancement of ACC top-down facilitation

Spinal sensory transmission occurs under descending biphasic modulation, and descending facilitation is believed to contribute to chronic pain [39]. In addition to descending modulation from the brainstem rostral ventromedial medulla (RVM), top-down facilitatory modulation of spinal nociceptive transmission has been reported from the ACC [40]. Stimulation of the ACC-potentiated spinal excitatory synaptic transmission and this modulation is independent of the RVM. Peripheral nerve injury enhanced the spinal synaptic transmission and occluded the ACC-spinal cord facilitation. Inhibition of ACC reduced the enhanced spinal synaptic transmission caused by nerve injury [40]. These findings indicate that ACC-spinal top-down facilitation is enhanced after the injury in an RVM-independent manner and that such top-down facilitation may contribute to the process of chronic neuropathic pain.

Conclusions

Cortical synapses activated by painful signals are highly plastic and their potentiation likely encodes the unpleasantness of pain and pain-related emotional sufferings. Calcium-dependent signaling pathways play important roles in the induction and expression of such potentiation.

References

[1] Florence, S.L. *et al.* (1998) Large-scale sprouting of cortical connections after peripheral injury in adult macaque monkeys. *Science* 282, 1117–1121. 10.1126/science.282.5391.1117.

[2] Jain, N. *et al.* (1998) Reorganization of Somatosensory Cortex After Nerve and Spinal Cord Injury. *News Physiol Sci* 13, 143–149. 10.1152/physiologyonline.1998.13.3.143.

[3] Flor, H. *et al.* (1995) Phantom-limb pain as a perceptual correlate of cortical reorganization following arm amputation. *Nature* 375, 482–484. 10.1038/375482a0.

[4] Metz, A.E. *et al.* (2009) Morphological and functional reorganization of rat medial prefrontal cortex in neuropathic pain. *Proc Natl Acad Sci USA* 106, 2423–2428. 10.1073/pnas.0809897106.

[5] Zhuo, M. (2004) Central plasticity in pathological pain. *Novartis Found Symp* 261, 132–145; discussion 145–154.

[6] Ko, H.G. *et al.* (2018) Rapid turnover of cortical NCAM1 regulates synaptic reorganization after peripheral nerve injury. *Cell Rep* 22, 748–759. 10.1016/j.celrep.2017.12.059.

[7] Bliss, T.V. *et al.* (2016) Synaptic plasticity in the anterior cingulate cortex in acute and chronic pain. *Nat Rev Neurosci* 17, 485–496. 10.1038/nrn.2016.68.

[8] Chen, T. *et al.* (2014) Postsynaptic potentiation of corticospinal projecting neurons in the anterior cingulate cortex after nerve injury. *Mol Pain* 10, 33. 10.1186/1744-8069-10-33.

[9] Wei, F. *et al.* (2000) Role of EGR1 in hippocampal synaptic enhancement induced by tetanic stimulation and amputation. *J Cell Biol* 149, 1325–1334. 10.1083/jcb.149.7.1325.

[10] Wei, F. *et al.* (1999) Loss of synaptic depression in mammalian anterior cingulate cortex after amputation. *J Neurosci* 19, 9346–9354.

[11] Ko, S.W. *et al.* (2005) Selective contribution of Egr1 (zif/268) to persistent inflammatory pain. *J Pain* 6, 12–20. 10.1016/j.jpain.2004.10.001.

[12] Wei, F. and Zhuo, M. (2008) Activation of Erk in the anterior cingulate cortex during the induction and expression of chronic pain. *Mol Pain* 4, 28. 10.1186/1744-8069-4-28.

[13] Wei, F. and Zhuo, M. (2001) Potentiation of sensory responses in the anterior cingulate cortex following digit amputation in the anaesthetised rat. *J Physiol* 532, 823–833. 10.1111/j.1469-7793.2001.0823e.x.

[14] Kang, S.J. *et al.* (2012) Plasticity of metabotropic glutamate receptor-dependent long-term depression in the anterior cingulate cortex after amputation. *J Neurosci* 32, 11318–11329. 10.1523/JNEUROSCI.0146-12.2012.

[15] Zhao, M.G. *et al.* (2006) Enhanced presynaptic neurotransmitter release in the anterior cingulate cortex of mice with chronic pain. *J Neurosci* 26, 8923–8930. 10.1523/JNEUROSCI.2103-06.2006.

[16] Bie, B. *et al.* (2011) Increased synaptic GluR1 subunits in the anterior cingulate cortex of rats with peripheral inflammation. *Eur J Pharmacol* 653, 26–31. 10.1016/j.ejphar.2010.11.027.

[17] Niikura, K. *et al.* (2011) Enhancement of glutamatergic transmission in the cingulate cortex in response to mild noxious stimuli under a neuropathic pain-like state. *Synapse* 65, 424–432. 10.1002/syn.20859.

[18] Xu, H. *et al.* (2008) Presynaptic and postsynaptic amplifications of neuropathic pain in the anterior cingulate cortex. *J Neurosci* 28, 7445–7453. 10.1523/JNEUROSCI.1812-08.2008

[19] Chen, T. *et al.* (2014) Postsynaptic insertion of AMPA receptor onto cortical pyramidal neurons in the anterior cingulate cortex after peripheral nerve injury. *Mol Brain* 7, 76. 10.1186/s13041-014-0076-8.

[20] Li, H.S. *et al.* (2019) Protein kinase C lambda mediates acid-sensing ion channel 1a-dependent cortical synaptic plasticity and pain hypersensitivity. *J Neurosci* 39, 5773–5793. 10.1523/JNEUROSCI.0213-19.2019.

[21] Descalzi, G. *et al.* (2009) Presynaptic and postsynaptic cortical mechanisms of chronic pain. *Mol Neurobiol* 40, 253–259. 10.1007/s12035-009-8085-9.

[22] Hartmann, B. *et al.* (2004) The AMPA receptor subunits GluR-A and GluR-B reciprocally modulate spinal synaptic plasticity and inflammatory pain. *Neuron* 44, 637–650. 10.1016/j.neuron.2004.10.029.

[23] Zhuo, M. (2009) Plasticity of NMDA receptor NR2B subunit in memory and chronic pain. *Mol Brain* 2, 4. 10.1186/1756-6606-2-4.

[24] Zhuo, M. (2008) Cortical excitation and chronic pain. *Trends Neurosci* 31, 199–207. 10.1016/j.tins.2008.01.003.

[25] Zhuo, M. (2018) Potentiation of cortical excitatory transmission in chronic pain. *Pain* 159, 212–213. 10.1097/j.pain.0000000000001115.

[26] Zhuo, M. (2016) Neural mechanisms underlying anxiety-chronic pain interactions. *Trends Neurosci* 39, 136–145. 10.1016/j.tins.2016.01.006.

[27] Liu, S.B. *et al.* (2015) Long-term upregulation of cortical glutamatergic AMPA receptors in a mouse model of chronic visceral pain. *Mol Brain* 8, 76. 10.1186/s13041-015-0169-z.

[28] Liu, S.B. *et al.* (2020) Cyclic AMP-dependent positive feedback signaling pathways in the cortex contributes to visceral pain. *J Neurochem* 153, 252–263. 10.1111/jnc.14903.

[29] Li, X.Y. *et al.* (2010) Alleviating neuropathic pain hypersensitivity by inhibiting PKMzeta in the anterior cingulate cortex. *Science* 330, 1400–1404. 10.1126/science.1191792.

[30] Cao, X.Y. *et al.* (2009) Characterization of intrinsic properties of cingulate pyramidal neurons in adult mice after nerve injury. *Mol Pain* 5, 73. 10.1186/1744-8069-5-73.

[31] Li, X.Y. *et al.* (2014) Long-term temporal imprecision of information coding in the anterior cingulate cortex of mice with peripheral inflammation or nerve injury. *J Neurosci* 34, 10675-10687. 10.1523/JNEUROSCI.5166-13.2014.

[32] Blom, S.M. *et al.* (2014) Nerve injury-induced neuropathic pain causes disinhibition of the anterior cingulate cortex. *J Neurosci* 34, 5754–5764. 10.1523/JNEUROSCI.3667-13.2014.

[33] Tsuda, M. *et al.* (2017) Neuronal and microglial mechanisms for neuropathic pain in the spinal dorsal horn and anterior cingulate cortex. *J Neurochem* 141, 486–498. 10.1111/jnc.14001.

[34] Zhang, F. *et al.* (2008) Selective activation of microglia in spinal cord but not higher cortical regions following nerve injury in adult mouse. *Mol Pain* 4, 15. 10.1186/1744-8069-4-15.

[35] Wu, L.J. *et al.* (2007) ATP-induced chemotaxis of microglial processes requires P2Y receptor-activated initiation of outward potassium currents. *Glia* 55, 810–821. 10.1002/glia.20500.

[36] Wu, L.J. and Zhuo, M. (2008) Resting microglial motility is independent of synaptic plasticity in mammalian brain. *J Neurophysiol* 99, 2026–2032. 10.1152/jn.01210.2007.

[37] Chen, T. *et al.* (2010) Spinal microglial motility is independent of neuronal activity and plasticity in adult mice. *Mol Pain* 6, 19. 10.1186/1744-8069-6-19.

[38] Song, Q. *et al.* (2015) Minocycline does not affect long-term potentiation in the anterior cingulate cortex of normal adult mice. *Mol Pain* 11, 25. 10.1186/s12990-015-0025-2.

[39] Zhuo, M. (2017) Descending facilitation. *Mol Pain* 13, 1744806917699212. 10.1177/1744806917699212.

[40] Chen, T. *et al.* (2018) Top-down descending facilitation of spinal sensory excitatory transmission from the anterior cingulate cortex. *Nat Commun* 9, 1886. 10.1038/s41467-018-04309-2.

Chapter 8

NMDA NR2B

Study: Rodents' Higher IQ May Come at Painful Price

By Rick Weiss, January 29, 2001

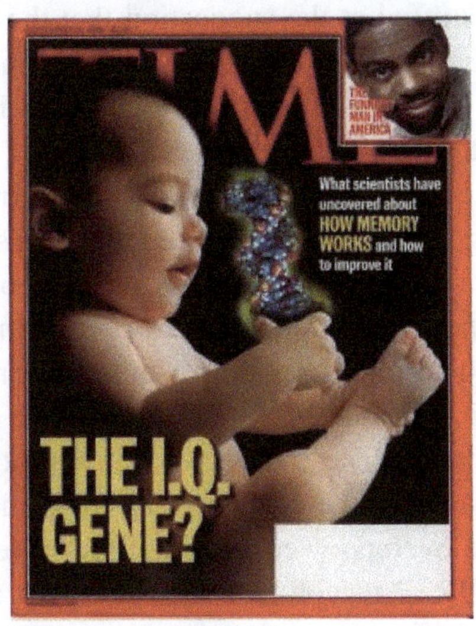

It hurts to be smart

That's one conclusion from the latest study of so-called Doogie mice —
"smart" rodents that are genetically engineered to have enhanced memory

and learning skills. Along with those extra IQ points, researchers have found, comes an added sensitivity to some kinds of pain.

The new work offers a sobering lesson about the difficulty of enhancing certain brain functions without simultaneously taking a toll on others. It also may temper whatever momentum there is to engineering genetic enhancements in people. "Beware what you ask for," said James L. McGaugh, a neuroscientist at the University of California at Irvine. "And when you get it, look carefully and see what else you got." Doogie mice — named after the precocious television character Doogie Howser, MD — made a big splash when they were introduced to the world in September 1999. Having been endowed with extra copies of a gene involved in memory formation, the animals outperformed their normal counterparts on a variety of tasks. They were better at recognizing objects they had seen before, remembered painful experiences longer, recalled with greater accuracy the location of submerged platforms in milky water, and were better at "unlearning" old associations that were no longer true.

Some scientists sniffed at the suggestion that the mice were particularly brainy, noting that intelligence is much more than a collection of four or five mental skills. Nonetheless, the work was surprising because it was the first to show that by adding a few extra copies of a single gene to an embryo, researchers could improve an animal's performance on a range of memory and learning tasks. Some suggested that drugs designed to mimic the gene's effects might help Alzheimer's patients or even make sharp people sharper. The new work hints that it won't be that easy.

Min Zhuo and his colleagues at Washington University School of Medicine in St. Louis assessed how the Doogie mice responded to tissue damage and inflammation. They suspected that sensations of pain caused by those types of injury may be controlled by the same "NR2B receptor" that Doogie mice are overendowed with and that gives the animals their superior memories. NR2B receptors are proteins that act as "coincidence detectors" in the brain. They recognize, for example, when a certain sound is linked to the arrival of food and help consolidate such coincidences into learned associations.

The researchers subjected the mice to stimuli that cause either short-term or long-term pain. They heated the animals' tails, poked their foot pads with stiff fibers, and injected their paws with irritating solutions.

Then they used molecular and neurological tests to see how the animals' brains responded and tracked the animals' behavior, measuring, for example, how long they licked the site of injury.

Those tests indicated that, compared with normal mice, Doogie mice are equally sensitive to short-term pain. But chronic inflammatory pain, such as that caused by the injected irritants, lasts significantly longer in Doogie mice. The team reported in today's issue of the journal of *Nature Neuroscience*: "Our results suggest that a genetic manipulation conferring enhanced cognitive abilities may also provide unintended traits, such as increased susceptibility to persistent pain." Other scientists conceded that it is difficult to know what mice are experiencing because they cannot talk. Even in people, the physical and cognitive components of pain are deeply integrated. Still, several scientists have now said that the new study offers strong substantiation that a Doogie mouse's pain is real. "This is very convincing evidence" that the mice have prolonged chronic pain responses, said McGaugh, who directs U.C. Irvine's center for the neurobiology of learning and memory. Moreover, he said, the finding makes sense. "Most of our brain regions are multipurpose. These things are all intertwined."

Indeed, as others have said, evolution rewards creatures that find more than one use for things, especially things as useful as a neural coincidence detector. "When nature finds an effective mechanism, it's used and reused," said Ira Black, chairman of neuroscience and cell biology at the Robert Wood Johnson Medical School in Piscataway, N.J. That's why so many drugs have unwanted side effects, and the same will probably prove to be true for new drugs that may someday take aim at NR2B, Black said. People who try to make smarter babies with NR2B genes, or boost their own memory with NR2B drugs, may have to accept some level of chronic pain. Alternatively, those who seek to kill their pain with NR2B-blocking drugs may also have to accept some cognitive side effects.

"You can't have it both ways," Black said.

Summary

The NMDA receptor is known to be critical for learning-related central plasticity. Recent evidence indicates that NMDA receptor-dependent plasticity is not unique to cognition-related plasticity. NMDA receptors, located in pain-related sensory synapses, are important for injury-related plasticity, from the spinal dorsal horn to the cortical synapse. Furthermore, injury triggered not only changes in glutamate AMPA receptor-mediated responses but also NMDA receptor-mediated responses, including NMDA NR2B receptors. Therefore, targeting NMDA NR2B receptors as a potential target for chronic pain may serve dual functions, first to prevent injury-induced new plasticity and, second, to reduce established plasticity mediated by NMDA NR2B receptors.

Keywords: NMDA receptor; NR2B (GluN2B); ACC; IC; LTP; smart mice; chronic pain; AC1

Introduction

The NMDA receptor is a major type of ionotropic glutamate receptor in the CNS and plays a fundamental role in both synaptic transmission and plasticity [1–4]. NMDA receptor-dependent synaptic plasticity is a key cellular model for studying central brain functions, including learning and memory, chronic pain, and drug addiction. For example, NMDA receptor-dependent LTP in the hippocampus contributes to the formation of long-term memory. In the ACC, NMDA receptor-dependent LTP plays an important role in injury-triggered cortical excitation and plastic changes. In this short review, we will review the progress made in NMDA receptor-dependent, and -independent, synaptic plasticity in the ACC, including LTP and LTD. The contribution of different types of NMDA receptors will be discussed. We believe that foundational knowledge of cortical plasticity will provide novel protein targets for better treatment of different forms of brain disease, including chronic pain, anxiety, and memory loss.

NMDA receptors and their subtypes

NMDA receptors are 5–10 times more permeable to Ca^{2+} than to Na^+ or K^+ and are a critical intracellular signaling molecule for triggering post-synaptic, and possible presynaptic, plastic changes. The NMDA receptor subunits consist of three families: GluN1, GluN2 (GluN2A, -2B, -2C, and -2D), and GluN3 (GluN3A and GluN3B). Structural studies have already reported that NMDA receptors are assembled in tetrameric form, in which two GluN1 subunits and two GluN2 and/or GluN3 subunits are combined together as di-heteromeric or tri-heteromeric receptors. The subunit composition varies during neurodevelopment and determines the channel properties of NMDA receptors [5,6]. In general, NMDA receptors demonstrate slower kinetics which enables relatively longer permeability of Ca^{2+} once activated in the presence of glutamate and the co-agonist glycine. Depolarization relieves the block of Mg^{2+} in a voltage-dependent manner and then Ca^{2+} influxes through the NMDA receptor channel, triggering a cascade of intracellular events that cause the change of synaptic plasticity. Aside from the postsynaptic NMDA receptors, presynaptic NMDA receptors have also been reported to modulate glutamate release, and extra-synaptic NMDA receptors are involved in synaptic plasticity [7–11].

NMDA receptor in the ACC neurons

Glutamate is the major excitatory neurotransmitter in the ACC. NMDA receptor plays a significant role in the ACC [4,12,13]. In the ACC, an NMDA receptor-mediated field potential was recorded after applying antagonists of AMPA and KA receptors. This potential was sensitive to the antagonist of NMDA receptor, AP-5 [14]. Similarly, an electrically evoked NMDA receptor-dependent calcium signal was recorded in the ACC neurons [12,15]. In addition, it was found that NMDARs act as key molecules in the synaptic plasticity of the ACC neurons. Application of AP-5 completely blocked the induction of pairing training and TBS-induced LTP in the ACC of adult mice [16]. Interestingly, after the LTP induction, NMDA receptor-mediated currents were decreased, which suggests that the expression of LTP in the ACC is NMDA receptor-independent. On the other hand, NMDA receptors in the ACC have also been reported to be involved in another form of synaptic plasticity: LTD. LTD induced by a 'pairing training' protocol can be blocked by applying AP-5 [17]. Therefore, NMDA receptors in the ACC neurons are a crucial part of the physiological process.

GluN2A vs. GluN2B, 2C, and 2D

The GluN2A and GluN2B subunits predominate in the ACC. GluN2A and GluN2B subunits determine distinct properties of NMDA receptors. For example, heteromers containing GluN1 plus GluN2B mediate a current that decays three to four times more slowly than receptors composed of GluN1 plus GluN2A. At birth, forebrain NMDA receptors are composed almost exclusively of GluN1 and GluN2B subunits, gradually incorporating more GluN2A subunits during postnatal development [18]. Major changes in the GluN2 subunits occur during the first two postnatal weeks. GluN2A expresses shortly after birth and rises steadily, becoming widely expressed in the adult. GluN2B expression is maintained at high levels following birth, peaks around the first postnatal week, and becomes progressively restricted to the forebrain [18,19]. Previous studies in the pyramidal neurons of the ACC demonstrated conclusively that by using a selective antagonist of GluN2A-containing NMDARs, NVP-AAM077, and a selective antagonist of GluN2B-containing NMDARs, Ro 25-6981, the GluN2A and GluN2B subunit-mediated currents were calculated after sequential application of those two antagonists. Application of NVP-AAM077 depressed the total NMDAR-mediated currents by 63.2% $\pm$ 1.7%, and the addition of Ro25-6981 to the same neuron further reduced currents by 13.7% $\pm$ 1.4% in the ACC. Reversing the order of applying antagonists resulted in total NMDAR-mediated currents being depressed by 18.0% $\pm$ 1.9% and 64.2% $\pm$ 3.0%, respectively [16]. Moreover, the rising time and decay time constant of GluN2B-mediated EPSCs were significantly longer than those of GluN2A-mediated currents in ACC pyramid neurons [20,21].

Among all the subunits, NMDA receptors with GluN2C/2D subunits displayed a lower sensitivity to Mg^{2+} ions and slower deactivation kinetics compared with GluN2A/2B containing NMDA receptors. GluN2C/2D-contaning NMDA receptors allowed for the selective modification of circuit properties in regions expressing GluN2C/2D subunits [22–25]. The GluN2C subunit appears on postnatal day 10 (P10), and its expression is mainly confined to the granule cells of the cerebellum and the olfactory bulb. The GluN2D subunit is expressed in the caudal region of the embryonic brain. Its expression drops during neurodevelopment and mainly distributes in the diencephalon and brainstem in the central nervous system in adult animals. Nevertheless, the expression of GluN2C/2D was also detected in cerebral cortices [26–28]. One recent study has proved that the GluN2C/2D is expressed in the ACC of adult mice and comprises 24% of the rest of the NMDA receptor-mediated current in the pyramidal neurons in the ACC after blocking the GluN2A and 2B by NVP-AAM077 and Ro 25-6981. This was determined by the decrease in the amplitude of NMDA receptor-mediated EPSCs using the selective antagonist of GluN2C/2D, UBP145 [29]. According to this study, the proportion of GluN2C/2D is (100% − 63.7% − 15.9%) × 24% = 4.9%. In the ACC, GluN2C/2D has been suggested to be involved in presynaptic regulation of spontaneous glutamate release [33].

NMDA receptor-dependent post-LTP and -independent pre-LTP

There are two major forms of LTP in the ACC: pre-LTP and post-LTP. The induction of post-LTP depends on the postsynaptic NMDA receptors. The inhibition of NMDA receptors by bath application of AP-5 reliably blocks the induction of post-LTP no matter which protocol the post-LTP was induced by [16]. The key mechanism for the involvement of NMDA receptor in the induction of post-LTP is its voltage dependence. In order to activate postsynaptic NMDA receptors, two events need to happen simultaneously. First, glutamate needs to be released under the help of, and binds to, postsynaptic NMDA receptors. Second, the postsynaptic membrane needs to be depolarized so that the block by extracellular Mg^{2+} can be removed. Thus, calcium influxes from the extracellular space into the postsynaptic cells via opened NMDA receptors which then activates a series of signaling molecules within the postsynaptic cells, including CaM, PKA, AC1, cAMP, immediate early genes, as well as enzymes producing diffusible retrograde messengers [1,4,13,30]. In the ACC, it has been shown that bath application of selective GluN2A or GluN2B antagonists significantly reduced but did not block the induction of post-LTP by the pairing protocol, as detected by whole-cell patch-clamp recording. This suggests that tri-heteromers of the NMDA receptor may also be the dominant form of the receptor that contributes to post-LTP at these synapses [16].

In addition to NMDA receptor-dependent post-LTP, there is non-NMDA receptor-dependent pre-LTP. With the antagonist of NMDA receptor, AP-5 perfused in the bath solution, low-frequency stimulation (2 Hz) induces the pre-LTP in the brain slices of the ACC [31]. This form of LTP cannot be induced in the ACC slices obtained from mice lacking the GluK1 subunit but can be induced in slices from mice lacking the GluK2 subunit and blocked by a potent GluK1-selective KAR antagonist: UBP310. The L-type voltage-gated calcium channels have also been found to contribute to the induction of pre-LTP since nimodipine reduced the amplitude of pre-LTP. The transduction of pre-LTP in the ACC also involves the activation of the cAMP signaling pathway, FMRP signaling pathway, and requires AC1 [32]. The expression of pre-LTP involves the modulation of hyperpolarization-activated cyclic nucleotide-gated (HCN) channels [3,31]. These key molecules may become potential drug targets for the treatment of chronic pain and its related anxiety.

Presynaptic roles of NMDA receptors

There are cumulative studies investigating the NMDA receptors in the presynaptic terminals. Presynaptic NMDA receptors may regulate neurotransmitter release by (1) directly or indirectly causing Ca^{2+} influx to trigger vesicle exocytosis or (2) modulating downstream intracellular signaling cascades through metabotropic effects that are independent of Ca^{2+} flux. The roles of presynaptic NMDA receptors are mostly investigated in the somatosensory cortex and visual cortex [33,34]. Presynaptic NMDA receptors have synapse-specific expression in several brain regions, including layer IV-layer II/III synapses of the somatosensory cortex and visual cortex in rats and mice, and local layer IV or V of the visual cortex in rats. Presynaptic NMDA receptors are required for the induction of spike timing-dependent LTD in the layer IV-layer II/III synapses of the mouse visual cortex during early life [35]. Additionally, some forms of presynaptic NMDA receptor-mediated plasticity may involve changes in PKA phosphorylation of Rab3-interacting molecule 1 (RIM1) [36].

Interestingly, presynaptic NMDA receptors were found to facilitate spontaneous neurotransmitter release in the absence of depolarization, which suggests that they are activated by ambient glutamate, and may not require the relief of the Mg^{2+} block [37]. This may result from the incorporation of presynaptic NMDA receptor subunits with reduced Mg^{2+} sensitivity, such as GluN2C, GluN2D, or GluN3A subunits. Our recent study has found that the GluN2C/2D modulates the presynaptic glutamate release in the ACC. The role of presynaptic NMDA receptors in spontaneous release also requires activation of the c-JunN-terminal kinase 2 (JNK2) pathway [38]. The presynaptic NMDA receptors in spontaneous and evoked release work in a non-overlapping manner to modulate the presynaptic release.

Genetically overexpression of NR2B enhanced LTP

The genetic approach has been used to examine if enhancing the NMDA receptor functions also increases LTP [39]. The NR2B subunits were overexpressed postnatally in the mouse forebrain using the CaM-kinase-II promoter. To assess bidirectional synaptic plasticity in the 1–100 Hz range, a series of LTP experiments in the Schaffer collateral CA1 pathway were conducted. A single tetanic stimulation (100 Hz, 1 s) typically evoked smaller but reliable potentiation in 4–6-month-old control slices, in comparison to that evoked in younger adult slices. However, the same stimulus evoked significantly larger potentiation in transgenic slices. The enhancement of potentiation was not due to changes in inhibitory GABA-mediated mechanisms as it was also observed in the presence of 100 μM picrotoxin. Enhanced LTP is NMDA receptor-dependent, and the enhanced LTP was completely blocked by AP-5 (100 μM).

Enhancement of LTP in NR2B overexpressed mice is frequency-dependent. Enhanced long-term potentiation in the transgenic slices was also observed when prolonged, repetitive stimulation (10 Hz), to the Schaffer–CA1 path was applied. Although 10-Hz stimulation for 1.5 min (900 pulses) did not induce reliable synaptic potentiation in wild-type animals, it evoked robust synaptic potentiation in transgenic slices. However, repetitive stimulation delivered at 5 Hz for 3 min (also giving 900 pulses) did not produce significant synaptic potentiation in either group of mice.

Behavioral enhancement of pain responses in NR2B transgenic mice

In situ hybridization with a probe for NMDA NR2B showed that NR2B expression was significantly enhanced in two forebrain structures that are important for pain perception: the ACC and IC (Figure 1) [40].

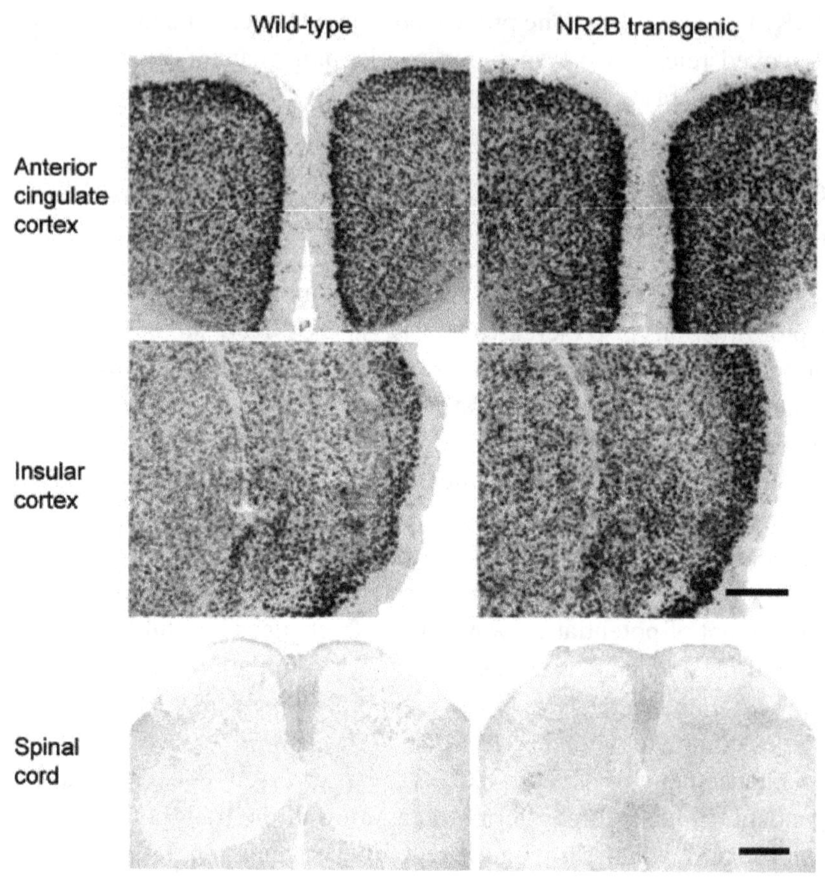

Figure 1.   Enhanced NR2B expression in the ACC and insular cortex but not the spinal dorsal horn in NR2B transgenic mice. Expression of NR2B was examined by *in situ* hybridization in wild-type and NR2B transgenic mice. Whereas the two forebrain regions exhibited enhanced NR2B expression in transgenic compared to wild-type mice, no upregulation of NR2B expression was seen in the spinal cord dorsal horn. Scale bars: 300 m (adapted from Wei *et al.* [43]).

As predicted, no detectable NR2B expression was found in the dorsal horn of wild-type or transgenic mice. Electrophysiological experiments confirmed that NMDA receptor-mediated responses were also enhanced in the transgenic mice. Brain slices of these areas were prepared from adult mice, and excitatory postsynaptic field potentials (fEPSPs) were recorded upon local electrical stimulation (Figure 2). In each region, after a blockade of AMPA and KA receptors by CNQX (20 μM), a slow fEPSP was observed (Figure 2) that could be entirely, and reversibly, blocked by the NMDA receptor antagonist AP-5 (100 μM). Consistent with the high levels of NR2B transgene expression found throughout the cerebral cortex in transgenic mice, these mice — as compared to wild-type mice — exhibited enhanced NMDA receptor-mediated fEPSPs in both the ACC and IC.

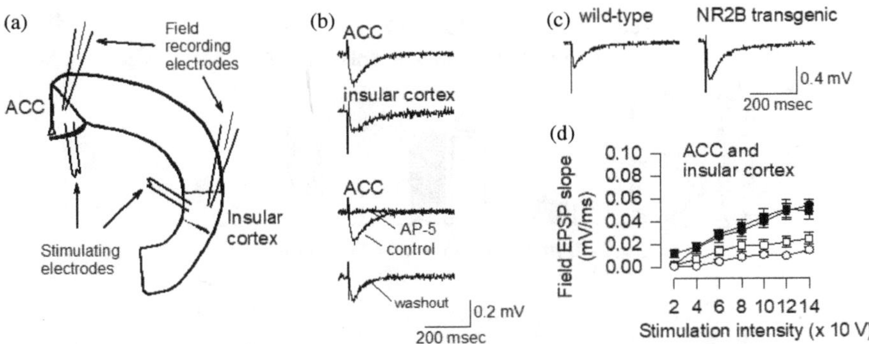

Figure 2. Forebrain-targeted NR2B overexpression enhanced NMDA receptor-mediated synaptic responses in the ACC and IC. (a, b) Traces of NMDA receptor-mediated fEPSPs recorded from the ACC and insular cortex in the presence of 20 μM CNQX. (c) Bath application of 100 μM AP-5 completely (top) and reversibly (bottom) blocked NMDA receptor-mediated fEPSPs in the ACC. Similar results were found in insular cortex (data not shown). (d) The input (stimulation intensity, 200-μs duration)–output (fEPSP slope) relationship in the ACC (wild type, $n = 8$; transgenic, $n = 12$) and insular cortex (wild-type, $n = 6$; transgenic, $n = 9$) reveals enhanced NMDA receptor-mediated responses in transgenic relative to wild-type mice ($P < 0.001$). Above are examples of fEPSP traces recorded from the ACC in wild-type and transgenic animals. We also integrated fEPSPs in the ACC and found an increased area under the curve for transgenic (33.2 ± 4.3 mV·ms) relative to wild-type mice (19.1 ± 3.3 mV·ms, $P < 0.05$). Similar results were found in the insular cortex (transgenic, 33.1 ± 6.1 mV·ms versus wild type, 17.0 ± 2.7 mV·ms; $P < 0.05$) (adapted from Wei *et al.* [43]).

NR2B overexpression affected the behavioral responses of mice to tissue injury and inflammation using two models: peripheral injection of formalin and peripheral injection of complete Freund's adjuvant (CFA) [40]. Within the first hour after hind paw formalin injection (5%, 10 μl), nociceptive behavioral responses were similar between wild-type and transgenic mice (Figure 3). However, transgenic mice exhibited more pronounced phase 3 behavioral responses (during the second hour after

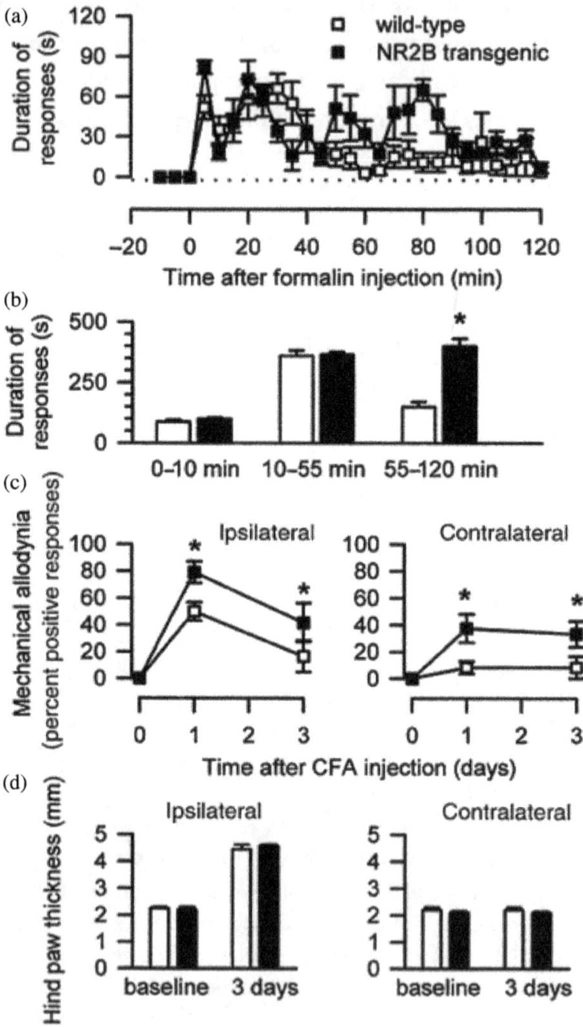

injection), as compared to wild-type mice (Figure 3). No obvious difference in the degree of hind paw oedema was found between wild-type and transgenic mice, consistent with the fact that both Phase 1 and Phase 2 responses were not affected in NR2B transgenic mice. Moreover, NR2B overexpression also enhanced allodynic responses to non-noxious stimuli in an inflammatory pain model. After hind paw injection of CFA, behavioral nociceptive responses to a previously non-noxious mechanical stimulus were significantly enhanced in NR2B transgenic mice (Figure 3).

Physiological pain is not affected

Experiments were performed to examine whether NR2B overexpression affected acute nociceptive processing. No significant difference in TF response latency was observed, indicating that spinal nociceptive transmission was not significantly altered in transgenic mice. Likewise, no differences were found in response latencies to an HP at three different plate temperatures (50, 52.5, and 55°C) or to a cold plate (0°C). Thus, acute behavioral responses to noxious thermal stimuli were indistinguishable between NR2B transgenic and wild-type mice.

Figure 3. (*Continued*)   Enhanced behavioral responses to formalin or CFA injection in NR2B transgenic mice. (a) The number of seconds during which wild-type ($n = 16$) and NR2B transgenic mice ($n = 9$) were engaged in nociceptive behavioral responses to hind paw formalin injection was plotted in 2-min intervals. Among nine NR2B transgenic mice, six of them were from transgenic line 1, and three were from line 2. Because no significant difference was found between them, the data were pooled. (b) Results from (a) were grouped into three phases (see Results). *Significant difference from wild-type mice. (c) The responses of animals to a mechanical stimulus (a 0.4-mN von Frey fiber applied to the dorsum of a hind paw) that elicited no responses before dorsal hind paw CFA injection were recorded 1 and 3 days after injection. The data were plotted as percent positive responses to stimulation of the ipsilateral or contralateral hind paw (relative to side of injection) for wild-type ($n = 4$) and NR2B transgenic ($n = 4$) mice. *Significant difference between wild-type and transgenic mice in the indicated conditions. (d) Hind paw edema was measured with a fine caliper in wild-type and NR2B transgenic mice (white bars, wild type, $n = 4$; black bars, NR2B transgenic, $n = 4$). No significant difference was found between wild-type and transgenic mice (adapted from Wei *et al.* [43]).

Injury-induced enhancement of NMDA receptor functions

Does the genetic overexpression of NR2B mimic physiological or pathological conditions? In a set of experimental models of chronic pain, forebrain NMDA NR2B receptors were found to be increased, and such enhancement of NMDA receptor functions is known to contribute to chronic pain [20]. After persistent inflammation (the CFA animal model for chronic inflammation), the expression of NMDA NR2B receptors in the ACC was increased over a long period of time, thereby increasing the NR2B component in NMDA receptor-mediated EPSCs (Figure 4). In the behavioral allodynia test, microinjection into the ACC or systemic administration of NMDA NR2B receptor-selective antagonists inhibited behavioral responses to peripheral inflammation (Figure 7). These results are consistent with genetic studies that mice with NMDA receptor NR2B subunit forebrain overexpression selectively enhanced inflammation-related

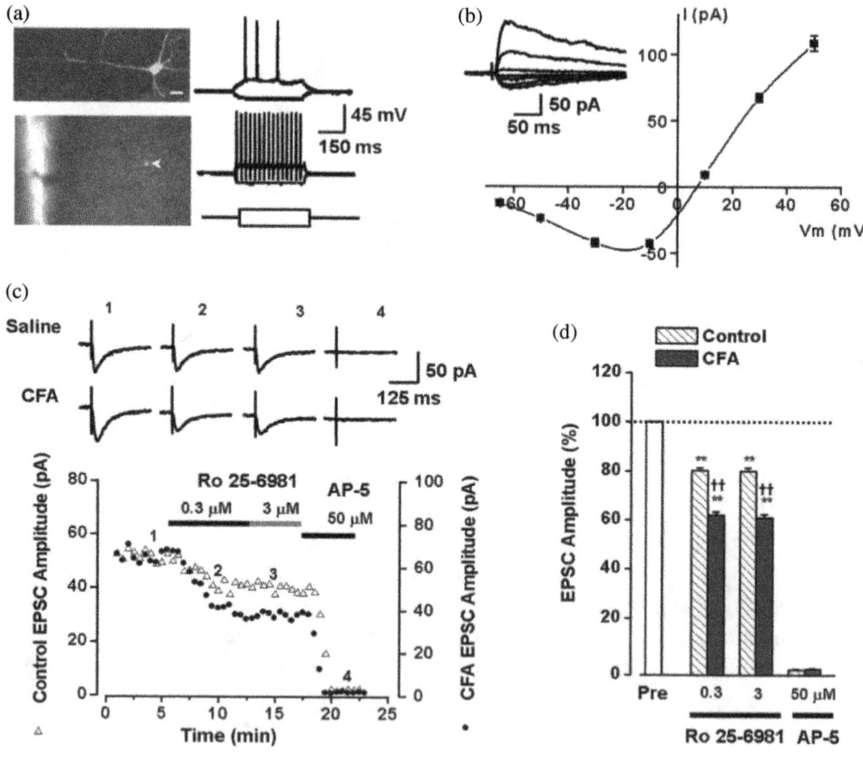

persistent pain without significant changes in acute pain. The anti-allodynic effects of NMDA NR2B receptor antagonists have been also reported in other animal models of chronic pain. These findings provide direct evidence that NMDA NR2B receptors undergo long-term plastic changes in the brain after injury.

Biochemical evidence for the upregulation of NR2B expression in the ACC after injury

To examine long-term changes in NR2B expression in the ACC, we performed *in situ* hybridization in saline- and CFA-injected mice [24]. We found that mRNA for the NMDA NR2B receptor was increased in the ACC at one and three days after peripheral CFA injection. The increase in

Figure 4. (*Continued*)   Increased NMDA NR2B receptor-mediated postsynaptic currents in adult ACC neurons are shown after inflammation. (a) Confocal image of a pyramidal neuron in an adult ACC slice loaded with Lucifer yellow (top left). The arrow indicates the location of the neuron in the ACC slice (bottom left). Scale bar: 20 μm. When injected with current steps from -100 to 100 pA in 400 ms (bottom right), the neuron shown at the left fired repetitive action potentials with frequency adaptation (top right). A minority of neurons recorded ($n = 4$) showed fast-spike properties (middle right). (b) Current–voltage plot of the NMDA receptor EPSCs. The currents exhibited strong outward rectification and the reversal potential of the linear part was 4.4 ± 2.0 mV ($n = 6$ neurons/5 mice). Inset shows NMDA receptor-mediated EPSCs recorded at holding potentials from -65 to 50 mV. Each trace represents the average of five consecutive recordings. (c) A selective NR2B antagonist, Ro 25-6981, partially inhibited NMDA receptor-mediated EPSCs. The time course of changes in EPSC amplitude before and during the application of Ro 25-6981 (0.3 and 3 μM) and AP-5 (50 μM) in ACC neurons from both saline (control) and CFA-injected mice is shown. Traces show the currents at different time points during application of drugs. Ro 25-6981 produced its maximal effect at 3 min after bath application, and a higher dose of Ro 25-6981 (3 μM) had no additional effects. The remaining currents can be totally blocked by AP-5 (50 μM). (d) Summary data of the effects of Ro 25-6981 (0.3 and 3 μM) in ACC neurons of control (saline-injected) and CFA-injected mice. Ro 25-6981 produced significantly greater inhibitory effects on NMDA receptor-mediated EPSCs in ACC neurons in CFA-injected mice ($n = 9$ neurons/8 mice) than those in control adult mice ($n = 8$ neurons/7 mice). AP-5 (50 μM) completely blocked the currents, confirming the currents are mediated NMDA receptors ($n = 5$ slices/5 mice). The double asterisks indicate significant difference from control, and the double daggers indicate significant difference from control NMDA receptor-mediated EPSCs. Error bars represent SEM (adapted from Wu *et al.* [20]).

NR2B was bilateral since the ACC ipsilateral and contralateral to the injected hind paw showed similar changes. In contrast, no obvious change in NR2A was found in the ACC, indicating that the upregulation is selective for NR2B subtype receptors. This result is in contrast to experience-dependent plasticity of NMDA receptors in the visual cortex. An increase in NR2A subtype receptors was reported during experience-related modification, providing strong evidence for different contributions of NR2B and NR2A in physiological vs. pathological processes. We also performed experiments for the NR1 subtype receptor, an important subtype for NMDA receptor function. Our results showed that the NR1 subtype receptor was unaffected by CFA inflammation.

To determine if NR2B was upregulated in lower pain-related central regions, we performed similar experiments in the midbrain PAG and the spinal cord dorsal horn of the same mice. Unlike in the ACC, we found no significant change in the level of NR2B mRNA in the PAG or spinal cord dorsal horn. These findings indicate that changes in NR2B are relatively regionally selective and that neurons involved in pain transmission (i.e., spinal dorsal horn neurons), and endogenous analgesia modulation (PAG), did not undergo similar changes as ACC neurons.

To better quantify the changes in NMDA receptor subunits, we performed biochemical analyses using the western blot method. The upregulation of NR2B proteins in the ACC after the CFA injection was also confirmed by western immunoblotting [24]. The level of NR2B expression in the ACC was significantly increased at 1, 3, and 7 days after the CFA injection. The increase in NR2B expression was time-dependent, as there was no significant change in NR2B protein level at 1 or 6 hrs after CFA injection as compared with saline-injected mice. We also performed a western blot analysis for the other NMDA receptor subunits, NR1 and NR2A, and could not see any significant increase at 3 and 7 days after CFA injection. These findings indicate that changes in NMDA receptors are selective for NR2B subunits in the ACC. Similar to the changes in NR2B mRNA, changes in NR2B protein expression were bilateral. Quantitative analysis of NR2B expression in the ACC showed that NR2B in the ACC contralateral and ipsilateral to the injected hind paw had similar increases. Furthermore, baseline expression levels of NR2B in the left ACC and right ACC were identical.

Injury triggered upregulation of NMDA NR2B functions in the IC

Similar to these findings in the ACC, the upregulation of NMDA receptors including NR2B after nerve injury has been found in the IC [41]. Similarly, peripheral injury significantly enhanced NMDA receptor-mediated currents in IC neurons (Figure 5). These enhanced currents are likely due to upregulation of synaptic NMDA receptors (Figure 6).

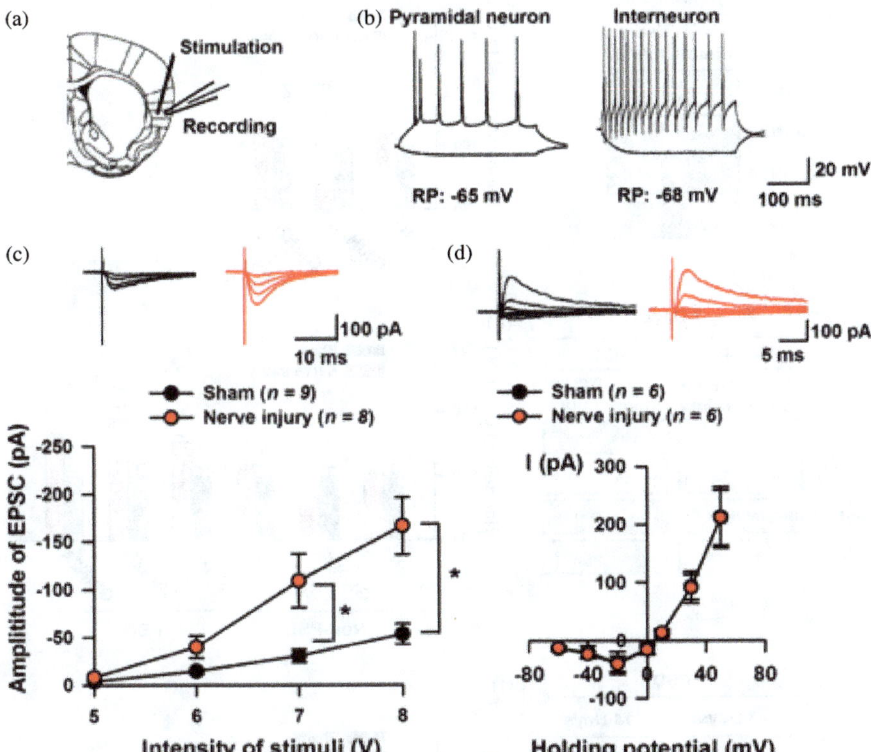

Figure 5. Enhanced NMDAR current in layer II/III in the IC after peripheral nerve injury. (a) Diagram indicating the placement of the stimulating and recording electrodes in the IC. (b) When injected with current steps from −100 to 100 pA in 400 ms, the pyramidal neuron fired repetitive action potentials with frequency adaptation (left). The interneurons showed fast-spike properties (right). (c) Synaptic input–output curves in slices from sham control mice ($n = 9$ neurons) and nerve injury mice ($n = 8$ neurons). (d) The I–V curve in slices from sham control mice ($n = 6$ neurons) and nerve injury mice ($n = 6$ neurons). *$P < 0.05$ compared with sham control (adapted from Qiu *et al.* [41]).

Activation of cAMP-dependent signaling enhanced the amount of synaptic NMDA receptors in acutely isolated insular cortical slices and increased the surface localization of NMDA receptors in cultured cortical neurons. Furthermore, the increase in the amount of NMDARs required phosphorylation of the NMDA receptor subunit GluN2B at Tyr (1472) by a pathway involving AC1, PKA, and Src family kinases. Behavioral and pharmacological experiments found that injecting

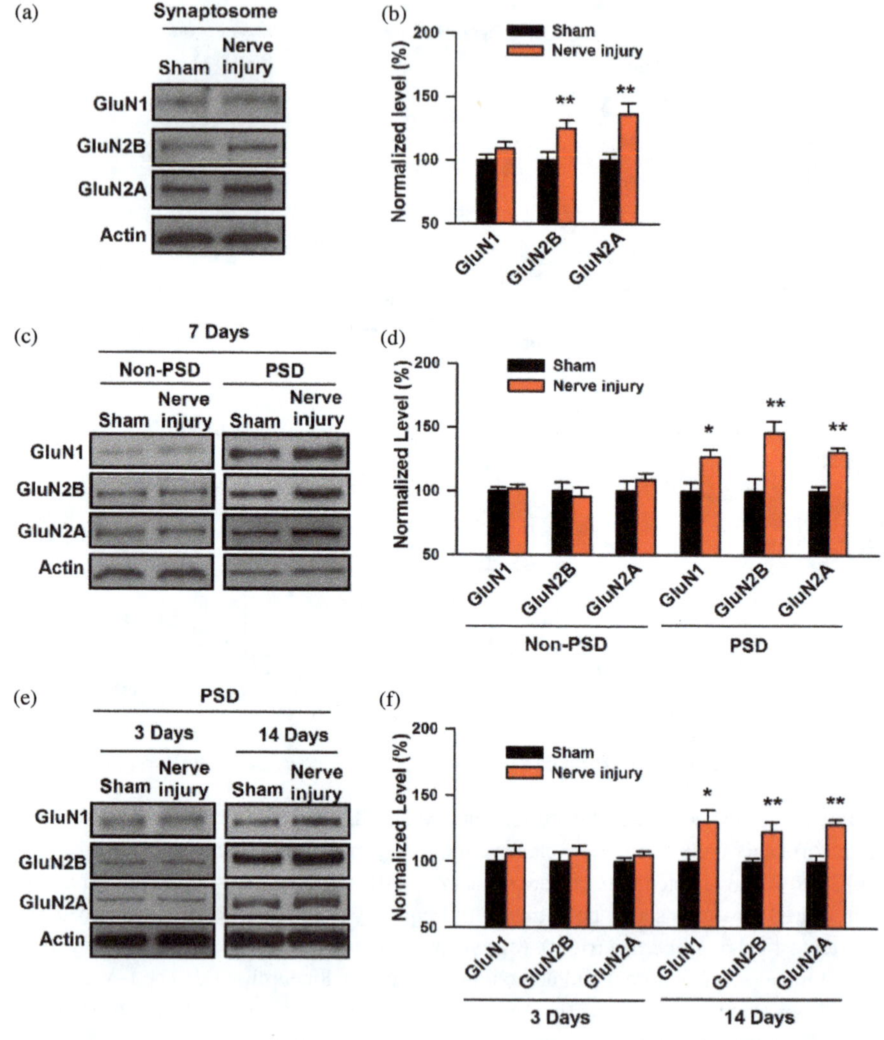

NMDAR or GluN2B-specific antagonists into the IC reduced behavioral responses to normally non-noxious stimuli in the mouse model of neuropathic pain, suggesting that activity-dependent plasticity takes place in the insular cortex after nerve injury and that inhibiting the increase in NMDAR function may help prevent or treat neuropathic pain.

GluN2B and its downstream AC1 as potential drug targets for treating chronic pain

In the genetic overexpression of NMDA GluN2B subunit mice, we found that chronic pain — not acute or physiological pain — was selectively enhanced [39,40], providing the first genetic evidence that forebrain GluN2B receptor is critical for chronic pain. In the behavioral allodynia test, microinjection into the ACC, or systemic administration of GluN2B receptor-selective antagonist, inhibited behavioral responses to peripheral inflammation. The use of GluN2B antagonists to control chronic pain has been reported by many groups. The action off-site for GluN2B antagonists is mainly due to cortical GluN2B receptors, although it may also exert some of its effects at the spinal cord level. Understanding the basic mechanisms of GluN2B-involved long-term plasticity may help us to design better medicines for treating chronic pain.

Figure 6. (*Continued*)   Synaptic NMDARs are upregulated in the IC after peripheral nerve injury. (a) Western blots for the GluN1, GluN2A, and GluN2B subunit in the synaptosomal fraction of the IC at 7 days after nerve injury or sham control. (b) The levels of GluN1, GluN2A, and GluN2B in the synaptosome fraction of the IC on day 7 after nerve injury or sham control ($n = 5$ for each group). (c) Western blots for GluN1, GluN2A, and GluN2B in the non-PSD-enriched fraction and the PSD-enriched fraction of the IC obtained on day 7 after nerve injury or sham control. (d) The levels of GluN1, GluN2A, and GluN2B in the non-PSD fraction and PSD-enriched fraction of the IC on day 7 after nerve injury or sham control ($n = 7$ for each group). (e) Western blots for GluN1, GluN2A, and GluN2B in the PSD-enriched fraction of the IC obtained 3 days or 14 days after nerve injury or sham control. (f) The levels of GluN1, GluN2A, and GluN2B in the PSD-enriched fraction of the IC obtained 3 days or 14 days after nerve injury or sham control ($n = 4$–6 for each group). *$P < 0.05$; **$P < 0.01$ compared with sham control; error bars indicate SEMs (adapted from Qiu *et al.* [41]).

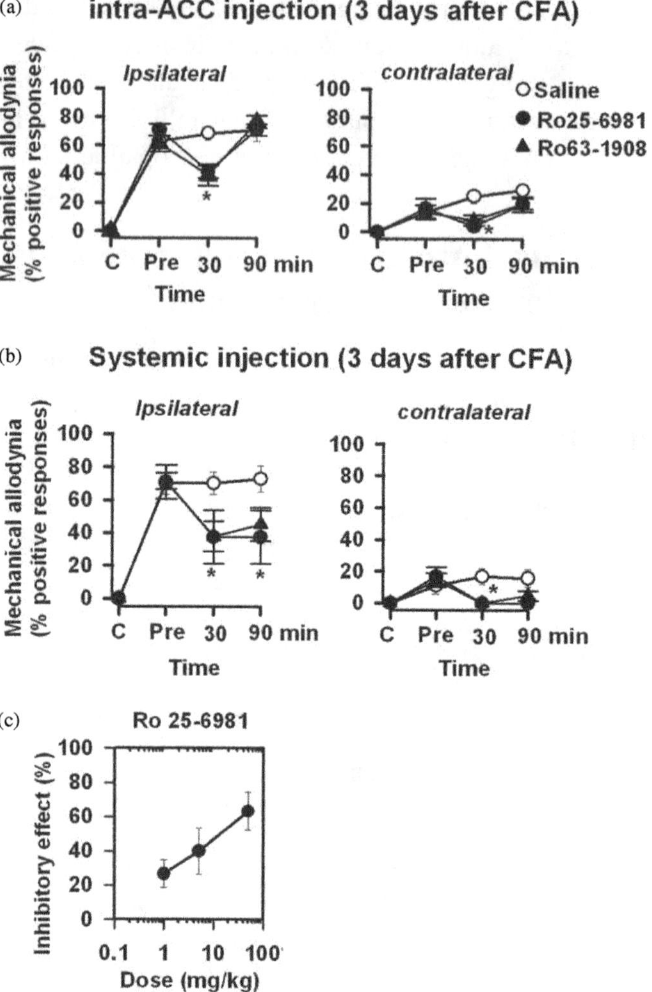

Figure 7. Effects of selective NMDA NR2B receptor antagonists on inflammation-induced behavioral allodynia, acute pain, and motor function. (a) The NR2B receptor antagonists Ro 25-6981 ($n = 6$ mice) or Ro 63-1908 ($n = 5$ mice) microinjected bilaterally into the ACC (1 μg/0.5 μl for each side) significantly reduced mechanical allodynia at 3 days after CFA injection. Responses were recorded before CFA injection, previous to drug or saline injection (Pre), and for 30–90 min afterward. Allodynia detected from the uninjected hind paw was also reduced. *$P < 0.05$. (b) Systemic administration of Ro 25-6981 (5 mg/kg, i.p.; $n = 6$ mice) or Ro 63-1908 (5 mg/kg, i.p.; $n = 5$ mice) significantly reduced mechanical allodynia bilaterally. (c) Dose-dependent response curve for 1–50 mg/kg Ro 25-6981 showing the percentage of the inhibitory effect on behavioral allodynia ($n = 5$–8 mice). Error bars represent SEM (adapted from Wu *et al.* [20]).

Previous investigations have revealed that AC1 is likely acting downstream of NMDA receptors, including GluN2B receptors [4,13,42]. In central neurons, activation of NMDA receptors can activate several key intracellular pathways, including CaMKII, tyrosine kinase, and PKC. For cognitive functions, genetic deletion of AC1 failed to impair key memory functions, indicating that such functions can be taken over by other signaling pathways [43]. Thus, targeting AC1 is likely to cause fewer cognitive side effects as compared with GluN2B receptors. Cumulative studies have shown that AC1 activity is required for both GluN2B-containing NMDAR-dependent LTP, as well as NMDAR-independent pre-LTP [31,43,44]. Wang *et al.* further demonstrated that LTP was also blocked by a selective AC1 inhibitor, NB001, which had an analgesic effect in animal models of neuropathic pain when administered intraperitoneally or orally, without apparent side effects [43]. Meanwhile, NB001 was reported to have analgesic effects on other pain models. Intraperitoneally application of NB001 markedly decreased the amount of spontaneous lifting and increased the mechanical paw withdrawal threshold in mice suffering from bone cancer pain [45]. NB001 also produced a significant analgesic effect in both acute persistent and chronic inflammatory, muscle pain [46]. Additionally, in a visceral pain model of irritable bowel syndrome, NB001 produced inhibition of injury-induced behavioral anxiety and spontaneous pain [47,48]. Therefore, AC1 has a better potential to become a drug target for the treatment of chronic pain compared with GluN2B.

Molecular mechanism for triggering synaptic changes

Recent studies in other central synapses suggest that NMDA NR2B receptor upregulation is likely to be reliant on activity-dependent mechanisms. The molecular motor protein KIF 17 has been shown to be involved in the active transport of NMDA receptor NR2B subunits. NR2B contains a cAMP response element-binding protein (CREB) binding domain which may couple increases in intracellular calcium with the increase in NR2B expression. Since NMDA receptors play an important role in activity-dependent plasticity in the ACC, we suggest that NMDA NR2B subunit may be regulated through NMDA-calcium-CaM-dependent signaling pathways. The activation of NMDA receptors triggers postsynaptic calcium, leading to the activation of calcium-stimulated CREB in the ACC after peripheral or central injury. Experiments using AC1 knockout mice also demonstrated that AC1 activity is critical for NR2B upregulation after the injury. Changes in NMDA NR2B antagonist-sensitive EPSCs caused by the nerve injury were abolished in mice lacking AC1. It is possible that NMDA NR2B receptor-AC1-cAMP-CREB-NR2B forms a positive cellular feedback to reinforce the NMDA receptor functions in the ACC neurons. Together with NMDA receptor-mediated AMPA receptor potentiation, these synaptic changes greatly enhance excitatory synaptic transmission in the ACC after the injury and thus contribute to cortical pain (Figure 8).

We found that AC1, a key enzyme for pain-related cortical plasticity, was significantly increased in the ACC in an animal model of irritable bowel syndrome. Inhibiting AC1 activity by selective AC1 inhibitor NB001 significantly reduced the upregulation of AC1 proteins in the ACC. Furthermore, it has been found that AC1 is required for NMDA GluN2B receptor upregulation and increases of NMDA receptor-mediated currents. These results suggest that AC1 may form a positive regulation in the cortex during chronic visceral pain.

Conclusion and future directions

In conclusion, NMDARs play an important role in the synaptic plasticity of the ACC. Among all of the subtypes of NMDAR, the GluN2 subfamily determines the crucial properties of the NMDAR. In pyramidal neurons in the ACC, post- and pre-LTP can be induced with or without NMDARs, respectively. Both the GluN2A and GluN2B are required for the induction of post-LTP, while the other two GluN2 subunits, GluN2C and GluN2D, may be involved in the presynaptic modulation of spontaneous glutamate release. The NMDAR-dependent LTD in the ACC also requires GluN2A and -2B. Since GluN2B and its downstream AC1 act as critical molecules in chronic pain and memory, and inhibiting AC1 has less cognitive side

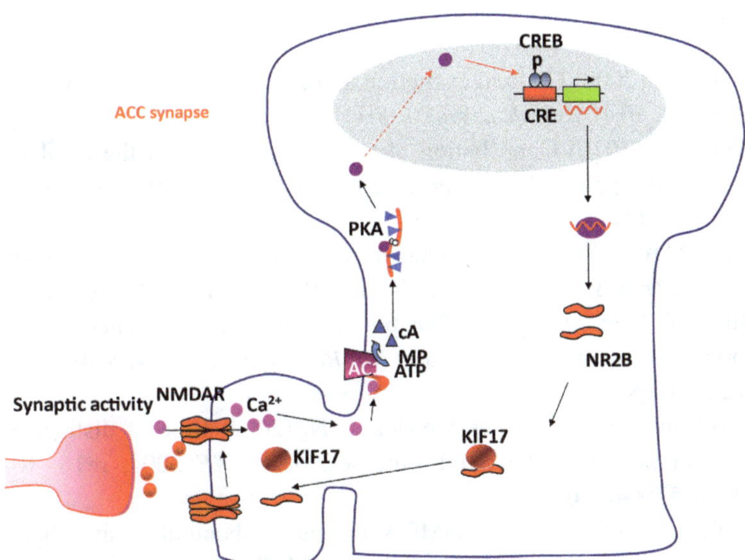

Figure 8. Central signaling pathways contribute to chronic pain. In the ACC, glutamate is the major fast excitatory transmitter between input fibers and pyramidal cells. Peripheral injury such as tissue inflammation or nerve injury triggers a burst of abnormal activity in the ACC circuits and subsequently activates postsynaptic NMDA receptors on cingulate pyramidal cells located in layers II-III. Activation of NMDA receptor triggers calcium influx. In adult ACC pyramidal cells, most of the NMDA receptors are the combination of NR1-NR2A and NR1-NR2B with possible minor components of currents made of NR1-NR2A-NR2B. Postsynaptic increases in Ca^{2+} lead to activation of Ca^{2+}-calmodulin (CaM)-dependent pathways. Among them, Ca^{2+} and CaM-stimulated AC1 is activated, and this activation leads to the generation of the key second messenger cAMP. Subsequently, cAMP activates PKA. PKA then translocates to the nucleus and phosphorylates CREB. NR2B contains a CREB binding domain which may couple increases in intracellular calcium with the increase in NR2B expression. Subsequently, postsynaptic synthesis of NMDA NR2B is increased, and together with endogenous motor protein KIF17, these new NR2B subunits are added to postsynaptic NMDA receptors. Such positive feedback control may further enhance neuronal excitability within the ACC and contribute to chronic pain.

effects, selective antagonists of AC1 may become a potent drug target for treating chronic pain. Future studies will focus on the molecular mechanisms of the different NMDAR subunits to understand how they work under physiological and pathological conditions.

References

[1] Zhuo, M. (2008) Cortical excitation and chronic pain. *Trends Neurosci* 31, 199–207. 10.1016/j.tins.2008.01.003.

[2] Zhuo, M. (2016) Contribution of synaptic plasticity in the insular cortex to chronic pain. *Neuroscience* 338, 220–229. 10.1016/j.neuroscience. 2016.08.014.

[3] Zhuo, M. (2016) Neural mechanisms underlying anxiety-chronic pain interactions. *Trends Neurosci* 39, 136–145. 10.1016/j.tins.2016.01.006.

[4] Bliss, T.V. *et al.* (2016) Synaptic plasticity in the anterior cingulate cortex in acute and chronic pain. *Nat Rev Neurosci* 17, 485–496. 10.1038/nrn.2016.68.

[5] Cull-Candy, S.G. and Leszkiewicz, D.N. (2004) Role of distinct NMDA receptor subtypes at central synapses. *Sci STKE* 2004, re16. 10.1126/stke.2552004re16.

[6] Paoletti, P. *et al.* (2013) NMDA receptor subunit diversity: Impact on receptor properties, synaptic plasticity and disease. *Nat Rev Neurosci* 14, 383–400. 10.1038/nrn3504.

[7] Berretta, N. and Jones, R.S. (1996) Tonic facilitation of glutamate release by presynaptic N-methyl-D-aspartate autoreceptors in the entorhinal cortex. *Neuroscience* 75, 339–344. 10.1016/0306-4522(96)00301-6.

[8] Liu, H. *et al.* (1994) Evidence for presynaptic N-methyl-D-aspartate autoreceptors in the spinal cord dorsal horn. *Proc Natl Acad Sci USA* 91, 8383–8387. 10.1073/pnas.91.18.8383.

[9] Siegel, S.J. *et al.* (1994) Regional, cellular, and ultrastructural distribution of N-methyl-D-aspartate receptor subunit 1 in monkey hippocampus. *Proc Natl Acad Sci USA* 91, 564–568. 10.1073/pnas.91.2.564.

[10] Rumbaugh, G. and Vicini, S. (1999) Distinct synaptic and extrasynaptic NMDA receptors in developing cerebellar granule neurons. *J Neurosci* 19, 10603–10610.

[11] Momiyama, A. *et al.* (1996) Identification of a native low-conductance NMDA channel with reduced sensitivity to Mg2+ in rat central neurones. *J Physiol* 494 (Pt 2), 479–492. 10.1113/jphysiol.1996.sp021507.

[12] Li, X.H. *et al.* (2019) NMDA receptor dependent long-term potentiation in chronic pain. *Neurochem Res* 44, 531–538. 10.1007/s11064-018-2614-8.

[13] Zhuo, M. (2009) Plasticity of NMDA receptor NR2B subunit in memory and chronic pain. *Mol Brain* 2, 4. 10.1186/1756-6606-2-4.

[14] Liauw, J. *et al.* (2003) NMDA receptors contribute to synaptic transmission in anterior cingulate cortex of adult mice. *Sheng Li Xue Bao* 55, 373–380.

[15] Li, X.H. *et al.* (2017) Characterization of postsynaptic calcium signals in the pyramidal neurons of anterior cingulate cortex. *Mol Pain* 13, 1744806917719847. 10.1177/1744806917719847.

[16] Zhao, M.G. *et al.* (2005) Roles of NMDA NR2B subtype receptor in prefrontal long-term potentiation and contextual fear memory. *Neuron* 47, 859–872. 10.1016/j.neuron.2005.08.014.

[17] Toyoda, H. *et al.* (2005) Roles of NMDA receptor NR2A and NR2B subtypes for long-term depression in the anterior cingulate cortex. *Eur J Neurosci* 22, 485–494. 10.1111/j.1460-9568.2005.04236.x.

[18] Sheng, M. *et al.* (1994) Changing subunit composition of heteromeric NMDA receptors during development of rat cortex. *Nature* 368, 144–147. 10.1038/368144a0.

[19] Monyer, H. *et al.* (1994) Developmental and regional expression in the rat brain and functional properties of four NMDA receptors. *Neuron* 12, 529–540. 10.1016/0896-6273(94)90210-0.

[20] Wu, L.J. *et al.* (2005) Upregulation of forebrain NMDA NR2B receptors contributes to behavioral sensitization after inflammation. *J Neurosci* 25, 11107–11116. 10.1523/JNEUROSCI.1678-05.2005.

[21] Wu, L.J. and Zhuo, M. (2009) Targeting the NMDA receptor subunit NR2B for the treatment of neuropathic pain. *Neurotherapeutics* 6, 693–702. 10.1016/j.nurt.2009.07.008.

[22] Mullasseril, P. *et al.* (2010) A subunit-selective potentiator of NR2C- and NR2D-containing NMDA receptors. *Nat Commun* 1, 90. 10.1038/ncomms1085.

[23] Ogden, K.K. *et al.* (2014) Potentiation of GluN2C/D NMDA receptor subtypes in the amygdala facilitates the retention of fear and extinction learning in mice. *Neuropsychopharmacology* 39, 625-637. 10.1038/npp.2013.241.

[24] Ravikrishnan, A. *et al.* (2018) Region-specific expression of NMDA receptor GluN2C subunit in parvalbumin-positive neurons and astrocytes: Analysis of GluN2C expression using a novel reporter model. *Neuroscience* 380, 49–62. 10.1016/j.neuroscience.2018.03.011.

[25] Zhang, Y. *et al.* (2012) NR2C in the thalamic reticular nucleus; effects of the NR2C knockout. *PLoS One* 7, e41908. 10.1371/journal.pone.0041908.

[26] Sun, L. *et al.* (2000) Expression of NR1, NR2A-D, and NR3 subunits of the NMDA receptor in the cerebral cortex and olfactory bulb of adult rat. *Synapse* 35, 212–221. 10.1002/(SICI)1098-2396(20000301)35:3<212:: AID-SYN6>3.0.CO;2-O.

[27] Scherzer, C.R. *et al.* (1998) Expression of N-methyl-D-aspartate receptor subunit mRNAs in the human brain: Hippocampus and cortex.

J Comp Neurol 390, 75–90. 10.1002/(sici)1096-9861(19980105)390:1<75::aid-cne7>3.0.co;2-n.

[28] Akbarian, S. *et al.* (1996) Selective alterations in gene expression for NMDA receptor subunits in prefrontal cortex of schizophrenics. *J Neurosci* 16, 19–30.

[29] Chen, Q.Y. *et al.* (2021) NMDA GluN2C/2D receptors contribute to synaptic regulation and plasticity in the anterior cingulate cortex of adult mice. *Mol Brain* 14, 60. 10.1186/s13041-021-00744-3.

[30] Zhuo, M. (2014) Long-term potentiation in the anterior cingulate cortex and chronic pain. *Philos Trans R Soc Lond B Biol Sci* 369, 20130146. 10.1098/rstb.2013.0146.

[31] Koga, K. *et al.* (2015) Coexistence of two forms of LTP in ACC provides a synaptic mechanism for the interactions between anxiety and chronic pain. *Neuron* 85, 377–389. 10.1016/j.neuron.2014.12.021.

[32] Koga, K. *et al.* (2015) Impaired presynaptic long-term potentiation in the anterior cingulate cortex of Fmr1 knock-out mice. *J Neurosci* 35, 2033–2043. 10.1523/JNEUROSCI.2644-14.2015.

[33] Banerjee, A. *et al.* (2016) Roles of presynaptic NMDA receptors in neurotransmission and plasticity. *Trends Neurosci* 39, 26–39. 10.1016/j.tins.2015.11.001.

[34] Dore, K. *et al.* (2017) Unconventional NMDA receptor signaling. *J Neurosci* 37, 10800–10807. 10.1523/JNEUROSCI.1825-17.2017.

[35] Larsen, R.S. *et al.* (2014) Synapse-specific control of experience-dependent plasticity by presynaptic NMDA receptors. *Neuron* 83, 879–893. 10.1016/j.neuron.2014.07.039.

[36] Lonart, G. *et al.* (2003) Phosphorylation of RIM1alpha by PKA triggers presynaptic long-term potentiation at cerebellar parallel fiber synapses. *Cell* 115, 49–60. 10.1016/s0092-8674(03)00727-x.

[37] Kavalali, E.T. (2015) The mechanisms and functions of spontaneous neurotransmitter release. *Nat Rev Neurosci* 16, 5–16. 10.1038/nrn3875.

[38] Nistico, R. *et al.* (2015) Presynaptic c-Jun N-terminal kinase 2 regulates NMDA receptor-dependent glutamate release. *Sci Rep* 5, 9035. 10.1038/srep09035.

[39] Tang, Y.P. *et al.* (1999) Genetic enhancement of learning and memory in mice. *Nature* 401, 63–69. 10.1038/43432.

[40] Wei, F. *et al.* (2001) Genetic enhancement of inflammatory pain by forebrain NR2B overexpression. *Nat Neurosci* 4, 164–169. 10.1038/83993.

[41] Qiu, S. *et al.* (2013) An increase in synaptic NMDA receptors in the insular cortex contributes to neuropathic pain. *Sci Signal* 6, ra34. 10.1126/scisignal.2003778.

[42] Li, X.H. *et al.* (2020) Neuronal adenylyl cyclase targeting central plasticity for the treatment of chronic pain. *Neurotherapeutics* 17, 861–873. 10.1007/s13311-020-00927-1.

[43] Wang, H. *et al.* (2011) Identification of an adenylyl cyclase inhibitor for treating neuropathic and inflammatory pain. *Sci Transl Med* 3, 65ra63. 10.1126/scitranslmed.3001269.

[44] Liauw, J. *et al.* (2005) Calcium-stimulated adenylyl cyclases required for long-term potentiation in the anterior cingulate cortex. *J Neurophysiol* 94, 878–882. 10.1152/jn.01205.2004.

[45] Kang, W.B. *et al.* (2016) Analgesic effects of adenylyl cyclase inhibitor NB001 on bone cancer pain in a mouse model. *Mol Pain* 12. 10.1177/1744806916652409.

[46] Vadakkan, K.I. *et al.* (2006) Genetic reduction of chronic muscle pain in mice lacking calcium/calmodulin-stimulated adenylyl cyclases. *Mol Pain* 2, 7. 10.1186/1744-8069-2-7.

[47] Zhang, M.M. *et al.* (2014) Effects of NB001 and gabapentin on irritable bowel syndrome-induced behavioral anxiety and spontaneous pain. *Mol Brain* 7, 47. 10.1186/1756-6606-7-47.

[48] Liu, S.B. *et al.* (2020) Cyclic AMP-dependent positive feedback signaling pathways in the cortex contributes to visceral pain. *J Neurochem* 153, 252–263. 10.1111/jnc.14903.

Chapter 9

Pain and Fear

A team of researchers, led by the University of Toronto, has discovered how, and where, a painful event becomes permanently etched in the brain. This discovery has significant and far-reaching implications for pain-related emotional disorders, such as anxiety and post-traumatic stress. University of Toronto's professor of physiology Dr. Min Zhuo and his colleagues Dr. Bong-Kiun Kaang (Seoul National University) and Dr. Bao-Ming Li (Fudan University) identified where emotional fear memory and pain were located in the brain by studying the biochemical processes in an entirely different part of the brain. In a paper published in *Neuron* [7], the researchers used mice to show how receptors, activated in the pre-frontal cortex (the area of the brain believed to be involved with higher intellectual functions), played a critical role in the development of fear. Up until this point, researchers had pointed to activation in the hippocampus as the origin of fear memory.

"This is critical, as it changes how and where scientists thought fear was developed," says Zhuo, the EJLB-CIHR Michael Smith Chair in Neurosciences and Mental Health. "By understanding the biomolecular mechanisms behind fear, we could potentially create therapeutic ways to ease emotional pain in people. Imagine reducing the ability of distressing events, such as amputations, to be permanently imprinted in the brain." Zhuo says that fear memory does not occur immediately after a painful event; rather, it takes time for the memory to become part of our consciousness. The initial event activates NMDA receptors (molecules on cells that receive messages and then produce specific physiological effects in the cell), which are normally quiet but are triggered when the brain receives a shock. Over time, these receptors leave their imprint on brain cells.

By delivering shocks to mice, the researchers activated the NMDA receptors and traced a subunit of the molecule — a protein called GluN2B — long believed to be associated with fear memory in the hippocampus and the amygdala, an almond-shaped structure in front of the hippocampus. To further test the protein's influence, researchers reduced the amount in mice and found they were less hesitant to avoid shocks. "We tested the animals using both spatial and auditory cues," Zhuo says. "In one experiment, the mice received small shocks when entering a chamber and they developed fear memory. In another experiment, we used sound tones to be associated with shocks. When GluN2B was blocked, they no longer avoided the chamber or reacted to the tone." Zhuo and his team then studied the mouse brain slices and discovered traces of GluN2B in the pre-frontal cortex, supporting their theory that fear memory develops in that region. "By identifying GluN2B in the prefrontal cortex of the brain, we propose that fear memory originates from a network of receptors, rather than one simple area," Zhuo says. "It is more complex than previously thought."

The next step, according to Zhuo, is to determine how GluN2B directly affects memory formation and storage in the brain. "While we know it exists in the hippocampus, amygdala, and the pre-frontal cortex — we don't know exactly how it alters them," Zhuo says. "Once we understand the implications for each part, we will be able to reduce levels of GluN2B accordingly and effectively reduce fear memory. In the future, people may be able to take therapeutic measures before experiencing a particularly uncomfortable situation." The University of Toronto Innovations Foundation is currently working with Zhuo to push for the translation of this finding into treatments.

Summary

Pain affects our emotions and our decision-making. Noxious stimuli that give us painful sensations trigger fear by activating neurons in the amygdala. Many neurons located in the central amygdala areas, thought to be important for anxiety, are also activated. At the cortical level, prefrontal cortical neurons are activated, and excitatory glutamatergic synapses undergo long-term plastic changes that may contribute to long-term fear and anxiety and affect other emotional responses. Molecular studies reveal that different signaling proteins play distinct roles in these processes, although they are all distributed in similar regions/neurons. It is clear that the specific central mechanisms for pain, fear, or anxiety may be explained by the functions of these selective signaling proteins within the circuits and that such findings may help us to identify new drugs for the treatment of these emotional disorders.

Keywords: Pain; Fear memory; ACC; amygdala; hippocampus; trace fear; AC1; CaMKIV; CREB; LTP

Introduction

It is well known that pain causes fear and anxiety in patients. Several successful pain medicines were actually originally used for the treatment of mood disorders. Recent studies using modern neurobiological approaches and human brain imaging techniques have revealed that pain, fear, and anxiety share many common brain structures, such as the anterior cingulate cortex, prefrontal cortex, amygdala, and hippocampus. At the molecular level, many signaling molecules that contribute to the expression of synaptic changes are also indicated in pain, fear, and anxiety. This chapter will review recent progress made in our understanding of the molecular aspects of fear and anxiety and explore the possible relationship between acute and chronic pain.

Pain triggers classic fear memory

Fear is a reaction to a specific, and immediately present, stimulus. Fear memory requires the involvement of higher brain structures, including the hippocampus, amygdala, and related cortical structures. In animal models of fear memory, noxious stimuli are most commonly used for triggering long-term fear memory in experimental animals, such as rats and mice. It has been found that the duration and magnitude of fear are intensity-related. Greater pain stimulation induced greater, and longer, fearful responses.

Brain circuits that are critical for fear

The amygdala has consistently been shown to be a key structure underlying fear memory formation in both animal and human studies. The amygdala is a well-organized brain structure with intensive internucleus connections. Pathways projecting from the thalamus and cerebral cortex terminate at the lateral amygdala (LA) and basolateral amygdala (BLA). Pyramidal cells, which are primarily glutamatergic neurons, relay information to the central amygdala (CeA). Efferents from the CeA then project to the periaqueductal gray, brainstem, and hypothalamus and induce fear and anxiety-related behavioral and autonomic responses.

The hippocampus is important for contextual fear memory. Electrolytic lesions of the dorsal hippocampus have been shown to block the acquisition and expression of contextual fear memory. Interestingly, auditory fear memory is largely unaffected by the dorsal hippocampal lesions. Thus, it is believed that contextual fear memory requires both the hippocampus and amygdala, and the projection from the hippocampus to the amygdala is thought to contribute to contextual conditioning [1]. Recent studies have indicated that cortical region ACC also plays important roles in both the induction and expression of fear memory (Figure 1).

Cortical contribution to fear

In addition to the amygdala and hippocampus, our recent studies have demonstrated that the ACC can contribute to the formation of fear

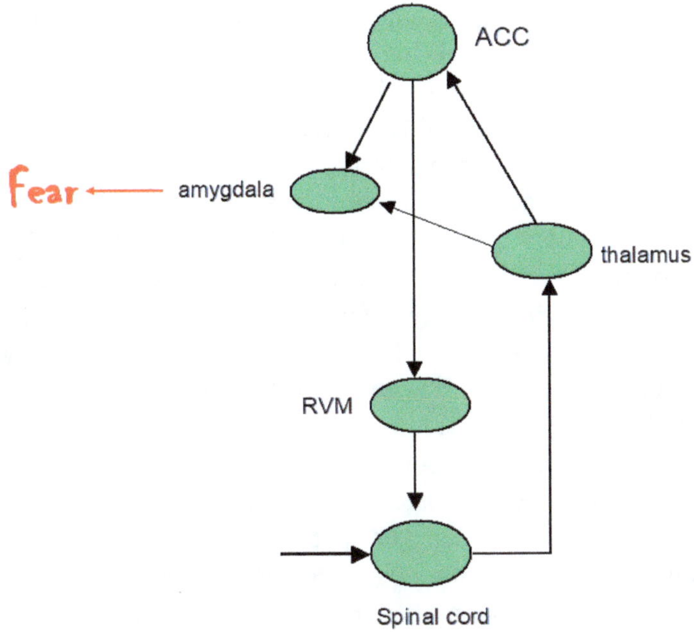

Figure 1. Amygdala and hippocampus in fear memory.

memory [2–4]. Brain imaging studies on humans have demonstrated that both the amygdala and the ACC are activated during fear learning. Activity within the amygdala may signal danger, whereas the ACC establishes necessary neural activity to sustain attention to the threat [5]. In fear conditioning paradigms that introduce a time interval between a visual cue and shock [6], ACC activity has been found to decrease between the cue and shock presentations. This may represent a "coincidence detection" mechanism, whereby ACC activity during the visual cue and shock may signal that the two are indeed related.

Animal studies have also identified the ACC as a critical area involved in the acquisition and storage of fear memory. For example, the expression of pCREB in the ACC was significantly increased by auditory fear conditioning [2]. Consistent with this finding, the c-fos mRNA expression in the ACC increases by 50% by trace fear conditioning [29]. Infusion of NMDA into the ACC reduced freezing in trace-fear-conditioned mice. Additionally, blocking of GluN2B activities in the

ACC by pharmacological, or small interfering RNA, could impair early memory of contextual fear [7]. Furthermore, experiments showed that electrical stimulation of the ACC induced an increase in ultrasonic activity (mostly between 30 and 50 kHz) during ACC stimulation at 0.3 mA (Figure 2). This frequency range of ultrasonic activity was also increased by a noxious foot shock. When paired with ACC electrical stimulation with a tone, presented in a conditioning chamber (Figure 3), long-term fear memory was detected in most of the mice, both to the tone presentation in a novel context (auditory memory) (Figure 3(b)) and to the conditioning environment (contextual memory) (Figure 3(b)). In the other animals, there is no specific freezing either to the conditioning context or to the ton [3]. Biochemical studies found that ACC showed a 20% rapid upregulation of AMPA receptor GluA1 membrane subunits that is evident immediately after conditioning [8]. Additionally, an electromyogram (EMG)-based study found that ACC putative pyramidal and non-pyramidal neurons were involved in the termination of fear behavior ("unfreezing"), and the spike activity of these neurons was reduced during freezing. Some of the neurons were also found to acquire unfreezing locked activity and change their tuning [9]. These results suggest that the ACC is involved in the acquisition of fear memory.

The contributions of the ACC to remote fear memory have been evaluated using contextual fear conditioning. Frankland *et al.* found that the recall of remote (36 days after conditioning), but not recent, contextual fear memory (1 day after conditioning) elevated the expression of Zif268 and c-fos in the ACC. Lidocaine infusion into the ACC also disrupted remote fear memory [4]. The formation of remote fear memory is accompanied by a neuronal structural change in the ACC [10]. Inhibition of protein synthesis in the ACC blocked the formation of long-term inhibitory avoidance memory [11]. These data suggest that the ACC also plays a critical role in the storage of remote fear memory.

Inhibition of ACC activity reduced classical fear memory

Both animal and human studies indicate that neurons in the ACC respond to peripheral painful stimuli, and injury triggers the activation of

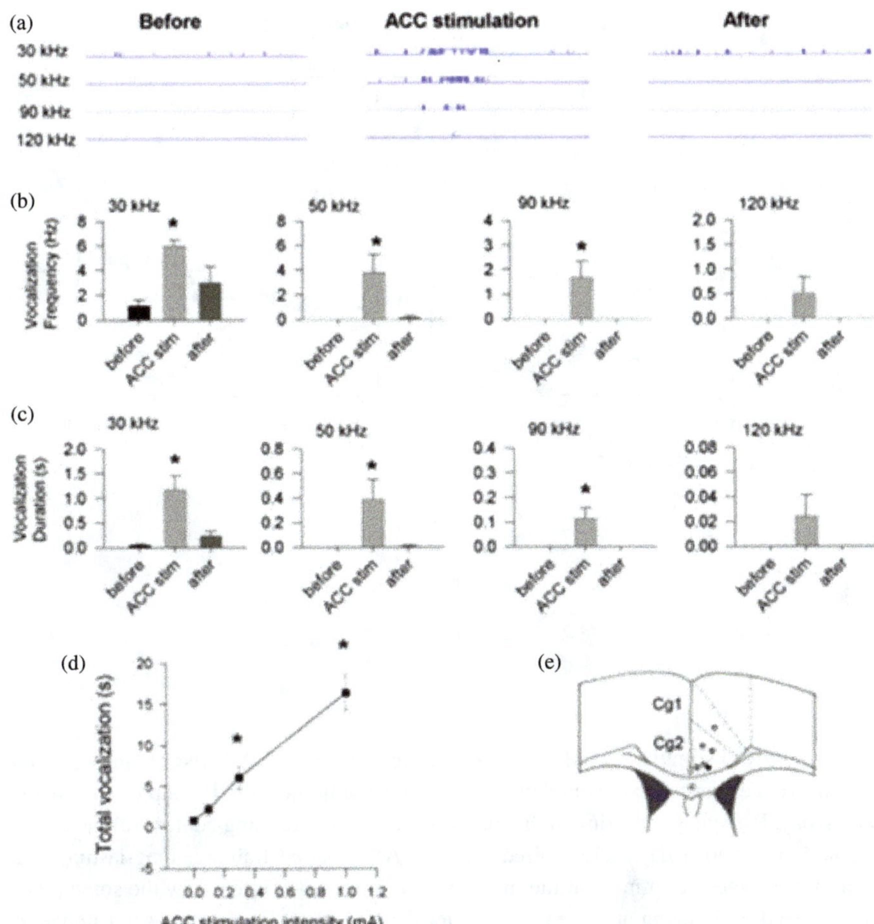

Figure 2. ACC stimulation induces ultrasonic vocalization in freely moving mice. (a) An example of ultrasonic responses from a single mouse at four different frequencies before, during, and after ACC stimulation (at 0.3 mA). 1 min duration; see filled circle for the stimulation site within the ACC in (e). (b) ACC stimulation (0.3 mA; $n = 6$) increased the frequency of individual ultrasonic responses. *P <0.05, comparing the frequency during ACC stimulation with baseline response before the stimulation. (c) ACC stimulation (0.3 mA; $n = 6$) also increased the duration of single ultrasonic response. *$P < 0.05$, comparing the duration during ACC stimulation with baseline duration before the stimulation. (d) Summarized data of ACC stimulation ($n = 6$) produced ultrasonic responses at different intensities. Total vocalization responses (sec) within 2 min ACC stimulation were plotted against the intensity of stimulation. (e) Stimulating sites in ACC on the schematic representation of coronal section 0.62 mm anterior to the Bregma. Filled circle, for data shown in (a); open circles, other sites for data shown in (b–d) (adapted from Tang *et al.* [3]).

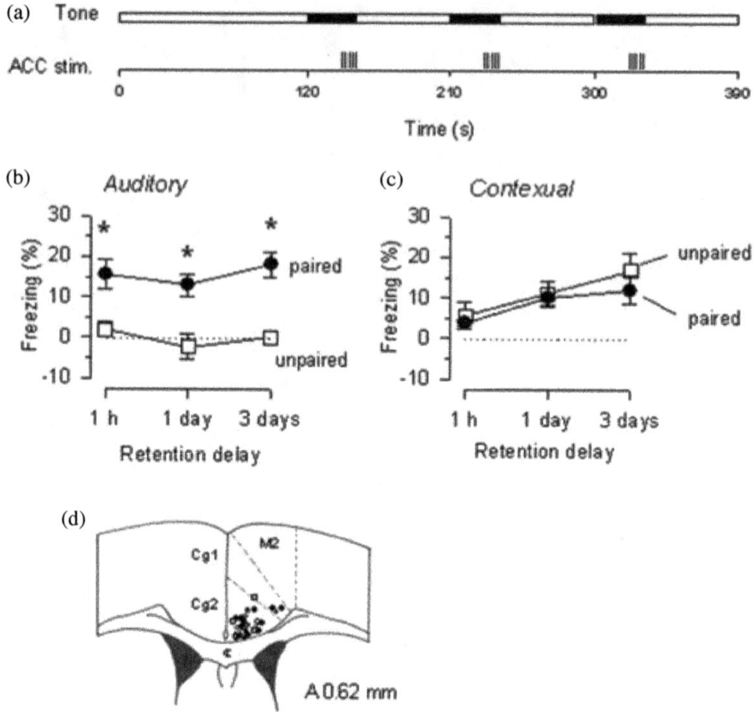

Figure 3.   ACC stimulation induces long-term fear memory. (a) Three pairings of 30 s tone and 10 s electrical train stimulation were delivered in the paired group on the conditioning day. Percentage freezing to the tone (b) and the conditioning context (c) measured at 1 h, 1 day, and 3 days after paired training. After paired training, long-lasting fear memory was detected in most of the mice ($n = 16$ mice, filled circles), while some other mice showed no freezing across the test periods ($n = 5$, data not shown). After unpaired training of tone and ACC stimulation, mice showed no freezing to the tone ($n = 6$, open squares) but clear memory to the training environment. *$P < 0.05$ compared with the unpaired group. (d) Stimulating sites in ACC on the schematic representation of coronal section 0.62 mm anterior to the Bregma. Filled circles, effect sites of ACC paired group; open circles, no effect sites of ACC paired group; open squares, ACC unpaired group (adapted from Tang *et al.* [3]).

immediate early genes in the ACC. Unlike the somatosensory cortex, neurons in the ACC have wide and diffuse receptive fields, and often cover the whole body of an animal, supporting its role in coding unpleasantness of pain. Supporting this hypothesis, lesions to the ACC or inhibition of excitatory synaptic transmission in the ACC produces antinociceptive effects or analgesia and blocks formalin-induced conditioned place avoidance. These results suggest that activation of the ACC, or increased excitatory activity within the ACC, is unpleasant or painful, acting directly or indirectly through other related brain regions. However, a recent imaging study reported that the activity in the ACC was increased during placebo analgesia. We think that the activation of endogenous analgesia system is unlikely to account for placebo analgesia. Previous studies have shown that stimulation of ACC did not trigger the endogenous analgesia system. Rather, we propose that enhanced inhibitory activity may contribute to the placebo effect. In a previous study, to test whether increasing inhibitory transmission in the ACC might relieve the unpleasantness resulting from foot shock, researchers microinjected a $GABA_A$ receptor agonist, muscimol, into the ACC prior to the fear conditioning. Muscimol significantly reduced the fear memory score 1 day and 3 days after classical fear conditioning (Figure 4) [3]. Similar research using ibotenic acid to cause bilateral lesions in the rostral ACC before training produced deficits in the acquisition of tone–shock associative learning but not contextual fear conditioning. Transient inactivation of ACC by muscimol microinjection also impaired the acquisition of cued fear learning. A recent study using an optogenetic strategy to inactivate the ACC after training showed that it had no effect on tone–shock-paired fear learning. However, when the ACC was active, both contextual fear

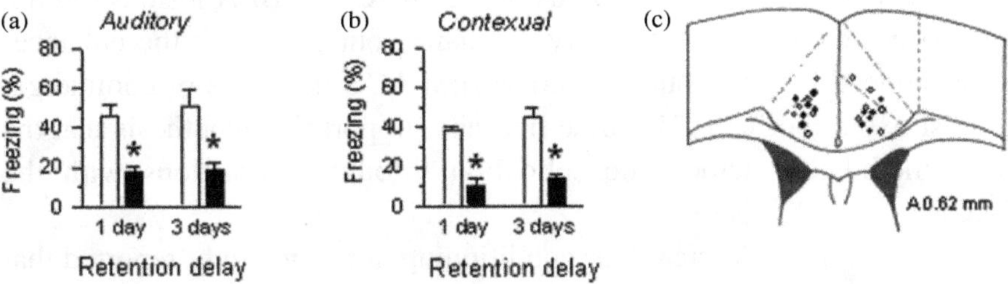

Figure 4.   ACC inactivation by a $GABA_A$ receptor agonist impairs fear memory by foot shock. (a, b) Mice receiving muscimol microinjection (1 $\mu g/\mu l$, 0.5 μl/each side, $n = 6$ mice, black bars) into bilateral ACC 15 min before conditioning showed reduced auditory (a) and contextual (b) fear memory induced by classic foot shock conditioning. *$P < 0.05$ compared with the saline-treated group ($n = 6$, open bars). (c) Microinjection sites in the ACC. Filled circles, muscimol group; open circles, saline group (adapted from Tang *et al.* [3]).

and cued fear learning were impaired [12]. The effect of ACC inhibition depends on the inhibition sites in the ACC as well as the acquisition time of fear memory.

Interconnection between amygdala and ACC

The amygdala and its related structures are well known for the induction of fear memory. Previous research has demonstrated that neurons in the ACC send projection fibers to the amygdala. While the prefrontal cingulate cortex has recently been shown to be involved in the extinction of fear memory, it is unclear whether the ACC inputs to the amygdala potentially contribute to the formation of fear memory. To test if fear-like long-term memory induced by ACC conditioning requires the involvement of amygdala, one study selectively blocked the NMDA receptors by bilaterally injecting the NMDA receptor antagonist AP-5 into the basolateral amygdala (BLA) during training (ACC stimulation paired with the environment or a tone). AP-5 significantly reduced ACC stimulation-induced fear memory to the tone 1 day after conditioning, indicating that the NMDA receptor function in the amygdala is required for the formation of auditory fear memory. Importantly, the contextual memory was not affected (Figure 5(a)), supporting the notion that other structures such as the hippocampus may contribute to contextual fear memory [3]. Another experiment used NMDA-induced lesions of the BLA. The BLA lesions did not impair the retention of inhibitory avoidance but did block the enhancement induced by post-training intra-rostral ACC infusions of cholinergic agonist oxotremorine [13]. These findings support the hypothesis that the BLA modulates memory consolidation through interactions with the ACC.

In addition to classical fear conditioning, a recent study reported that BLA-projecting ACC (ACC-BLA) neurons preferentially encode socially derived aversive cue information. Inhibition of ACC-BLA alters real-time amygdala representation of the aversive cue during observational conditioning. Selective inhibition of the ACC-BLA projection impaired acquisition, but not expression, of observational fear conditioning [14]. The monosynaptic glutamatergic connectivity of ACC-BLA was found in another study, but there was very little, to no, ACC input to the central

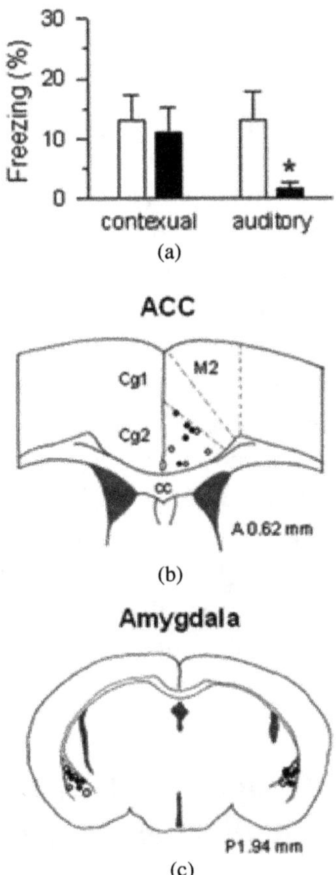

Figure 5.   NMDA receptors in the amygdala are required for auditory fear memory induced by ACC stimulation. (a) Bilateral microinfusions of AP-5 (2 μg/μl, 0.5 μl/side) in the BLAC 15 min before conditioning impaired auditory fear memory but no effect on contextual fear memory when tested 1 day later. Filled bars, AP-5 group ($n = 5$); open bars, saline group ($n = 4$). *$P < 0.05$ compared with saline-injected group. (b) Stimulation sites in ACC. Filled circles, AP-5 group; open circles, saline group. (c) Microinjection sites in the BLAC on the schematic representation of coronal section 1.94 mm posterior to the Bregma. Filled circles, AP-5 group; open circles, saline group (adapted from Tang *et al.* [3]).

amygdala. The ACC-BLA projection suggests its inhibitory control of innate freezing response to predator odors [12]. The role of ACC-amygdala interaction depends on the type of fear and the time when the projection is activated/inhibited.

Interconnection between hippocampus and ACC

Anatomic evidence has proved that there are reciprocal projections between the dorsal hippocampus (dHPC, posterior hippocampus in primates) and ACC, and the ACC-hippocampus projection is capable of eliciting contextual memory retrieval [15]. Precise context fear memory depends on the dHPC, regardless of the animal age [16]. Different from the dHPC, the ventral hippocampus (vHPC anterior in primates) relates to stress, emotion, and mood [17]. A report showed that inactivation of ACC and vHPC individually by local lidocaine infusion attenuates fear generalization, as inactivation of these regions returned the memory to a contextually precise form [18]. Another study found that ACC neurons projected to the vHPC unidirectionally. Using chemogenetic or optogenetic manipulations to inhibit the ACC-vHPC projecting neurons, reduced contextual fear generalization at the remote time point, whereas activating ACC-vHPC projecting neurons facilitated contextual fear generalization at a recent time point [19]. These results indicate that the interconnection between hippocampus and ACC modulates the contextual fear memory retrieval as well as generalization.

CaMKIV contributes to activation of learning-related immediate early genes

Activity-dependent immediate early genes are commonly used to map brain regions involved in fear conditioning. CREB is a fundamental immediate early gene that contributes to long-term memory. We measured the activation of CREB by fear conditioning in adult wild-type and CaMKIV$^{-/-}$ mice through immunocytochemistry. To demonstrate the associative specificity of pCREB expression in the various brain areas, we carried out a series of experiments in five different groups of wild-type and CaMKIV$^{-/-}$ mice: control (no stimulus), those conditioned with tone alone, shock alone, shock–tone (unpaired training), and tone–shock (paired training). At 45 min after paired tone–shock presentations, greater pCREB immunoreactivity was found in the hippocampal CA1 and the basolateral part of the amygdala of wild-type mice than in that of wild-type mice receiving unpaired training (Figures 6 and 7). Auditory

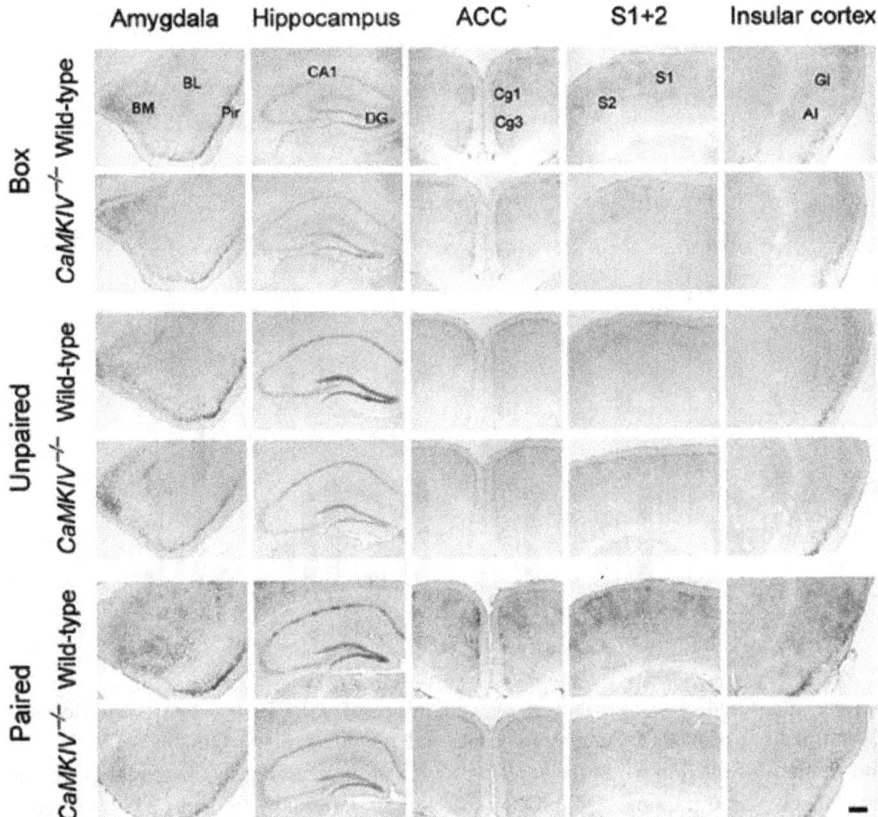

Figure 6. CaMKIV contributes to activation of CREB by fear conditioning. pCREB expression in the amygdala (BM, BL), hippocampus (CA1, DG), ACC (Cg1, Cg3), S1 and S2, and insular cortices (GI, AI) from wild-type and *CaMKIV*⁻/⁻ mice in a control situation (Box), after shock–tone unpaired training or after tone–shock paired training. Notation as in Fig. 2. BM, basomedial amygdala. Scale bar: 300 μm (adapted from Wei *et al.* [2]).

stimulation alone did not produce a significant change in pCREB levels in the hippocampus or amygdala, whereas shock alone caused modest increases in pCREB in the hippocampus and amygdala. In both wild-type and CaMKIV⁻/⁻ mice, similar basal levels of pCREB were found in different areas of the hippocampus, including CA1, CA3, dentate gyrus (DG), and amygdala (Figures 6 and 7). This suggests that the deletion of CaMKIV does not affect the basal expression of pCREB in neurons. In the

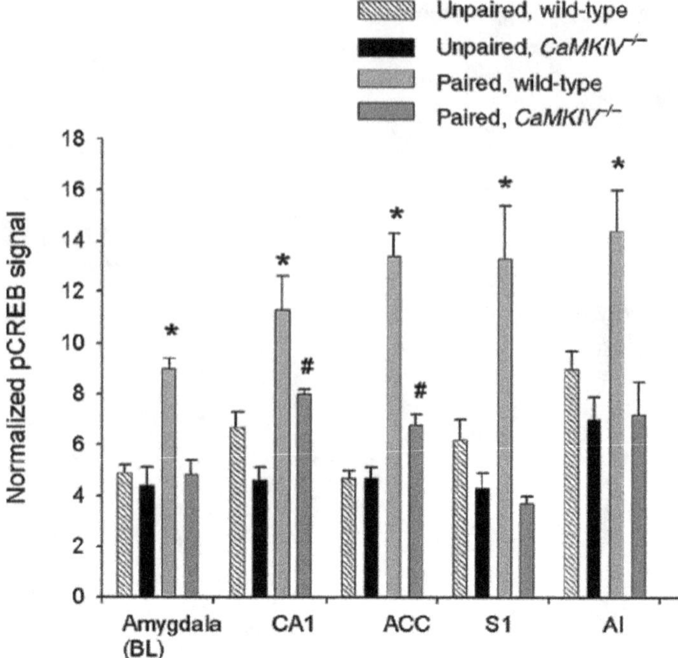

Figure 7. CaMKIV contributes to activation of CREB by fear conditioning. Quantifications of pCREB staining in basolateral amygdala, hippocampal CA1, ACC, S1, and AI are illustrated in wild-type and *CaMKIV*⁻/⁻ mice receiving shock–tone unpaired or tone–shock paired training. * significant difference from wild-type and *CaMKIV*⁻/⁻ controls; # significant difference from unpaired *CaMKIV*⁻/⁻ mice (adapted from Wei *et al.* [2]).

groups of mice receiving paired fear conditioning, pCREB levels in the CA1 region were significantly lower, but not absent, in CaMKIV⁻/⁻ mice as compared to wild-type mice (Figures 6 and 7). This finding indicates that CaMKIV contributes only partially to the activation of CREB in the CA1 areas. However, paired conditioning activated pCREB in the basolateral nucleus of the amygdala (as compared to wild-type mice receiving paired training) and was completely abolished in CaMKIV⁻/⁻ mice (Figure 7). This suggests that CaMKIV activity is essential to fear conditioning-induced CREB activation.

We also examined three cortical areas involved in the processing of sensory information and possibly fear-related memory: the ACC, primary somatosensory cortex, and the agranular insular cortex (IC). In wild-type mice, paired fear conditioning induced significantly larger increases in pCREB levels in all three cortical regions than those with unpaired training. Tone or shock alone increased pCREB expression in the somatosensory and IC of wild-type mice. However, CaMKIV seems to contribute differently to the activation of CREB by fear conditioning in the different cortical areas examined. In CaMKIV$^{-/-}$ mice, fear conditioning-induced pCREB was not seen in the somatosensory and IC but was seen (albeit at a lesser level than in wild-type mice) in the ACC (Figures 6 and 7). Basal expression of pCREB in these regions was similar in wild-type and CaMKIV$^{-/-}$ mice. We also examined the expression of pCREB in auditory pathways, including the cochlear nucleus, lateral superior olivary complex, inferior colliculus, and medial nucleus of the trapezoid body. Similar levels of pCREB after tone alone, unpaired shock–tone, or paired tone–shock were found in wild-type and CaMKIV$^{-/-}$ mice, indicating that the auditory sensory transmission is normal. This conclusion is supported by behavioral data indicating that immediate responses (such as freezing response measured after fear conditioning) are normal in CaMKIV$^{-/-}$ mice. These results indicate that CaMKIV is crucial for the activation of CREB in areas related to fear memory *in vivo* and that its contribution is region specific.

Learning-related activation in the insular cortex

In addition to the ACC, it has been found that fear conditioning activates immediate early genes in the IC. The IC is another cortical area that is important for pain and aversive memory. Glutamate has been found to be an excitatory neurotransmitter in the IC. Furthermore, it was found that learning-like TBS induced LTP in the insular slices of adult mice. Physiological studies have proved that fear conditioning occludes electrically induced postsynaptic-LTP in the agranular IC. Injecting GluK1-specific antagonists into the agranular IC, which blocked the induction of presynaptic LTP in the IC, reduced behavioral responses to fear stimuli in mice [20]. Moreover, using a non-selective synapse blocker, $CoCl_2$, to inhibit the local IC neurotransmission, freezing behavior was attenuated in the groups that received the blocker either immediately after conditioning or before re-exposure to the aversive context but not before the conditioning session [21]. A recent study reported that the IC acts as a state-dependent regulator of fear that is necessary to establish an equilibrium between the extinction and maintenance of fear memories in mice [22]. Therefore, the IC also participates in the modulation of fear memory.

Synaptic LTP as a cellular model for fear

How is fear information learned and stored in amygdala circuits? One leading hypothesis is that synaptic transmission undergoes long-term plastic changes in the amygdala following fear conditioning [23–27]. LTP has been found to be the most likely synaptic mechanism underlying fear memory in the amygdala. First, electrophysiological studies using *in vitro* amygdala slices or *in vivo* recordings have shown that the auditory afferent pathways, including auditory thalamus-amygdala and cortex-amygdala pathways, undergo potentiation after LTP-inducing stimulation. Second, the associative nature of LTP in the amygdala supports the fact that fear conditioning requires the convergence of auditory and nociceptive inputs onto single neurons in the LA. Third, neuronal activity in the LA has been shown to be modified during auditory fear conditioning in a manner similar to that observed after artificial LTP induction. Fourth, drugs that inhibit the induction and/or consolidation of LTP in the amygdala also inhibit fear memory. Lastly, fear conditioning occluded LTP-induced presynaptic enhancement of synaptic transmission in the cortical pathway to the lateral amygdala.

ACC-ACC connections unlikely undergo LTP *in vivo*

What about the possibility that synaptic transmission within the ACC may contribute to memory formation as those reported in the amygdala? We suspect that synaptic changes between two sides of the ACC connected by callosal projection fibers may serve as an ideal candidate for the storage of such information since the thalamic-cortical connections are important for ascending nociceptive sensory transmission. In order to detect possible changes in the ACC, we performed recordings of synaptic responses in the ACC before and after fear conditioning. As shown in Figure 8, we found that synaptic responses to stimulation of the callosal projection fibers from the other side of the ACC were not affected by fear conditioning.

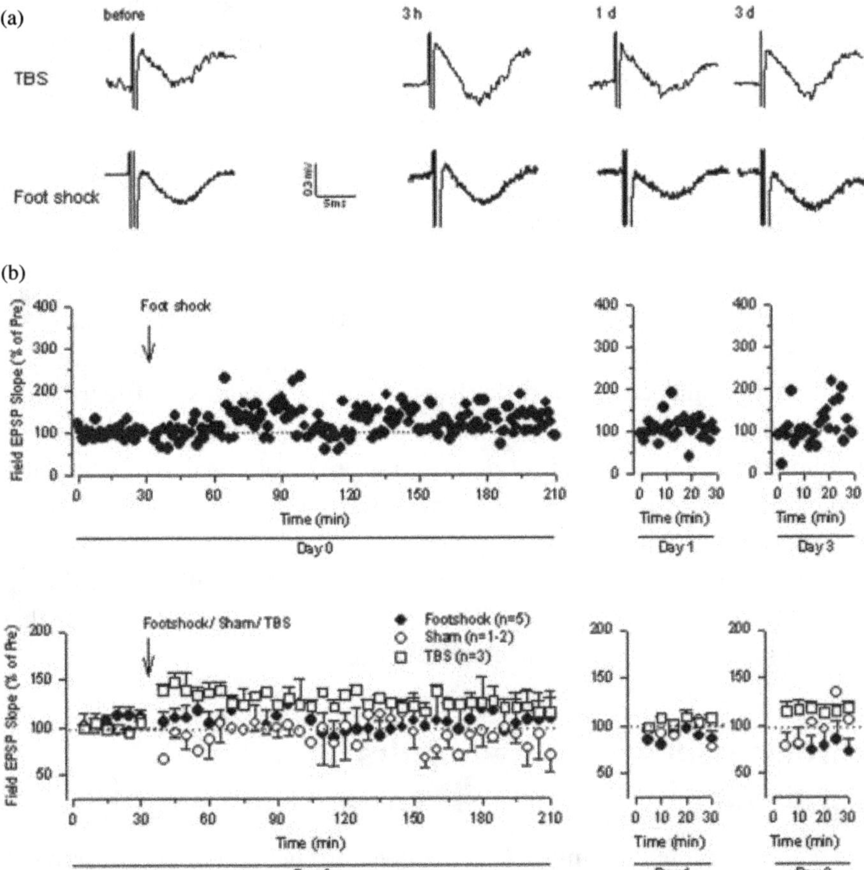

Figure 8. LTP in cortico-cortical synapses after fear conditioning. Fear conditioning did not cause long-term plasticity in ACC-ACC synapses. (a) Evoked fast responses in the ACC by stimulation applied to the other side of ACC were not affected by single foot shock that induced classic fear memory. However, TBS induced synaptic potentiation in the ACC. Insets show traces of evoked responses before, 3 h, 1 day, and 3 days after fear conditioning. (b) Summarized data of experiments shown in (a) and mice receiving the treatment without the foot shock ($n = 2$ mice) (adapted from Tang *et al.* [3]).

Furthermore, measurements of the same responses at 1 and 3 days after fear conditioning also did not reveal any significant changes (Figure 8), indicating that synaptic transmission between the ACCs did not undergo synaptic potentiation after fear conditioning. To test if synaptic transmission may undergo potentiation by theta burst stimulation, we also tested responses in freely moving mice after applying TBS locally into the ACC. As shown in Figure 8, we found that synaptic transmission was enhanced for the initial 3 hours.

NMDA GluN2B enhancement: Both chronic pain and fear are enhanced

In a previous study, we demonstrated that forebrain GluN2B overexpression significantly enhanced synaptic potentiation in the hippocampal CA1 region. In the same mice, behavioral responses to inflammation were also significantly enhanced. Interestingly, in the behavioral tests of fear memory, transgenic mice showed faster learning than wild-type littermates, and the enhancement lasted for days. Electrophysiological and behavioral studies have shown that activation of NMDA GluN2B receptors in the amygdala is critical for learning-related LTP and behavioral memory. A recent review has found enhanced memory, along with fear, in more than 30 lines of transgenic mice. Thus, it is likely that our brains code this information using the same mechanisms.

Chronic pain selectively impairs trace fear memory

It is widely reported in clinic literature that chronic pain affects or impairs cognitive functions. However, in animal models of chronic pain, it has been less reported. Our preliminary studies using classic fear memory tests found that chronic inflammation did not significantly affect the establishment of classic fear memory (induced by signal electrical stimulation). To test whether CFA-induced inflammatory pain influences the acquisition of attention-demanding associative memory, we tested CFA- and saline-injected mice in the trace fear conditioning paradigm.

Mice were injected with either saline or CFA in one hind paw 3 days before trace fear conditioning. After a 1 min baseline (ITI-0), mice received 10 CS–US pairings. Trace fear memory was tested the following day with the presentation of the CS in a novel chamber. There was no significant effect of CFA on freezing behavior during the training session, and freezing significantly increased in both saline- and CFA-injected mice throughout training (Figure 9(a)). When tested the next day, however, there was a significant effect of CFA treatment on freezing behavior (Figure 9(b)). CFA-injected mice froze significantly less during ITI-3,

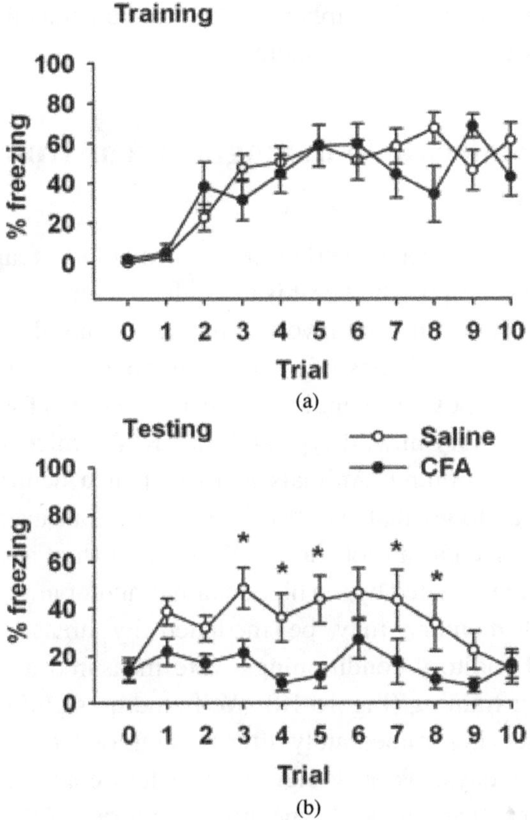

Figure 9. Trace fear memory deficit in CFA-injected mice. Freezing during trace training (a) and testing (b) over 10 trials 3 days after injection of CFA ($n = 8$) or saline ($n = 7$) in wild-type mice. The asterisks indicate significantly different from saline-injected mice; in all cases, $P < 0.05$ is considered significant (adapted from Wu *et al.* [28]).

ITI-4, ITI-5, ITI-7, and ITI-8, as compared with the mice injected with saline. This difference was not attributable to a change in the sensitivity to the foot shock because shock reactivity was not significantly altered between CFA-injected and saline-injected mice. To show that the CFA-induced fear memory deficit was specific for trace fear memory, CFA-injected and saline-injected mice were tested in the traditional delay conditioning paradigm. Mice were trained with a single CS–US pairing and tested for contextual and cued fear memory the following day. There was no difference in the freezing behavior between CFA- and saline-injected mice for either contextual or cued conditioning. These results suggest that CFA-induced peripheral inflammation selectively interferes with the expression of trace fear memory.

CaMKIV: Molecule to distinguish fear from chronic pain

Considering the significant overlap between memory, fear, and chronic pain, one key question is whether it is possible to identify selective molecular signaling proteins that are involved in one, but not the other, process. We have screened many lines of mice throughout years and have found CaMKIV to be one possible candidate. We assessed two forms of associative emotional memory in wild-type and CaMKIV$^{-/-}$ mice: contextual and auditory fear conditioning. Animals learn to fear a neutral conditioned stimulus (such as tone) that has been paired with an aversive unconditioned stimulus (such as a foot shock). Early contextual and auditory fear conditionings are mediated by the hippocampus and/or amygdala, whereas late contextual memory may be mediated by areas of the cortex. Contextual and auditory conditionings were measured at 1 hour, 1 day, and 7 days, after training (Figure 10). We found no significant difference in contextual freezing immediately after training or 1 hour later. By contrast, at 1 and 7 days, we saw significantly less contextual freezing in CaMKIV$^{-/-}$ mice than in wild-type mice (Figure 10(a)). Furthermore, when we tested auditory fear conditioning 1 day or 7 days after training,

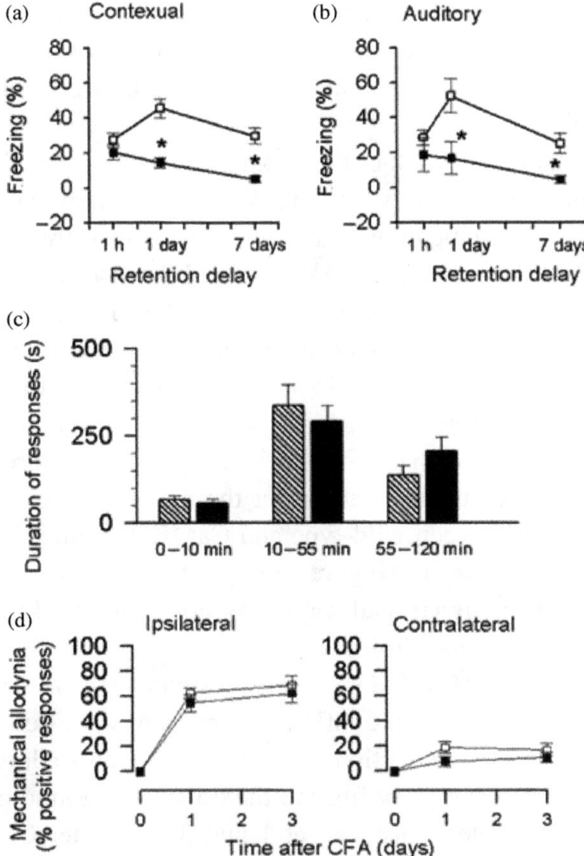

Figure 10. CaMKIV is required for fear memory but not behavioral responses to tissue injury. (a, b) Contextual and auditory fear conditioning at 1 h, 1 days, and 7 days after training (wild-type mice, $n = 8$; $CaMKIV^{-/-}$ (KO), $n = 7$). (c) Nociceptive behavioral responses to hind paw formalin injection during three different phases. No significant difference was found between wild-type (hatched) and $CaMKIV^{-/-}$ mice (black). (d) The responses of animals to a mechanical stimulus that elicited no responses before dorsal hind paw CFA injection were recorded 1 and 3 days after CFA injection. The data are plotted as percentage positive responses to stimulation of the ipsilateral or contralateral hind paw for wild-type and $CaMKIV^{-/-}$ mice. *$P < 0.05$ as compared with wild-type mice (adapted from Wei *et al.* [2]).

we saw significantly less freezing in response to the tone in CaMKIV$^{-/-}$ mice than in wild-type mice (Figure 10(b)). No obvious difference in the behavioral response to the foot shock was found between wild-type and CaMKIV$^{-/-}$ mice.

Next, we examined whether the lack of CaMKIV affected acute nociceptive transmission or persistent pain. We saw no significant difference in tail-flick latency, indicating that spinal nociceptive transmission was not significantly altered in CaMKIV$^{-/-}$ mice. Likewise, we found no differences in latencies to a hot-plate test (55°C). Thus, the behavioral responses of CaMKIV$^{-/-}$ and wild-type mice to noxious thermal stimuli were indistinguishable. We then looked at the response to a more prolonged nociceptive stimulus via hind paw formalin injections. We had previously demonstrated that formalin injections caused three phases of licking responses 120 plus minutes after the injection. We found no significant difference between wild-type and CaMKIV$^{-/-}$ mice in any of the three phases of response (Figure 9(c)). These results suggest that CaMKIV-dependent signal pathways are not required for behavioral responses to formalin injection.

To explore whether CaMKIV-dependent signal pathways contribute to behavioral responses in chronic inflammatory pain, we tested mechanical allodynia after hind paw injection of complete Freund's adjuvant (CFA). Application of a fine von Frey fiber to the dorsum of a hind paw elicited no response in untreated mice, but at 1 and 3 days after CFA injection (50%, 10 μl) into the dorsum of a single hind paw, mice responded to stimulation of either the same (ipsilateral) or, to a lesser extent, the contralateral hind paw by hind paw withdrawal. Again, we found that mechanical allodynia was similar in wild-type and CaMKIV$^{-/-}$ mice (Figure 10(d)) and that the degree of local inflammation in wild-type and CaMKIV$^{-/-}$ mice was similar. Together, these results indicate that CaMKIV is preferentially involved in fear memory induced by a noxious shock but not in the behavioral responses to acute noxious stimuli or tissue inflammation. These results also suggest that behavioral responses to peripheral tissue inflammation commonly used in pain research do not reflect fear memory.

Enhanced trace fear memory in CaMKIV transgenic mice

Synaptic activities in the ACC are important for trace fear memory [28,29]. Therefore, the enhanced LTP in the ACC may lead to improved trace fear memory. To test this idea, wild-type and CaMKIV overexpressed mice were examined in a trace fear conditioning paradigm with a weak electric shock (0.5 mA) [28]. Trace fear memory conditioning differs from the classic delay paradigm as animals must sustain attention during the trace interval to learn the CS–US association. Before conditioning, CaMKIV transgenic mice displayed similar movement in the conditioning chamber as compared with wild-type mice. During 10 CS–US pairing training, both mice showed increasing freezing throughout the training session. However, CaMKIV overexpressed mice displayed dramatically increased freezing response starting from the third trial compared with wild-type mice. Moreover, CaMKIV overexpressed mice reached the highest level of freezing at the fourth trace, while wild-type mice did at the fifth trace (Figure 11(a)). These data show that CaMKIV

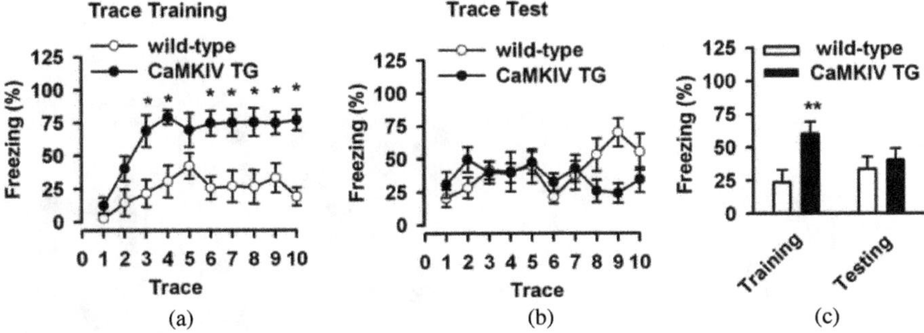

Figure 11. Enhanced learning of trace fear conditioning at weak shock intensity in CaMKI transgenic mice. (a) CaMKIV transgenic mice ($n = 8$) showed significantly increased freezing compared with wild-type mice ($n = 7$) during trace training at 0.5-mA intensity electric shock. (b) At 0.5-mA intensity electric shock, CaMKIV transgenic mice showed comparable freezing within the traces of the 24-h retention test compared with wild-type mice. (c) Statistical results showed that CaMKIV transgenic mice showed increased average freezing during trace training at 0.5-mA intensity electric shock compared with wild-type mice. *$P < 0.05$ and **$P < 0.01$ (adapted from Wu. *et al.* [28]).

overexpressed mice can learn faster in trace fear conditioning. In the test session (24 h post-training) however, the transgenic mice showed no difference in freezing as compared with wild-type mice (Figures 11(b) and 11(c)).

To test whether the unchanged 24 h trace fear memory was a result of the electric shock (0.5 mA) being too weak to be consolidated, we then tested the trace fear memory with a strong electric shock (0.8 mA). Similar to the 0.5-mA electrical shock, both mice showed increasing freezing throughout the training session during 10 CS–US pairing training (Figure 12(a)). Interestingly, CaMKIV transgenic mice displayed a similar freezing response compared with wild-type mice during 10 CS–US pairing training (Figure 12). However, on the test session 24 h post-training, the transgenic mice showed a significantly increased freezing response compared with wild-type mice (Figures 12(b) and 12(c)). Taken together, these results suggest that mice overexpressing CaMKIV are enhanced in fear memory. However, the enhancement in acquisition or 24 h post-training retention of trace fear memory is dependent on the shock intensity.

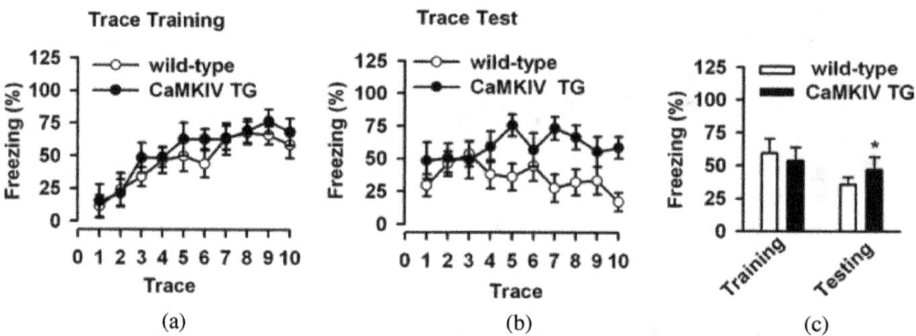

Figure 12. Enhanced memory of trace fear conditioning at strong shock intensity in CaMKIV transgenic mice. (a) At 0.8-mA intensity electric shock, CaMKIV transgenic mice ($n = 7$) showed similar freezing compared with wild-type mice ($n = 10$) during trace training. (b) At 0.8-mA intensity electric shock, CaMKIV transgenic mice showed increased freezing within the traces of the 24-h retention test compared with wild-type mice. (c) Statistical results showed that CaMKIV transgenic mice showed increased average freezing during the 24-h retention test compared with wild-type mice. $*P < 0.05$ (adapted from Wu *et al.* [28]).

Table 1.   Genetic separation of pain and fear.

Genetic manipulation	Pain		Memory	
	Acute pain	*Chronic pain*	*LTP-LTD*	*Spatial fear memory*
GluN2B forebrain overexpression	No change	enhanced	Enhanced LTP	enhanced enhanced
PDZ93 KO	No change	reduced	ND ND	reduced
Kainate GluR5 KO	No change	reduced	No change ND	No change
Kainate GluR6 KO	No change	No change	Reduced ND	reduced
CaMKIV KO	No change	No change	Reduced	No change/reduced reduced
CaMKIV forebrain overexpression	No change	No change	Altered	reduced reduced
AC1&8 KO	No change	reduced/ blocked	Reduced/blocked	reduced reduced
Eg1 (NGFIA) KO	No change	reduced	Altered/reduced ND	reduced

Conclusions and future directions

It is becoming clear that there are cortical and subcortical regions that are involved in both the induction and expression of fear. Physical and emotional painful stimuli are the most effective stimuli for generating long-term fear in the brain. Studies at the synaptic level provide strong evidence that synaptic LTP serves as a key mechanism for such long-term memory (see Table 1). Despite rapid progress made in recent animal studies, clinical treatment for fear and related diseases such as PTSD is still lacking. It is critical to understand the molecular and cellular mechanisms for pain and fear and identify selective molecular targets for the development of future treatments for chronic pain and the related fear, as well as emotional disorders.

References

[1] Frankland, P.W. and Bontempi, B. (2005) The organization of recent and remote memories. *Nat Rev Neurosci* 6, 119–130. 10.1038/nrn1607.

[2] Wei, F. *et al.* (2002) Calcium calmodulin-dependent protein kinase IV is required for fear memory. *Nat Neurosci* 5, 573–579. 10.1038/nn0602-855.

[3] Tang, J. *et al.* (2005) Pavlovian fear memory induced by activation in the anterior cingulate cortex. *Mol Pain* 1, 6. 10.1186/1744-8069-1-6.

[4] Frankland, P.W. *et al.* (2006) Stability of recent and remote contextual fear memory. *Learn Mem* 13, 451–457. 10.1101/lm.183406.

[5] Bishop, S.J. (2008) Neural mechanisms underlying selective attention to threat. *Ann NY Acad Sci* 1129, 141–152. 10.1196/annals.1417.016.

[6] Knight, D.C. *et al.* (2004) Neural substrates mediating human delay and trace fear conditioning. *J Neurosci* 24, 218–228. 10.1523/JNEUROSCI.0433-03.2004.

[7] Zhao, M.G. *et al.* (2005) Roles of NMDA NR2B subtype receptor in prefrontal long-term potentiation and contextual fear memory. *Neuron* 47, 859–872. 10.1016/j.neuron.2005.08.014.

[8] Descalzi, G. *et al.* (2012) Rapid synaptic potentiation within the anterior cingulate cortex mediates trace fear learning. *Mol Brain* 5, 6. 10.1186/1756-6606-5-6.

[9] Steenland, H.W. *et al.* (2012) Predicting aversive events and terminating fear in the mouse anterior cingulate cortex during trace fear conditioning. *J Neurosci* 32, 1082–1095. 10.1523/JNEUROSCI.5566-11.2012.

[10] Restivo, L. *et al.* (2009) The formation of recent and remote memory is associated with time-dependent formation of dendritic spines in the hippocampus and anterior cingulate cortex. *J Neurosci* 29, 8206–8214. 10.1523/JNEUROSCI.0966-09.2009.

[11] Zhang, Y. *et al.* (2011) Induction and requirement of gene expression in the anterior cingulate cortex and medial prefrontal cortex for the consolidation of inhibitory avoidance memory. *Mol Brain* 4, 4. 10.1186/1756-6606-4-4.

[12] Jhang, J. *et al.* (2018) Anterior cingulate cortex and its input to the basolateral amygdala control innate fear response. *Nat Commun* 9, 2744. 10.1038/s41467-018-05090-y.

[13] Malin, E.L. *et al.* (2007) Involvement of the rostral anterior cingulate cortex in consolidation of inhibitory avoidance memory: interaction with the basolateral amygdala. *Neurobiol Learn Mem* 87, 295–302. 10.1016/j.nlm.2006.09.004.

[14] Allsop, S.A. *et al.* (2018) Corticoamygdala transfer of socially derived information gates observational learning. *Cell* 173, 1329–1342 e1318. 10.1016/j.cell.2018.04.004.

[15] Rajasethupathy, P. *et al.* (2015) Projections from neocortex mediate top-down control of memory retrieval. *Nature* 526, 653–659. 10.1038/nature15389.

[16] Gafford, G.M. *et al.* (2013) Memory accuracy predicts hippocampal mTOR pathway activation following retrieval of contextual fear memory. *Hippocampus* 23, 842–847. 10.1002/hipo.22140.

[17] Fanselow, M.S. and Dong, H.W. (2010) Are the dorsal and ventral hippocampus functionally distinct structures? *Neuron* 65, 7–19. 10.1016/j.neuron.2009.11.031.

[18] Cullen, P.K. *et al.* (2015) Activity of the anterior cingulate cortex and ventral hippocampus underlie increases in contextual fear generalization. *Neurobiol Learn Mem* 124, 19–27. 10.1016/j.nlm.2015.07.001.

[19] Bian, X.L. *et al.* (2019) Anterior cingulate cortex to ventral hippocampus circuit mediates contextual fear generalization. *J Neurosci* 39, 5728–5739. 10.1523/JNEUROSCI.2739-18.2019.

[20] Shi, T.Y. *et al.* (2018) Kainate receptor mediated presynaptic LTP in agranular insular cortex contributes to fear and anxiety in mice. *Neuropharmacology* 128, 388–400. 10.1016/j.neuropharm.2017.10.037.

[21] Alves, F.H. *et al.* (2013) Involvement of the insular cortex in the consolidation and expression of contextual fear conditioning. *Eur J Neurosci* 38, 2300–2307. 10.1111/ejn.12210.

[22] Klein, A.S. *et al.* (2021) Fear balance is maintained by bodily feedback to the insular cortex in mice. *Science* 374, 1010–1015. 10.1126/science.abj8817.

[23] McKernan, M.G. and Shinnick-Gallagher, P. (1997) Fear conditioning induces a lasting potentiation of synaptic currents in vitro. *Nature* 390, 607–611. 10.1038/37605.

[24] Rumpel, S. *et al.* (2005) Postsynaptic receptor trafficking underlying a form of associative learning. *Science* 308, 83–88. 10.1126/science.1103944.

[25] Rogan, M.T. *et al.* (1997) Fear conditioning induces associative long-term potentiation in the amygdala. *Nature* 390, 604–607. 10.1038/37601.

[26] Clem, R.L. *et al.* (2008) Ongoing in vivo experience triggers synaptic metaplasticity in the neocortex. *Science* 319, 101–104. 10.1126/science.1143808.

[27] Clem, R.L. and Huganir, R.L. (2010) Calcium-permeable AMPA receptor dynamics mediate fear memory erasure. *Science* 330, 1108–1112. 10.1126/science.1195298.

[28] Wu, L.J. *et al.* (2008) Genetic enhancement of trace fear memory and cingulate potentiation in mice overexpressing Ca2+/calmodulin-dependent protein kinase IV. *Eur J Neurosci* 27, 1923–1932. 10.1111/j.1460-9568.2008.06183.x.

[29] Han, C. J. *et al.* (2003) Trace but not delay fear conditioning requires attention and the anterior cingulate cortex. *Proc Natl Acad Scie USA* 100(22), 13087–13092.

Chapter 10

Pain and Anxiety

News and Views

Anxiety and pain: Two sides of the same synapse?

Researchers have known for a decade what patients have suspected for much longer: chronic pain produces anxiety, and anxiety makes the pain worse. But why? Scientists have found some tantalizing clues, including intense activity in the same part of the brain during both sensations. But an answer has been elusive. Now, neuroscientists at the University of Toronto have mapped a mechanism in the brain's anterior cingulate cortex, or ACC, that could explain the link between anxiety and chronic pain.

Dr. Min Zhuo is a professor in the department of physiology at University of Toronto and the Canada Research Chair in Pain and Cognition. Dr. Zhuo and his lab recently published a paper in the journal *Neuron* that showed how neuroplasticity — the brain's ability to physically re-organize itself in response to experience — can spur the interplay between chronic pain and anxiety. They also showed that a drug they developed for chronic pain can limit anxiety. Dr. Zhuo spoke with the Faculty of Medicine writer Jim Oldfield about his findings and what they could mean for patients:

Oldfield: "What's the link between pain and cognition in your work?"
Zhuo: "Most chronic pain researchers study pain as a sensory experience. They think that if we reduce the physical pain, the patient will be better. But I think we need to look at both the mental and physical in the study of pain. Recently we've seen more acknowledgment that patients with chronic pain are often anxious, and that if this pain lasts a long time many will suffer depression and lose hope. As well, more researchers are moving away from the idea that mental disorders are just genetic — i.e., you have a bad gene and that's why you have disease. In my view, if chronic pain causes enough physical change in the brain, anyone can behave like a mentally ill patient. So, to better understand and treat pain we need to look at its physical cause but also cognition and mental health, or emotional well-being.

Oldfield: "What findings does the Neuron paper describe?"
Zhuo: "We found a possible explanation for why chronic anxiety and pain make each other worse. Neurons communicate through synapses, which are structures that convey nerve impulses from one neuron to another. These synapses are plastic, meaning they change physically in response to stimuli. Repeated experiences can reinforce these changes and over time the physical changes can become permanent. It's like a learned memory. And when that happens, you can experience a sensation without an underlying cause. You've probably heard of amputees who have pain in a limb that doesn't exist. Well, we found a similar "learned memory" process for anxiety, in a different part of the same synapse where we see activity linked to chronic pain. So, when these two functions become partners, they make each other worse. When you have chronic pain, you develop anxiety. Because you have anxiety, you suffer more chronic pain. This is the first evidence that anxiety is linked to long-lasting changes at synapses."

Oldfield: Is there a way to short-circuit this process?
Zhuo*: "We identified a channel or genetic process called HCN that maintains this anxious learning. So theoretically, if we find an inhibitor to erase the long-term potentiation for anxiety — the learned memory — we should be able to reduce anxiety through this channel. But a big challenge is that drugs in the central nervous system often produce side effects. The HCN channel is also expressed in heart and other types of cells, so any drug targeting this channel would have to be selective for neurons. Fortunately, a drug we've been developing for pain called NB001 inhibits a protein called AC1 that is upstream from HCN, and it does so in a way that is selective for neural cells. It's a very promising therapy for chronic pain and anxiety."*

Oldfield: "What are the challenges in bringing this drug to market?"
Zhuo*: "Our research is still at the pre-clinical stage. But it's very difficult to move drug discovery forward, in part due to the world economic situation. There are a few angel funders because of the recession. Big Pharma is risk-averse. NGOs support experimental research, but their funds are very tight. On the bright side, we have teamed up with investors and researchers internationally on clinical tests in cancer patients. Many cancer patients have pain that is badly managed; in the late stages they're almost asleep with high-dose opiates. Others don't get enough medicine and live with terrible pain and anxiety. My hope is that one day we can treat these patients with a drug that works."*

Summary

It is well known that anxiety is associated with pain and chronic pain. However, less is known about the possible cellular and molecular mechanisms of this association. Recent studies have suggested that many brain regions that are critically involved in chronic pain are also contributing to the regulation of behavioral anxiety. Furthermore, many receptors involved in anxiety are also affected by peripheral injury that causes chronic pain. This chapter will discuss these findings and seek to better understand the basic mechanisms for anxiety that is triggered and/or caused by chronic pain.

Keywords: Pain; ACC; Anxiety; oxytocin; pre-LTP; KA; amygdala; chronic pain; LTD

Introduction

It is well known that pain causes fear and anxiety in patients. In fact, several previous successful pain medicines were originally used for the treatment of mood disorders. Recent studies using modern neurobiological approaches and human brain imaging techniques reveal that pain, fear, and anxiety share many common brain structures. Brain structures in common include the anterior cingulate cortex, prefrontal cortex, amygdala, and hippocampus. At the molecular level, many signaling molecules that are contributing to the induction of expression of synaptic changes are also indicated in pain, fear, and anxiety. This chapter will review recent progress made in our molecular understanding of fear and anxiety and explore its possible relationship with pain and chronic pain.

Anxiety is a common human emotional experience that causes a decreased quality of life and increased social burden [1–3]. However, current treatment options for anxiety are limited as the molecular mechanisms of these complicated emotional disorders are not fully understood. With the development of integrative methods, including genetic manipulations, a variety of molecular targets involved in anxiety have been revealed. These targets vary from membrane receptors such as the 5-HT, $GABA_A$, and GluK1 KA receptors to intracellular signaling proteins such as CaMKIV and AC8, to transcription factors such as CREB and Egr-1. We propose that all these molecules act together to form a balance between excitatory and inhibitory transmissions that are critical for physiological anxiety, and that prolonged disturbance of any of them can promote pathological anxiety-like behavior. Studies on the interactions between these molecules will help elucidate the cellular mechanisms of anxiety and will provide molecular targets for treating the disorders.

Animal models for studying anxiety

As anxiety is a highly conserved behavior among all species, animal models are widely used to study the molecular mechanisms of anxiety disorders [4,5]. Numerous studies have been conducted on animal models of anxiety to help elucidate the neural mechanisms involved. As animals respond to the potential presence of a threat with characteristic responses and defensive behaviors, investigators use these physiological and behavioral parameters to characterize numerous models of anxiety-like behaviors in animals. The ethologically based animal models of anxiety attempt to create the natural conditions under which such emotional states are elicited and therefore are thought to minimize possible confounding effects of motivational or perceptual states. However, there still exists individual differences and variable behavioral baseline levels in these models. While there are many animal models of anxiety, only three will be discussed here in detail: the elevated plus maze (EPM), light/dark box, and open field.

One of the best characterized and pharmacologically validated animal models of anxiety is the elevated plus maze. The apparatus has four arms that extend from a central, elevated platform in the shape of a plus sign. Two opposing arms are enclosed by high walls, while the other two opposing arms are not enclosed and have a minimal lip. The maze creates a conflict between the rodent's natural tendency to explore new environments and their fear of the unfamiliar, openness, and elevation. The most common measure of anxiety, in this test, is quantified by the amount of time entering, or staying in, the open arm. The animal who enters the open arm more, and stays in longer, would be considered to have a lower level of anxiety, whereas an animal who enters less, and stays in for shorter periods of time, would be considered to have higher levels of anxiety. An example of an elevated plus maze test for GluK1 knockout mice is shown in Figure 1. The plus maze allows for a rapid screening of anxiety-modulating drugs, or mouse genotypes, without training, or complex schedules. However, it is possible for the behavior patterns to be influenced by variability in test conditions (e.g., age, sex, handling, time of testing, illumination, and method of scoring), which may contribute to discrepancies among results.

Similar to the EPM, the light/dark box test is also based on the conflict between a rodent's natural tendency to explore a novel environment and its inherent aversion to brightly lit areas. The light/dark test apparatus is divided into two compartments by a panel. One compartment is brightly

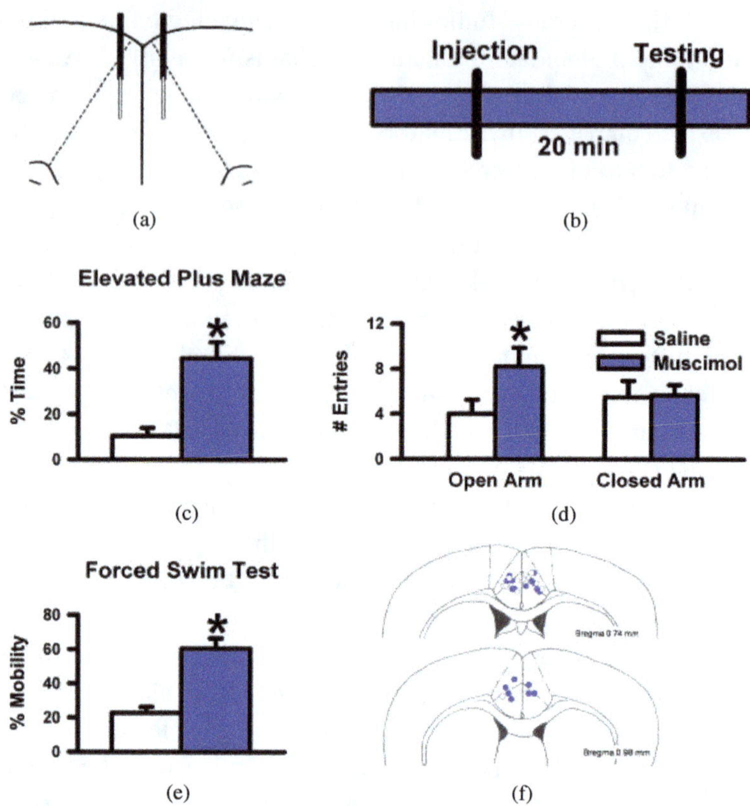

Figure 1. The ACC is critical for anxiety- and depression-like behaviors. (a) Model showing the cannulation and microinjection into the ACC. (b) Diagram showing the time-line of experiment. (c) Percent of time in the open arms of the elevated plus maze. (d) Number of entries into the open and closed arms of the elevated plus maze. (e) Percent of active mobility in the forced swim test. (f) Microinjection sites in the ACC (adapted from Kim *et al.* [13]).

illuminated, and another, slightly smaller, compartment is painted black. The animals are allowed to roam freely between both compartments. The most common measure of anxiety, in this test, is measured by the number of transitions between, and time spent in, the lit chambers. The latency to exit the dark chamber can also be used as an index of anxiety. An animal that spent a decreased amount of time in the light chamber would be inter-preted as possessing an increased level of anxiety.

The open field test consists of placing an animal in an unknown chamber with four surrounding walls and measuring a number of behaviors such as spontaneous exploratory locomotion and a tendency to prefer periphery rather than the center of the open field. Although this may seem like a simple task, there is no standardization between laboratories that use this model. Some open fields are square and others circular, some are clear and others opaque, others are bright and some are totally dark, and some have tops and others are open. Presence of objects within the arena, how an animal is placed into the open field, the length of the recording period, and what specifically is recorded may also vary across the literature.

The pharmacological validation for this task is fairly minimal. The effects of treatments on exploration are not measured directly — as anxiolytic treatments do not, by themselves, increase exploration — rather it is how the treatments decrease the stress-induced inhibition of exploratory behavior. Interestingly, compounds like benzodiazepines and 5-HT1A — commonly used for the treatment of anxiety disorders such as panic attacks, generalized anxiety, or obsessive-compulsive disorder — are ineffectual at this task. This suggests that this paradigm may not model all features of anxiety disorders.

Brain areas that involved in anxiety

The amygdala is a critical structure in our understanding of anxiety [6,7]. Although less is investigated at the synaptic level, anxiety is thought to be a concern about what might happen in the future. In both humans and animals, electrical stimulation of the amygdala elicits anxiety. Balance between excitatory and inhibitory transmissions is critical for the mediation of anxiety-like behaviors. For example, hyperexcitation due to enhanced excitatory transmission, or reduced inhibitory transmission, can promote anxiety. The most widely prescribed classes of anxiolytic drugs — the benzodiazepines and the selective serotonin reuptake inhibitors — modulate GABA- and serotonin-mediated neurotransmission to reduce neuronal excitability. Therefore, it is generally believed that the hyperexcitation of neural circuits — including the amygdala, septo-hippocampal area, and prefrontal cortex — are mainly responsible for anxiety disorders.

Since anxiety is regarded as an innate fear, the structural basis of anxiety resides in the neural circuitry related to fear response. The amygdala is a key structure for processing neuronal inputs from other parts of the brain, initiating output signals to responding nuclei and generating various physiological responses, including behavioral, autonomic, and hormonal responses related to anxiety. In both humans and animals, electrical stimulation of the amygdala elicits anxiety whereas lesion of the amygdala impairs the perception of fear. In addition to the amygdala, various studies using lesion, microinjection, electrophysiology, or imaging techniques have implicated other brain regions in anxiety, including cortex, hippocampus, hypothalamus, and brainstem.

The neural circuitry of anxiety is thought to share a common pathway underlying conditioned fear behavior (Figure 2). Anatomically, information from sensory modalities reaches the amygdala via projections from the thalamus, cortex, or hippocampus. There are internucleus interactions for the transmission and integration of the sensory input in the amygdala. Essentially, the information flows from the lateral amygdala to the basolateral amygdala before finally reaching the central amygdala. Then the efferents from central amygdala go to the periaqueductal gray, brainstem, and hypothalamus, which initiates fear-related behavioral, autonomic, and hormonal responses. Figure 2 shows this simplified information flow in the nuclei, from transmitting threat stimulus to initiating anxiety/fear behaviors. The normal function of this circuit is critical for physiological anxiety, while the dysfunction of this circuit will lead to pathological anxiety. We have to keep in mind that there are complex interactions between these brain regions underlying different aspects of anxiety, such as memory, aversion, or motivation. Moreover, we propose that the local interneurons will also affect the information flow between the different brain regions (Figure 2).

Cortical contribution to anxiety

In addition to the amygdala, animal and human research has demonstrated that different cortical regions have also been indicated in anxiety or

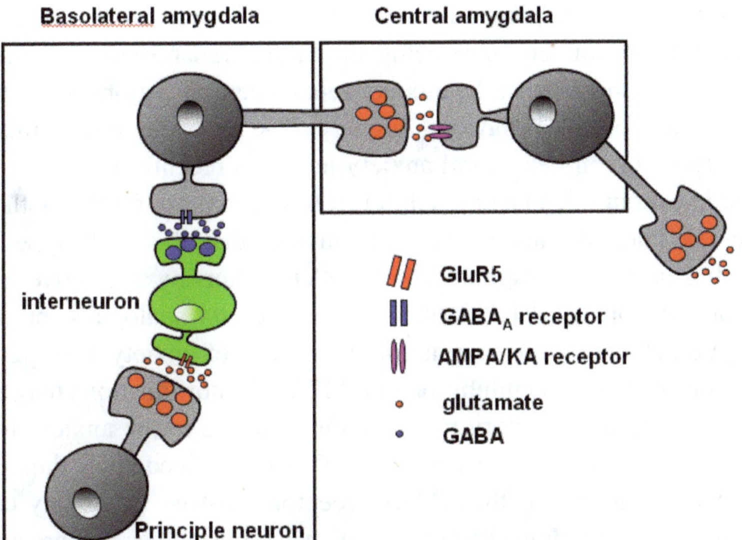

Figure 2. A model for KA receptor modulation in the amygdala. In the basolateral amygdala, GluR5 is mainly expressed in somatodendritic regions of interneurons. Activation of GluR5 initiates action potential firing that increases GABA release. The released GABA produces feed-forward inhibition on pyramidal neurons and the subsequent output to the central amygdala. Therefore, activation of GluR5 in the basolateral amygdala reduces the information flow transmitted to the central amygdala, which may explain the increased anxiety-like behaviors in GluR5$^{-/-}$ mice.

anxiety-related behaviors. For example, the ACC and IC, two major brain areas for unpleasantness and emotion, have been demonstrated to play important roles in various types of anxiety [8–11]. The ACC/IC has been implicated in anxiety and fear in both human and animal studies [2,12–14]. Human imaging studies observed increased ACC and/or IC activity in patients with anxiety disorders [15]. For example, the ACC shows greater activation in patients with social anxiety disorders [16], and activation of the IC has been reported in phobia subjects [17].

AMPA receptor

Glutamatergic synapses are major excitatory synapses in the central nervous system. AMPA receptors meditate most of the excitatory currents at the resting membrane potentials. According to the current view of cellular mechanisms for anxiety, enhancing inhibitory transmission or reducing excitatory transmission in the anxiety-related central regions is the major approach to control behavioral anxiety. The possible involvement of AMPA receptors in behavioral anxiety has been recently investigated. In genetic knockout mice lacking GluR1, it has been reported that both male and female knockout mice were more anxious than their wild-type litter-mates. Furthermore, antagonism of AMPA receptors by a selective AMPA receptor antagonist GYKI52466, and some of its analogues, produced anxiolytic behaviors in different animal models of anxiety. One possible interpretation is that the inhibition of AMPA receptor functions may affect the excitability of inhibitory neurons, thus leading to the anxiety imbalance of E/I. Clearly, more cellular analyses are needed in the future. Additional evidence for the AMPA receptor involved in anxiety comes from studies of withdrawal from chronic benzodiazepines that are associated with increased anxiety. Withdrawal anxiety, as measured in behavioral tests such as the EPM, is accompanied by the reduction of AMPA receptors in hippocampus and cortex.

In addition to AMPA receptors, the involvement of glutamate NMDA and mGluRs in anxiety has been also reported. Activation of certain types of mGluRs in the hippocampus produced anxiolytic behaviors, and thus the use of mGluR antagonists has been discovered to be possible new treatment for anxiety.

KA receptor-mediated synaptic transmission in the cortex and amygdala

In situ hybridization and immunostaining results show that KA receptor subunits are widely expressed in the cortex [18–20]. For example, GluK1-3 and GluK5 are both highly expressed, whereas GluK4 is either slightly detectable during postnatal days or not expressed at all. Results from *in situ* hybridization and immunostaining tests show that KA receptor subunits are expressed in the ACC [19]. Although KA receptors have been reported in many central synapse studies, research regarding the contribution of KA receptors to baseline synaptic transmission has been highly limited. In many brain regions, baseline synaptic transmission is mainly mediated or exclusively mediated by AMPA receptors. Since early reports of KA receptors in nociceptive transmission in the spinal cord dorsal horn [21], KA receptor-mediated synaptic responses have also been reported in the ACC and IC [22,23]. Single shock stimulation could induce small KA receptor-mediated excitatory postsynaptic currents (EPSCs) (see Figure 3). KA receptor EPSCs had a significantly slower rise time course and decay time constant compared with AMPA receptor-mediated EPSCs (Figure 3). High-frequency repetitive stimulation significantly facilitated KA receptor EPSCs. Studies using genetically modified mice with the deletion of GluK1 (GluR5) and/or GluK2 (GluR6) demonstrated that both GluK1 and GluK2 are involved in synaptic transmission in the adult ACC. KA EPSCs are ~5–10% of AMPA/KA EPSCs in all layers of the adult mouse and rat insular cortex [23]. Similar to the ACC, genetic deletion of GluK1 or GluK2 subunit partially reduced postsynaptic KA EPSCs, and exposure of GluK2 knockout mice to the selective GluK1 antagonist UBP 302 could significantly reduce the KA EPSCs. These data suggest that both GluK1 and GluK2 play functional roles in the IC.

In the amygdala, mRNA for GluK1,2 and GluK5 subunits are highly expressed. Among them, GluK1 is highly expressed. It has been reported that KA receptors contribute to postsynaptic excitatory responses of the basolateral amygdala induced by the stimulation of external capsule [24]. Interestingly, this excitatory response (about 30% of the total current) is sensitive to the inhibition of the GluK1 receptor antagonist. Due to the slow decay kinetics of KA-mediated currents, the postsynaptic responses summate in response to high-frequency stimulation.

KA receptor and synaptic regulation

In the basolateral amygdala, presynaptic GluK1 is shown to bidirectionally modulate the release of GABA, possibly due to different types of GluK1-containing KA receptors with different agonist affinities [25]. *In vitro* slice recordings showed that GluK1 is selectively expressed in interneurons, and its activation could largely depolarize those interneurons and increase synaptic GABA release, as well as GABA tonic currents. More importantly, the GluK1 activation in the basolateral amygdala

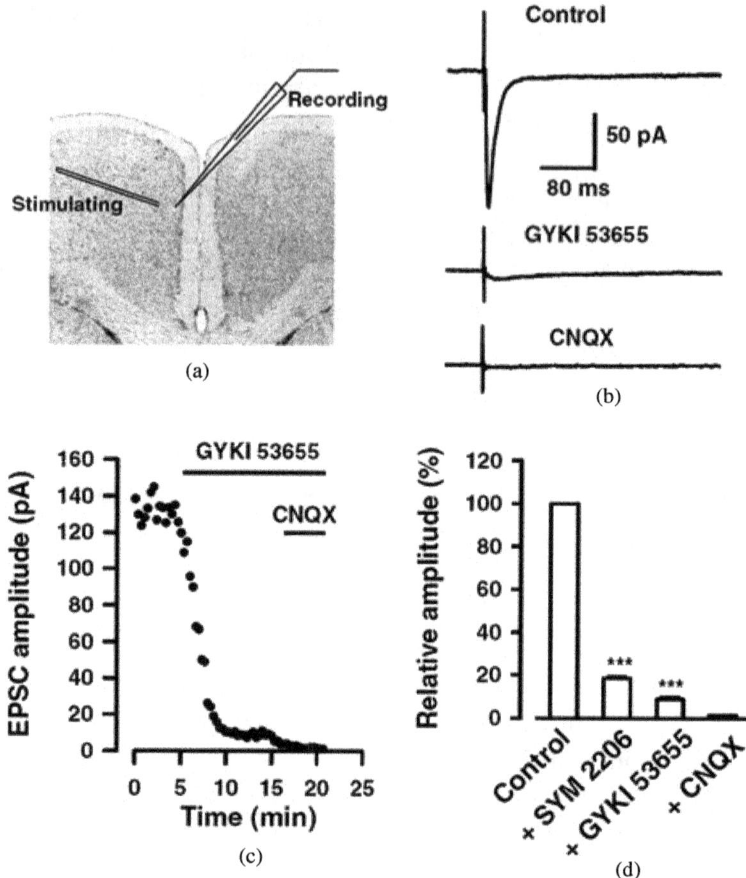

Figure 3. KA receptor-mediated EPSCs in adult ACC pyramidal neurons. (a) Diagram showing the placement of stimulating and recording electrodes in the ACC. (b) Control EPSCs were recorded in the presence of picrotoxin (PTX, 100 μM) and AP-5 (50 μM). After the perfusion of GYKI 53655 (100 μM), a small residual current remained, which could be totally blocked by CNQX (20 μM). In the following example of EPSCs, each trace represents an average of 5–10 consecutive recordings. (c) Sample points showing the time course of GYKI 53655 and CNQX effects on the neuron shown in b. (d) Statistical results showing the percentage of EPSCs in the presence of SYM 2206, GYKI 53655, or CNQX. Modified from Wu *et al.* [22] (adapted from Wu *et al.* [22].

reduced the output to the central amygdala, which may explain the increased anxiety phenotype in the GluK1 knockout mice [26]. In addition, the regulation of excitatory glutamate transmission by KA receptors has been also reported in the amygdala [27,28], and such regulation is also mixed (inhibition and facilitation).

In the ACC, activation of GluK1 also triggers action potential-dependent GABA release, which is also required for the activation of L-VGCCs and Ca^{2+} influx. The effect of GluK1 activation is selective to the GABAergic but not glutamatergic synaptic transmission [29]. Interestingly, we found that GluK2 knockout mice showed reduced fear memory but not anxiety, which may be due to its role in the induction of LTP in the lateral amygdala [30].

KA receptor and LTP

There are at least three major forms of LTP in central synapses [31,32]. The first type is post-LTP. The second one is pre-LTP. The last one is pre-LTP that requires diffusible messengers [31]. In general, NMDA receptor-dependent LTP takes place in most central synapses, in addition to the hippocampus [33]. In the ACC, there are at least two forms of LTP. The post-synaptic form of LTP requires the activation of NMDA receptors. This form of LTP is independent of KA receptors. The expression of LTP is likely due to increases in AMPA receptor-mediated functions, including phosphorylation of GluA1 AMPA receptors [34]. There is no involvement of KA receptor in post-LTP.

Another form of LTP, pre-LTP, has been recently identified [35–37] (see Figures 4 and 5). It requires the activation of KA GluK1 receptor. Activation of GluK1-containing KARs appears sufficient for the induction of pre-LTP in the ACC (Figures 5 and 6). Together, these results suggest that presynaptic GluK1 receptors are both necessary and sufficient for the induction of pre-LTP in the ACC.

In the amygdala, KA receptor-dependent LTP has been reported. Repetitive low-frequency stimulation induces LTP in the basolateral

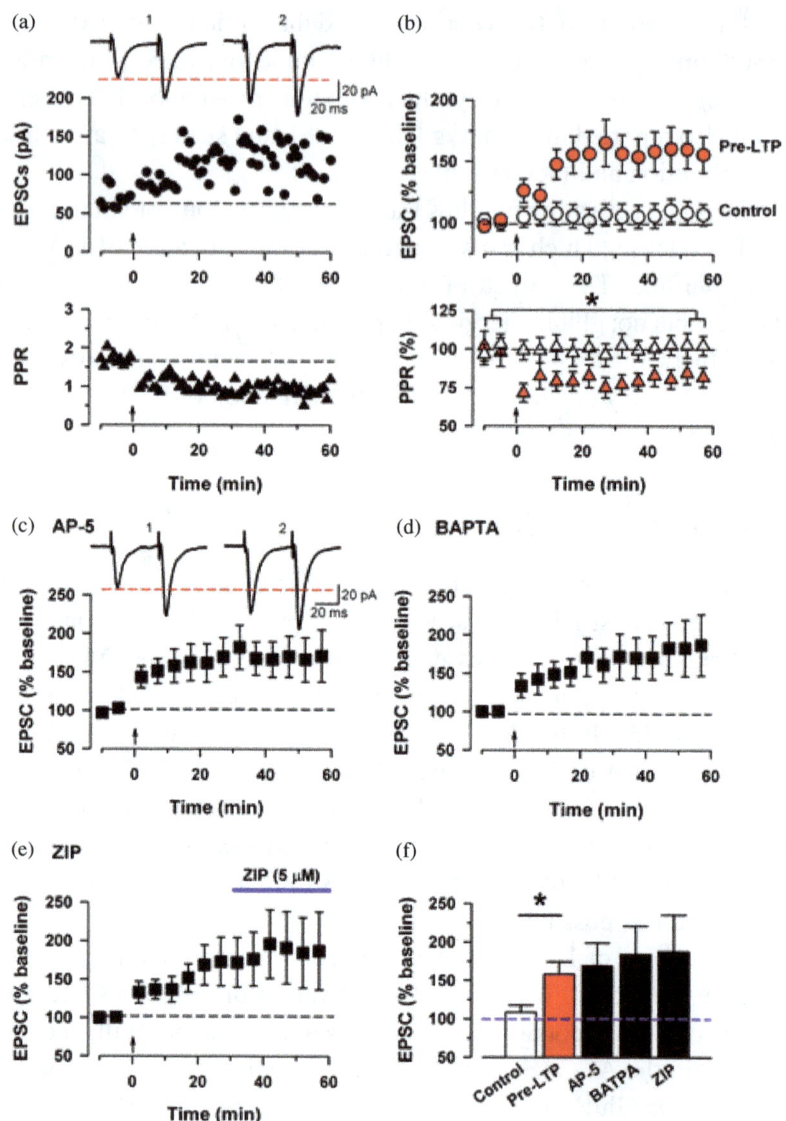

amygdala, and this form of LTP is NMDA receptor-independent and sensitive to inhibition of GluK1 receptor [38]. Moreover, this form of LTP is not synapse selective and can even be hetero-synaptic. The NMDA receptor-dependent form of LTP, however, can be used using the pairing protocol in the same regions. This form of LTP does not require KA receptors. However, there is evidence that presynaptic KA receptors may contribute to postsynaptic LTP [39]. Recently, a KA-dependent presynaptic form of LTP has also been reported. Low-frequency stimulation, paired with voltage clamps at postsynaptic sites, induces NMDA receptor-independent pre-LTP. KA GluK1 receptor is involved [39].

KA receptor and LTD

LTD is another major form of plasticity that contributes to synaptic functions. In the ACC and IC, the induction of LTD requires activation of NMDA receptor or L-type voltage-gated calcium channels or mGluRs [34,40,41]. There is no report of the involvement of KA receptors in the

---

Figure 4. (*Continued*)   Presynaptic NMDAR-independent LTP in the ACC. (a) Upper panel: Sample traces of eEPSCs with paired-pulse stimulation at 50 ms interstimulus interval during baseline (1) and 60 min after the pre-LTP (2) at a holding membrane potential of −60 mV. Middle panel: A time course plot of a representative single example. Bottom panel: Time course plot of PPR for this neuron. The arrow donates the time of LTP induction. (b) Pooled data to illustrate the time course of pre-LTP and changes in PPR. Upper panel: Pre-LTP (red circles: $n = 20$ neurons/16 mice) and control (white circles: $n = 13$ neurons/9 mice). Bottom panel: PPR for these neurons. $*P < 0.05$ for PPR change estimated for the 10 min intervals baseline and between 50 and 60 min following the induction stimulus. (c) An NMDA receptor antagonist, AP-5 (50 μM), did not affect pre-LTP ($n = 12$ neurons/9 mice). (d) BAPTA (20 mM) in the recording pipette did not block pre-LTP ($n = 11$ neurons/8 mice). (e) A PKMζ inhibitor, ZIP (5 μM), did not affect maintenance of pre-LTP, applied 30 min after the pre-LTP induction ($n = 10$ neurons/8 mice). (f) Summary of the effects of an NMDA receptor antagonist, postsynaptic injection of BATPA, or a PKMζ inhibitor on pre-LTP. There was statistical difference when comparing control with pre-LTP, BAPTA, AP-5, or ZIP groups (one-way ANOVA, $F_{4,61} = 3.45$, $*P < 0.05$). There was no difference comparing pre-LTP with BAPTA, AP-5, or ZIP groups (NS, $P > 0.05$). The mean amplitudes of eEPSCs were determined at 50–60 min after the pre-LTP induction stimulus (adapted from Koga *et al.* [35]).

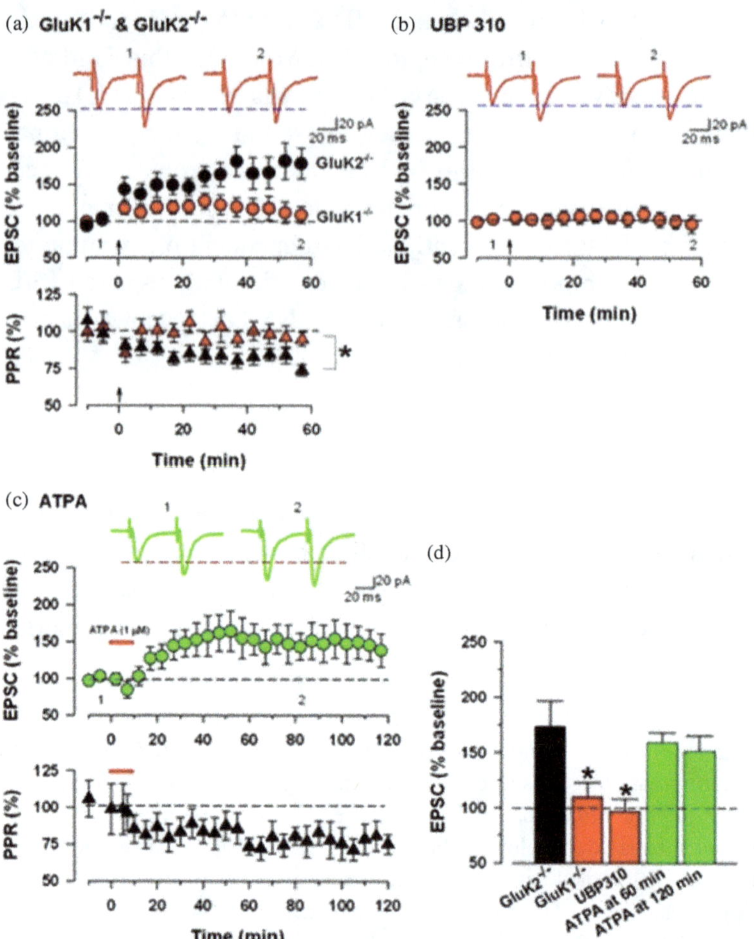

Figure 5. KA receptors mediate the induction of pre-LTP. (a) Upper panel: In GluK1$^{-/-}$ mice, sample traces of eEPSCs with paired-pulse stimulation at 50 ms during baseline (1) and 60 min after the induction stimulus (2) at a holding membrane potential of −60 mV. Middle panel: GluK2$^{-/-}$ mice showed normal pre-LTP (black circle). GluK1$^{-/-}$ mice showed greatly reduced pre-LTP (red circle). Bottom panel: PPR for the GluK2$^{-/-}$ (black) and GluK1$^{-/-}$ (red) groups. (b) A specific GluK1 antagonist, UBP310 (10 μM), completely blocked pre-LTP. (c) Upper panel: A GluK1 agonist, ATPA (1 μM for 10 min), induced long-lasting potentiation, recorded for 2 h. Bottom panel: Averaged data of PPR change before and after ATPA application. (d) Summary of the effects of GluK2$^{-/-}$, GluK1$^{-/-}$, a GluK1 antagonist, or a GluK1 agonist on pre-LTP. The amplitudes of eEPSCs in GluK1$^{-/-}$ or UBP310 groups were significantly decreased compared with control pre-LTP. There was no difference among control pre-LTP, GluK2$^{-/-}$, and ATPA. Modified from Koga *et al.* (2015) (adapted from Koga *et al.* [35]).

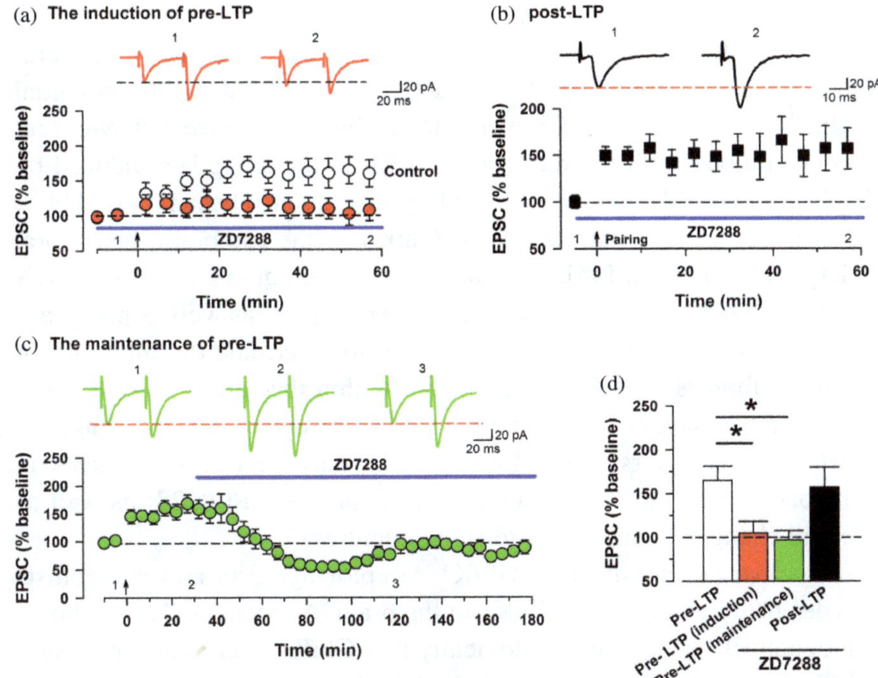

Figure 6.   HCN channels are involved in the maintenance of pre-LTP. (a) An HCN channel inhibitor, ZD7288 (10 μM), blocked the induction of pre-LTP (red circle, $n = 9$ neurons/6 mice). Open circles showed normal pre-LTP as a control ($n = 8$ neurons/6 mice). (b) Post-LTP was insensitive to ZD7288 (10 μM) ($n = 7$ neurons/7 mice). (c) ZD7288 (10 μM) blocked the maintenance of pre-LTP, applied 30 min after the pre-LTP induction ($n = 9$ neurons/9 mice). (d) Summary of pre-LTP and post-LTP experiments investigating HCN channels. The amplitudes of eEPSCs in the two ZD7288 pre-LTP groups were significantly decreased compared with pre-LTP without ZD7288 (one-way ANOVA, $F_{3,30} = 5.09$, *$P < 0.05$) (adapted from Koga *et al.* [35]).

induction of LTD. For the expression of LTD, it is likely mediated by reduced postsynaptic AMPA receptor functions, especially GluA2 subtype receptors in the ACC.

Roles of KA receptors in behavioral anxiety

Genetic studies in humans have suggested that certain types of KA subtype receptors may be linked to behavioral anxiety or depression [42]. Due to the limitations of selective compounds that either inhibit or activate selective subtypes of KA receptors, basic mechanisms of how KA receptors contribute to these mood disorders remain unclear. The generation of gene knockout mice helps us to better understand the potential functions of KA receptors. For example, in GluK2 KO mice, behavior fear is reduced along with the reduction of LTP in the amygdala induced by TBS[30]. In GluK4 KO mice, reduced anxiety has also been reported [43], and in mice with overexpression of GluK4 in the forebrain, behavioral anxiety was enhanced [44]. However, due to the complex effects of KA receptors on both excitatory and inhibitory synapses, as well as presynaptic vs. postsynaptic locations, it is difficult to determine the functions of selective subtypes of KA receptors even within the same central region. For example, gene deletion of the KA GluK1 receptor leads to changes in anxiety-like behaviors [30]. However, this may be due to changes in excitatory vs inhibitory transmission in the amygdala [26,28] as well as potential changes in amygdala-related structures.

Selective development of GluK1 receptor agonists and antagonists provides better evidence of GluK1 in the brain. A recent study has shown both genetically and pharmacologically that GluK1 is critical for presynaptic LTP in the ACC neurons [35]. This form of pre-LTP is absent in GluK1 KO mice as well as after inhibition of GluK1 receptor by UBP310. Furthermore, in mice exposed to a standard EPM or a raised open platform, the pre-LTP was partially reduced or completely blocked. These results provide strong evidence that GluK1-dependent pre-LTP in the ACC may be involved in behavioral anxiety. Erasing pre-LTP by a pharmacological inhibitor of HCN channels produced an inhibitory effect for injury-induced anxiety [35] (Figure 6).

Functional link between KA receptor-dependent LTP and behavioral anxiety

Anxiety is, unfortunately, often long-lasting as long as the environment or factors that contribute to the anxiety persist. It is thus expected that long-term changes in synaptic transmission of anxiety-related neurons or circuits may at least partially contribute to behavioral anxiety-like responses. There are at least two major mechanisms for KA-dependent LTP which may contribute to anxiety. One is to directly enhance neuronal responses to the same input signal, such as visual, auditory, or somatosensory. Consequently, neurons will fire more action potentials. Alternatively, synaptic LTP affects the jitter of neuronal action potentials [45,46]. As a result, neuronal activity will be altered. Such changes are not simply the increase or decrease of the number of spikes. It is also possible that both mechanisms may take place in a certain subset of neurons or neuronal circuits. KA-dependent plastic changes may provide a key synaptic basis for these neuronal circuit functions in the condition of anxiety.

For LTD, similar KA-dependent mechanisms may apply, including the depression of excitatory and inhibitory transmission. Future studies using integrative approaches to explore molecular basis of neuronal activity with anxiety-related neuronal circuits are clearly needed.

Intracellular signaling

CaMKIV and behavioral anxiety: The CaMKIV is a serine-threonine kinase that is activated by elevated intracellular Ca^{2+} during neuronal activity. The downstream proteins for CaMKIV include CREB and CRE modulator. CaMKIV is known to be important for gene expression, synaptic plasticity, and learning and memory. For example, deletion of CaMKIV affects LTP in the lateral amygdala and fear memory. Recently, we have found that CaMKIV knockout mice demonstrated a decrease in anxiety-like behaviors in both the elevated plus maze and light/dark box, suggesting CaMKIV plays a key role in anxiety. Microarray analysis has found downregulation of some anxiety-related genes in CaMKIV knockout mice, such as oxytocin and vasopressin, which may explain the anxiety phenotype for the knockout mice.

AC8: Calcium-stimulated AC8 is one of the AC isoforms important for the production of cAMP and activity-dependent gene regulations. Unlike AC1, AC8 is highly expressed in the thalamus, habenula, and hypothalamus brain regions involved in the neuroendocrine and behavioral responses to stress. Although naïve AC8 knockout mice exhibit normal anxiety, they show less stress-induced anxiety. The actions of AC8 in stress-induced anxiety may be due to its role in stress-induced CREB activation, gene expression, as well as long-term depression.

Activity-dependent immediate early genes: cAMP-responsive element binding protein (CREB) is well-known for its role in activity-dependent gene regulation and is involved in many neuronal processes, including synaptic plasticity and memory. Recently, it was found that CREB is also engaged in emotional behaviors, such as anxiety. CREB knockout mice exhibited an increase in anxiety-like behaviors in several paradigms, including the elevated plus maze, light/dark box, and open field. The mechanism by which CREB is involved in anxiety is not known, but it has been suggested that CREB could regulate the neuropeptide Y system (NPY) expression, and decreased concentrations of NPY are implicated in anxiety-like behaviors.

The zinc finger transcription factor Egr-1 is critical for coupling extracellular signals to changes in cellular gene expression. Using Egr-1 knockout mice, we have found that it is selectively required for synaptic potentiation in the amygdala, late auditory fear memory, and anxiety. Egr-1 knockout mice showed reduced anxiety behaviors as compared with wild-type mice in the elevated plus maze. The exact mechanism by which Egr-1 mediates anxiety is currently unknown, but it is likely some downstream gene expression regulated by Egr-1, such as synapsin I and II, plays a role.

Oxytocin and anxiety

Oxytocin is a well-known neurohypophysial hormone that plays an important role in behavioral anxiety and nociception. Microinjections of oxytocin into the ACC attenuate nociceptive responses and anxiety-like behavioral responses in animals with neuropathic pain [47] (Figure 7). Application of oxytocin selectively blocks the maintenance of pre-LTP but not post-LTP. In addition, oxytocin enhances inhibitory transmission and excites ACC interneurons. Similar results are obtained by using selective optical stimulation of oxytocin-containing projecting terminals in the ACC in animals with neuropathic pain. These results demonstrate that oxytocin acts on central synapses and reduces chronic-pain-induced anxiety by reducing pre-LTP.

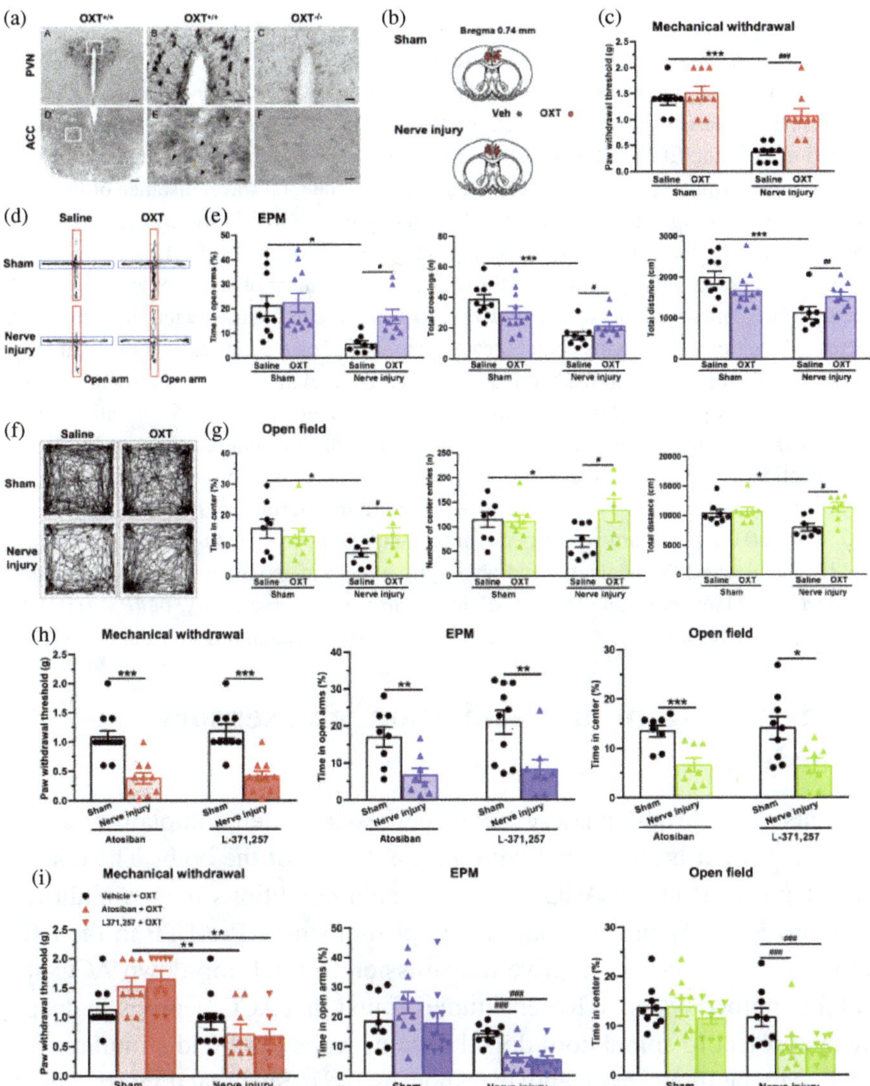

Figure 7. Microinjected oxytocin into the ACC attenuates chronic pain and anxiety-like behaviors. (a) Oxytocin (OXT)-immunoreactive (ir) cells in the PVN and projection fibers in the ACC. (b) and (e) were amplified from white inset on (a) and (d), respectively. The arrow indicates the expression of exhibits axonal oxytocin-ir in the ACC. No OXT-ir was observed in the PVN (c) and ACC (f) sections of OXT KO mice. Scale bars: 100 μm in (a) and (d); 20 μm in (b)–(f). (b) Bilateral microinjection sites for oxytocin in the ACC. (c) Oxytocin microinjected bilaterally into the ACC significantly increased paw

Figure 7. (*Continued*)   withdrawal threshold in nerve injured mice ($n = 9$ mice/group; ****P* < 0.001 for sham versus nerve injury; ###*P* < 0.001 for saline versus OXT). (d) Representative traces show the effects on movement of saline/oxytocin in sham/nerve injured mice in the EPM test. (e) Oxytocin microinjected into the ACC increased the time spent in open arms, the total number of crossings, and the total travel distance of the EPM in nerve injury group but had no effect on the sham group ($n = 8$–11 mice/group; **P* < 0.05 and ****P* < 0.001 for sham versus nerve injury; #*P* < 0.05 and ##*P* < 0.01 for saline versus OXT). (f) Representative traces show the effects on movement of saline/oxytocin in sham/ nerve injured mice in the open field test. (g) Oxytocin microinjected into the ACC increased the time spent in the center area, the number of center entries, and the total travel distance of the open field in the nerve injury group but had no effect on the sham group ($n = 7$–8 mice/group; **P* < 0.05 for sham versus nerve injury; #*P* < 0.05 for saline versus OXT). (h) Microinjected atosiban (peptide antagonist of oxytocin receptor) or L-371,257 (non-peptide antagonist oxytocin receptor) alone had no effect on the pain threshold and anxiety behaviors in the nerve injured mice ($n = 8$–12 mice/group; ***P* < 0.05, ***P* < 0.01, and ****P* < 0.001 for sham versus nerve injury, unpaired two-tailed *t*-test). (i) Atosiban or L-371,257 blocked the analgesic and anxiolytic effects of oxytocin in the nerve injured mice ($n = 8$–11 mice/group; ***P* < 0.01 for sham versus nerve injury, ###*P* < 0.001 for saline versus OXT). All error bars denote standard errors (adapted from Li *et al.* [47]).

Anxiety, descending facilitation, and sensory disorders

Patients with chronic anxiety and/or depression often complain of a pain experience that is associated with different parts of the body. The discovery of pre-LTP in the ACC in chronic pain conditions may explain this heretofore poorly understood clinical phenomenon. Pre-LTP in the ACC could affect spinal nociceptive transmission through top-down ACC spinal facilitatory systems. Recent studies found that ACC neurons in deeper layers project to spinal cord dorsal horn neurons [53], and stimulation of ACC can facilitate nociceptive responses [48]. Such facilitatory effects may be mediated by descending serotonergic projections from the RVM [48,49]. Activation of dorsal neurons thus triggers incoming pain-related somatosensory cortices and produces a pain experience as if it comes from certain regions of the body. Thus, the ACC-spinal network may provide the mechanism underlying anxiety-triggered body pain. By activation of ACC top-down modulation of the dorsal horn, abnormal activity triggered by anxiety or fear can initiate neuronal activity in the spinal cord that interacts with normal ongoing sensory inputs from the periphery. Such enhancement would then be sufficient to activate pain-related areas in the cortex, such as the somatosensory cortex, ACC, and insular cortex, triggering the sensation of pain. Some of these activities may further enhance top-down modulation by positive feedback activity. Thus, this direct cortico-spinal pathway may provide a novel means to link mood changes with the bodily sensations.

Conclusions

Anxiety is becoming a major medical problem in modern society [50,51,52]. Humans are suffering from the overload of information, real or not, which results in the experience of sadness and fear. Their brains are constantly exposed to all types of somatosensory, visual, and auditory insults. Altered emotional anxiety levels subsequently affect human decision and thus quality of life. In the case of injury, anxiety can also enhance the suffering of patients from initial physical injury.

References

[1] Davis, M. *et al.* (1994) Neurotransmission in the rat amygdala related to fear and anxiety. *Trends Neurosci* 17, 208–214. 10.1016/0166-2236 (94)90106-6.

[2] Gross, C. and Hen, R. (2004) The developmental origins of anxiety. *Nat Rev Neurosci* 5, 545–552. 10.1038/nrn1429.

[3] Gordon, J.A. and Hen, R. (2004) Genetic approaches to the study of anxiety. *Annu Rev Neurosci* 27, 193–222. 10.1146/annurev.neuro.27.070203. 144212.

[4] Finn, D.A. *et al.* (2003) Genetic animal models of anxiety. *Neurogenetics* 4, 109–135. 10.1007/s10048-003-0143-2.

[5] Holmes, A. (2001) Targeted gene mutation approaches to the study of anxiety-like behavior in mice. *Neurosci Biobehav Rev* 25, 261–273. 10.1016/s0149-7634(01)00012-4.

[6] Dityatev, A.E. and Bolshakov, V.Y. (2005) Amygdala, long-term potentiation, and fear conditioning. *Neuroscientist* 11, 75–88. 10.1177/1073858404270857.

[7]   Rodrigues, S.M. *et al.* (2004) Molecular mechanisms underlying emotional learning and memory in the lateral amygdala. *Neuron* 44, 75–91. 10.1016/j.neuron.2004.09.014.

[8]   Bushnell, M.C. *et al.* (2013) Cognitive and emotional control of pain and its disruption in chronic pain. *Nat Rev Neurosci* 14, 502–511. 10.1038/nrn3516.

[9]   Grupe, D.W. and Nitschke, J.B. (2013) Uncertainty and anticipation in anxiety: An integrated neurobiological and psychological perspective. *Nat Rev Neurosci* 14, 488–501. 10.1038/nrn3524.

[10]   Vogt, B.A. (2005) Pain and emotion interactions in subregions of the cingulate gyrus. *Nat Rev Neurosci* 6, 533–544. 10.1038/nrn1704.

[11]   Tovote, P. *et al.* (2015) Neuronal circuits for fear and anxiety. *Nat Rev Neurosci* 16, 317–331. 10.1038/nrn3945.

[12]   Tang, J. *et al.* (2005) Pavlovian fear memory induced by activation in the anterior cingulate cortex. *Mol Pain* 1, 6. 10.1186/1744-8069-1-6.

[13]   Kim, S.S. *et al.* (2011) Neurabin in the anterior cingulate cortex regulates anxiety-like behavior in adult mice. *Mol Brain* 4, 6. 10.1186/1756-6606-4-6.

[14]   Zhuo, M. (2016) Neural Mechanisms Underlying Anxiety-Chronic Pain Interactions. *Trends Neurosci* 39, 136–145. 10.1016/j.tins.2016.01.006.

[15]   Osuch, E.A. *et al.* (2000) Regional cerebral metabolism associated with anxiety symptoms in affective disorder patients. *Biol Psychiatry* 48, 1020–1023. 10.1016/s0006-3223(00)00920-3.

[16]   Boehme, S. *et al.* (2015) Neural correlates of emotional interference in social anxiety disorder. *PLoS One* 10, e0128608. 10.1371/journal.pone.0128608.

[17]   Ipser, J.C. *et al.* (2013) Meta-analysis of functional brain imaging in specific phobia. *Psychiatry Clin Neurosci* 67, 311–322. 10.1111/pcn.12055.

[18]   Porter, R.H. *et al.* (1997) Distribution of kainate receptor subunit mRNAs in human hippocampus, neocortex and cerebellum, and bilateral reduction of hippocampal GluR6 and KA2 transcripts in schizophrenia. *Brain Res* 751, 217–231. 10.1016/s0006-8993(96)01404-7.

[19]   Hadzic, M. *et al.* (2017) Ionotropic glutamate receptors: Which ones, when, and where in the mammalian neocortex. *J Comp Neurol* 525, 976–1033. 10.1002/cne.24103.

[20]   Ball, S.M. *et al.* (2010) Assembly and intracellular distribution of kainate receptors is determined by RNA editing and subunit composition. *J Neurochem* 114, 1805–1818. 10.1111/j.1471-4159.2010.06895.x.

[21] Li, P. *et al.* (1999) Kainate-receptor-mediated sensory synaptic transmission in mammalian spinal cord. *Nature* 397, 161–164. 10.1038/16469.

[22] Wu, L.J. *et al.* (2005) Kainate receptor-mediated synaptic transmission in the adult anterior cingulate cortex. *J Neurophysiol* 94, 1805–1813. 10.1152/jn.00091.2005.

[23] Koga, K. *et al.* (2012) Kainate receptor-mediated synaptic transmissions in the adult rodent insular cortex. *J Neurophysiol* 108, 1988–1998. 10.1152/jn.00453.2012.

[24] Li, H. and Rogawski, M.A. (1998) GluR5 kainate receptor mediated synaptic transmission in rat basolateral amygdala in vitro. *Neuropharmacology* 37, 1279–1286. 10.1016/s0028-3908(98)00109-9.

[25] Braga, M.F. *et al.* (2004) The physiological role of kainate receptors in the amygdala. *Mol Neurobiol* 30, 127–141. 10.1385/MN:30:2:127.

[26] Wu, L.J. *et al.* (2007) Increased anxiety-like behavior and enhanced synaptic efficacy in the amygdala of GluR5 knockout mice. *PLoS One* 2, e167. 10.1371/journal.pone.0000167.

[27] Negrete-Diaz, J.V. *et al.* (2012) Kainate receptor-mediated depression of glutamatergic transmission involving protein kinase A in the lateral amygdala. *J Neurochem* 121, 36–43. 10.1111/j.1471-4159.2012.07665.x.

[28] Aroniadou-Anderjaska, V. *et al.* (2012) Presynaptic facilitation of glutamate release in the basolateral amygdala: A mechanism for the anxiogenic and seizurogenic function of GluK1 receptors. *Neuroscience* 221, 157–169. 10.1016/j.neuroscience.2012.07.006.

[29] Wu, L.J. *et al.* (2007) Genetic and pharmacological studies of GluR5 modulation of inhibitory synaptic transmission in the anterior cingulate cortex of adult mice. *Dev Neurobiol* 67, 146–157. 10.1002/dneu.20331.

[30] Ko, S. *et al.* (2005) Altered behavioral responses to noxious stimuli and fear in glutamate receptor 5 (GluR5)- or GluR6-deficient mice. *J Neurosci* 25, 977–984. 10.1523/JNEUROSCI.4059-04.2005.

[31] Bliss, T.V. and Collingridge, G.L. (1993) A synaptic model of memory: Long-term potentiation in the hippocampus. *Nature* 361, 31–39. 10.1038/361031a0.

[32] Bliss, T.V. and Cooke, S.F. (2011) Long-term potentiation and long-term depression: A clinical perspective. *Clinics (Sao Paulo)* 66(Suppl 1), 3–17. 10.1590/s1807-59322011001300002.

[33] Bliss, T.V. and Collingridge, G.L. (2013) Expression of NMDA receptor-dependent LTP in the hippocampus: Bridging the divide. *Mol Brain* 6, 5. 10.1186/1756-6606-6-5.

[34] Bliss, T.V. *et al.* (2016) Synaptic plasticity in the anterior cingulate cortex in acute and chronic pain. *Nat Rev Neurosci* 17, 485–496. 10.1038/nrn.2016.68.

[35] Koga, K. *et al.* (2015) Coexistence of two forms of LTP in ACC provides a synaptic mechanism for the interactions between anxiety and chronic pain. *Neuron* 86, 1109. 10.1016/j.neuron.2015.05.016.

[36] Koga, K. *et al.* (2015) Impaired presynaptic long-term potentiation in the anterior cingulate cortex of Fmr1 knock-out mice. *J Neurosci* 35, 2033–2043. 10.1523/JNEUROSCI.2644-14.2015.

[37] Yamanaka, M. *et al.* (2016) Pre-LTP requires extracellular signal-regulated kinase in the ACC. *Mol Pain* 12. 10.1177/1744806916647373.

[38] Li, H. *et al.* (2001) Kainate receptor-mediated heterosynaptic facilitation in the amygdala. *Nat Neurosci* 4, 612–620. 10.1038/88432.

[39] Shin, R.M. *et al.* (2010) Hierarchical order of coexisting pre- and postsynaptic forms of long-term potentiation at synapses in amygdala. *Proc Natl Acad Sci USA* 107, 19073–19078. 10.1073/pnas.1009803107.

[40] Kang, S.J. *et al.* (2012) Plasticity of metabotropic glutamate receptor-dependent long-term depression in the anterior cingulate cortex after amputation. *J Neurosci* 32, 11318–11329. 10.1523/JNEUROSCI.0146-12.2012.

[41] Liu, M.G. *et al.* (2013) Long-term depression of synaptic transmission in the adult mouse insular cortex in vitro. *Eur J Neurosci* 38, 3128–3145. 10.1111/ejn.12330.

[42] Lerma, J. and Marques, J.M. (2013) Kainate receptors in health and disease. *Neuron* 80, 292–311. 10.1016/j.neuron.2013.09.045.

[43] Catches, J.S. *et al.* (2012) Genetic ablation of the GluK4 kainate receptor subunit causes anxiolytic and antidepressant-like behavior in mice. *Behav Brain Res* 228, 406–414. 10.1016/j.bbr.2011.12.026.

[44] Aller, M.I. *et al.* (2015) Increased dosage of high-affinity kainate receptor gene grik4 alters synaptic transmission and reproduces autism spectrum disorders features. *J Neurosci* 35, 13619–13628. 10.1523/JNEUROSCI.2217-15.2015.

[45] Li, X.Y. *et al.* (2014) Long-term temporal imprecision of information coding in the anterior cingulate cortex of mice with peripheral inflammation or nerve injury. *J Neurosci* 34, 10675–10687. 10.1523/JNEUROSCI.5166-13.2014.

[46] Tiesinga, P. *et al.* (2008) Regulation of spike timing in visual cortical circuits. *Nat Rev Neurosci* 9, 97–107. 10.1038/nrn2315.

[47] Li, X.H. *et al.* (2021) Oxytocin in the anterior cingulate cortex attenuates neuropathic pain and emotional anxiety by inhibiting presynaptic long-term potentiation. *Cell Rep* 36, 109411. 10.1016/j.celrep.2021.109411.

[48] Calejesan, A.A. *et al.* (2000) Descending facilitatory modulation of a behavioral nociceptive response by stimulation in the adult rat anterior cingulate cortex. *Eur J Pain* 4, 83–96. 10.1053/eujp.1999.0158.

[49] Li, P. *et al.* (1999) AMPA receptor-PDZ interactions in facilitation of spinal sensory synapses. *Nat Neurosci* 2, 972–977. 10.1038/14771.

[50] Tovote, P. *et al.* (2015) Neuronal circuit for fear and anxiety. *Nat Rev Neurosci* 11(1), 75–88.

[51] Janek, P.H. and Tye K.M. (2015) From circuits to behavious in the amygdala. *Nature* 517, 284–92.

[52] LeDoux, J.E. and Pine, D.S. (2016) Using neuroscience to help understand fear and anxiety: A two-system framework. *Am J Psychiatry* 173(11), 1083–1093.

[53] Chen, T. *et al.* (2014) Adenylyl cyclase subtype 1 is essential for late-phase long term potentiation and spatial propagation of synaptic responses in the anterior cingulate cortex of adult mice. *Molecular Pain* 10, 65.

Chapter 11

Descending Pain Facilitation

News and Views

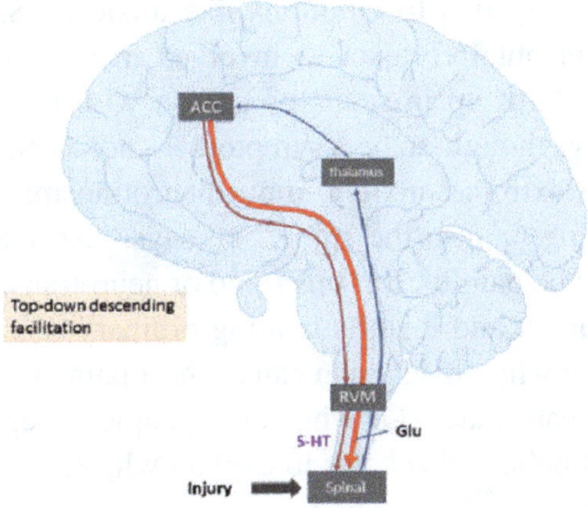

New research from a University of Toronto scientist could change the way we understand pain treatment. Dr. Min Zhuo discovered that the brain's frontal lobe is involved in pain transmission to the spine. If his findings in animals prove to be applicable to humans, this could potentially lead to a new class of non-addictive painkillers being developed.

For 20 years, Min Zhuo, a professor of physiology in the Faculty of Medicine at the University of Toronto, has been fascinated by 'invisible pain', in particular chronic invisible pain with no obvious cause. He has long suspected that the standard way of viewing spinal pain (resulting from injury or tissue inflammation), and the usual understanding of how to treat it (block it from entering the spinal cord), wasn't telling the whole story. Now, Zhuo has shown, in both mice and rats, that some spinal pain actually begins in the brain's frontal lobe — an area previously believed to be not involved. Zhuo has also demonstrated how treating pain in this area could be effective in preventing chronic pain.

Zhuo published his results on May 16, 2018, in the journal *Nature Communication*. "When doctors can't see anything wrong to cause chronic pain, often they think patients are making it up" says Zhuo, "but pain that originates in the frontal lobe would be very different from pain that comes from a physical injury, like a herniated disc. There wouldn't necessarily be any injury to 'see'. That's because our personality and emotions live in this region. If the frontal lobe can produce physical pain, that pain would be deeply tied to emotions like anxiety." Scientists already knew that the prefrontal cortex was involved in pain in some capacity because it would light up in scans of people with pain. However, that activity was always thought to be a symptom — not a cause — says Zhuo. "When you have extreme anxiety, more neurotransmitters are released that end up causing pain in the spine," he says. "Normal functions like walking shouldn't be painful. But this flood of neurotransmitters sends the spine into hyperdrive, and it starts treating ordinary sensations like pain. That could explain why anxiety can cause chest pain and make you think you're having a heart attack. Or why some people experience pain when you touch them. I believe this helps to explain why emotional pain causes physical pain."

The good news is that pain from the frontal lobe seems to be transmitted in a simple, more direct way to the spine making it relatively easy to shut down. Neurons in the frontal cortex send signals all the way down the spinal cord, says Zhuo, whereas pain signals from other areas of the brain are mediated by a complex network.

In animals, Zhuo found that pain was associated with an increase in neurotransmitters released from the frontal cortex. He was able to lessen pain by reducing the amount released. His next step is to test this process in people. For those who suffer from anxiety, along with neuropathic pain, a painkiller targeting the frontal lobe would be very beneficial, says Zhuo.

Summary

It is well known that pain transmission is under modulation of the descending inhibitory system (or called the endogenous analgesia system). Such inhibitory systems can be tonically active in physiological conditions. The discovery of descending facilitation in pain transmissions makes such modulation biphasic. Spinal pathways and spinal transmitters are different in descending facilitation vs inhibition. While descending inhibition can be triggered by various supraspinal structures, descending facilitation is mainly triggered by RVM as well as ACC. Furthermore, cumulative evidence shows that descending facilitation may become enhanced in chronic pain conditions.

Keywords: Pain; descending facilitation; ACC; spinal cord; RVM; descending inhibition; 5-HT

Introduction

Brain activity is able to affect sensory transmission through descending modulatory systems. For many years, it was believed that endogenous modulatory systems are mainly inhibitory or 'analgesic'. Cumulative studies, however, reveal that descending modulation of spinal sensory transmission is actually biphasic, including both inhibitory and facilitatory influences. Descending influences from the supraspinal central nuclei directly, or indirectly, modulates spinal sensory transmission and includes the ACC, amygdala, PAG, and RVM, which may function as the last relay between brain centers and the spinal cord [1–4]. Biphasic modulation of spinal nociceptive transmission from the RVM offers very fine regulation of spinal sensory thresholds and responses, perhaps reflecting the different types of neurons identified in this area. Integrative approaches have been used to investigate the mechanisms for descending facilitation, including electrophysiological, pharmacological, behavioral, and biochemical studies. In this review, we will summarize the data using whole animal preparations, *in vitro* spinal and brain slices, and genetically manipulated mice, to support the hypothesis that the positive feedback mechanism within the synapses, or between the different brain regions, is a key mechanism for persistent pain caused by injury.

Biphasic modulation of spinal pain transmission

The investigation of descending facilitatory systems has been carried out systematically in the brainstem RVM. At the whole animal level, electrophysiological, pharmacological, and behavioral experiments have been performed to help characterize the facilitation of responses of spinal sensory neurons to peripheral noxious stimuli, as well as the behavioral responses to noxious stimuli. Facilitation affects spinal nociceptive transmission from somatocutaneous areas as well as from visceral organs. Furthermore, facilitation is a common form of modulation of sensory transmission, affecting both noxious and non-noxious inputs. These unique features highlight the possibility that descending facilitation may serve as a key central mechanism that contributes to injury-related central pain or allodynia.

Biphasic modulation is intensity-dependent

A key feature of descending facilitation is that it is intensity-dependent. Whether facilitation or inhibition is observed depends, in part, on the intensity of the stimulation applied (Figures 1 and 2) [4,5]. According to effects on spinal sensory neuronal responses, we characterize sites within the brainstem into three different groups: biphasic, inhibitory, and facilitatory sites. At biphasic sites of stimulation, it is typical that electrical stimulation facilitates spinal nociceptive transmission at lesser intensities (5–25 μA) and inhibits responses of the same neurons at greater intensities (50–100 μA). At inhibitory sites, electrical stimulation only reduces and inhibits responses of spinal sensory neurons. At facilitatory sites, we found that electrical stimulation only caused increases in responses of spinal sensory neurons. To determine if facilitatory or inhibitory effects were simply due to different groups of spinal dorsal horn neurons recorded, we also investigated the effects of electrical stimulation at one intensity but at different sites in the RVM on the same spinal neuron. We found that the responses of the same spinal sensory neurons can be either inhibited or facilitated by electrical stimulation applied to different sites in the brainstem. Thus, spinal units receive both facilitatory and inhibitory influences descending from the brainstem. There is no clear anatomical separation between these different effects produced by stimulation in the

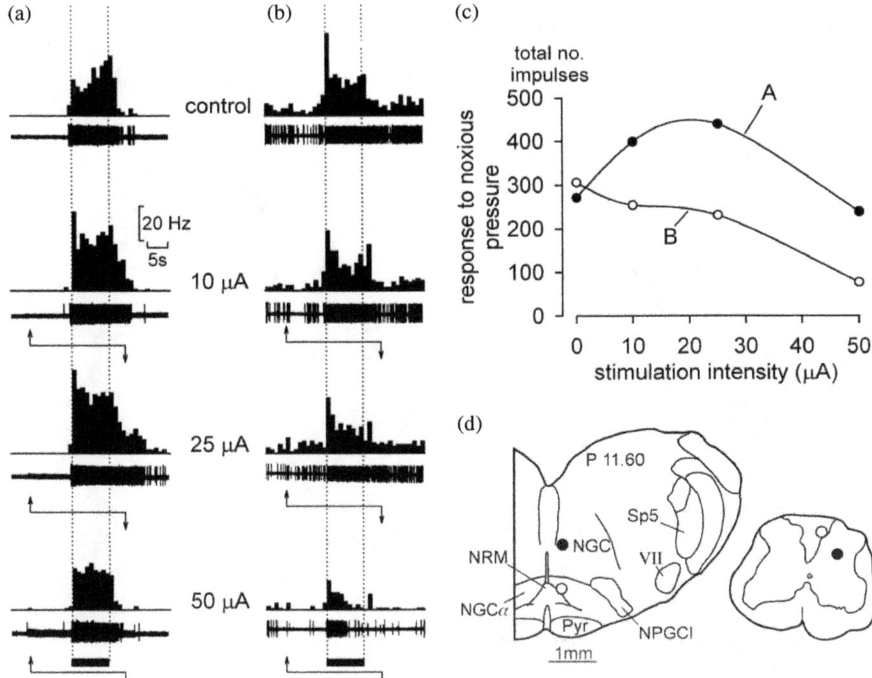

Figure 1. Examples (two different spinal neurons) of facilitation (a) and inhibition (b) of spinal nociceptive mechanical transmission produced by stimulation in the rostral medial medulla (RMM). (a, b) Peristimulus time histograms (1-s bin width) and corresponding oscillographic records illustrating a control response to noxious pressure (28.8 g) of the skin of the hind foot and the effect on responses of the same units during stimulation in the RMM (intensities given). (c) Graphic representation of the data in a and b; the point above 0 represents the response (total number of impulses in 10 s) in the absence of RMM stimulation. (d) Stimulation sites illustrated on a representative coronal brain section (Paxinos and Watson 1986) and spinal recording sites corresponding to the examples in a and b. Pyr, pyramidal tract; NGC, n. reticularis gigantocellularis; NGC, NGC pars; NPGCI, n. reticularis paragigantocellularis lateralis; NRM, n. raphe magnus; VII, facial nucleus; Sp5, spinal trigeminal tract (adapted from Zhuo and Gebhart [6]).

brainstem. Biphasic effects are often produced at sites of stimulation adjacent to those from which only inhibition is produced by similar intensities of stimulation. Further to this, inhibitory effects are produced at biphasic sites of stimulation adjacent to other biphasic sites from which facilitatory effects are produced. It is unlikely that these effects are simply due to activation of fibers passing through the RVM because microinjection of

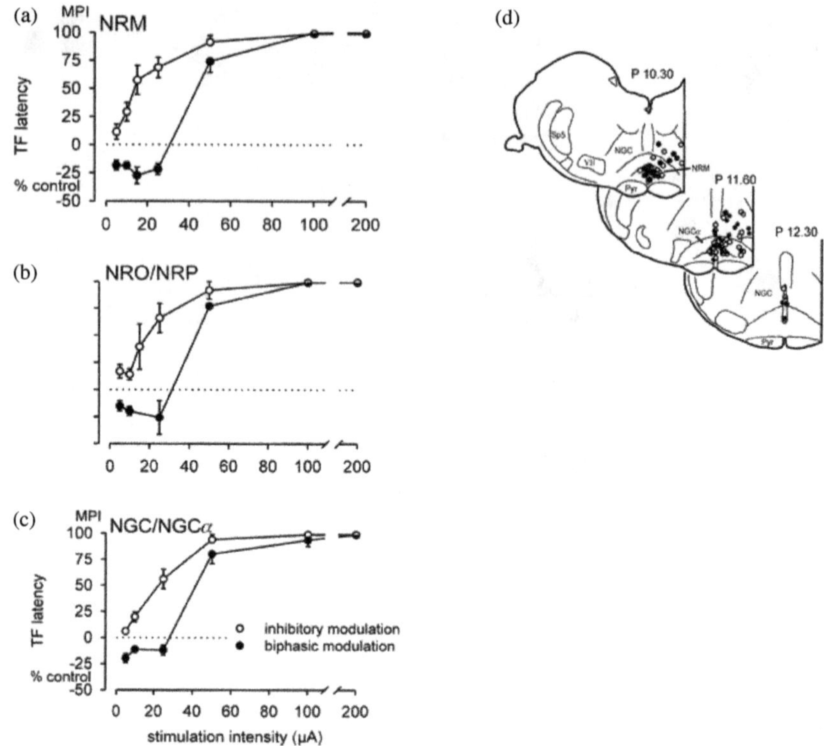

Figure 2. Summary of descending modulation of tail-flick (TF) reflex from the RVM. (a) TF latency presented as maximum possible inhibition (MPI; inhibitory effect) or % of control (facilitatory effect) against intensity of electrical stimulation in nucleus raphe magnus (NRM) for inhibitory modulation (○) and biphasic modulation (●). (b, c) Data from nucleus raphe obscurus (NRO)/nucleus raphe pallidus (NRP) and nucleus reticularis gigantocellularis (NGC)/nucleus reticularis gigantocellularis pars alpha (NGCα) presented as in A. (d) Sites of stimulation illustrated on representative coronal brain sections (Paxinos and Watson 1986[+]). Pyr, pyramidal tract; Sp5, spinal trigeminal tract; VII, facial nucleus (adapted from Zhuo and Gebhart [4]).

glutamate or selective receptor agonists into the RVM also produces similar biphasic effects.

Brainstem-spinal cord descending facilitation

For studying the effects of facilitation on stimulation-response functions (SRFs), responses of spinal neurons to graded, noxious cutaneous, or visceral stimuli were studied. In most cases, in particular within the range of noxious intensities of cutaneous or visceral stimulation, descending inhibition significantly reduced the slopes of SRFs. In contrast, descending facilitation enabled responses and caused a parallel shift of the SRF to the left without changing its slope. These different outcomes on the encoding properties of spinal neurons suggest that descending facilitation is likely to be mechanistically different from descending inhibition. The latency to stimulation-produced facilitation and inhibition is determined by employing a cumulative sum technique and bin-by-bin analysis of unit responses and further supports the theory that the mechanisms and pathways leading to inhibition and facilitation are different. The mean latency of stimulation-produced inhibition from the RVM is about 90 ms whereas the apparent mean latency to facilitation by electrical stimulation is greater than 200 ms. This suggests that descending facilitatory influences likely involve sites rostral to the RVM (e.g., ACC, see the following).

Facilitation of nociceptive visceral pain

Spinal visceral pain transmission is also under descending facilitatory modulation, in addition to the well-known descending inhibitory modulation [2,6,7]. Similar to biphasic modulation of spinal dorsal horn neurons' responses to cutaneous stimuli, spinal dorsal horn responses to colorectal distension (CRD) are also biphasic modulated (Figures 3 and 4). Such descending facilitatory effects can be induced either by electrical stimulation or L-glutamate microinjection into the RVM.

Facilitation of non-nociceptive transmission and possible implications

WDR neurons respond to both noxious and non-noxious stimuli and are implicated in the facilitation of ono-noxious responses in chronic pain. Consistent with these findings, it has been reported that activation of descending facilitation also increased the responses of spinal dorsal horn neurons to peripheral non-noxious stimuli, such as non-noxious mechanism brush of the hind paw skin [6]. Such facilitation of dorsal horn neurons' responses to non-noxious stimuli may contribute to pathological

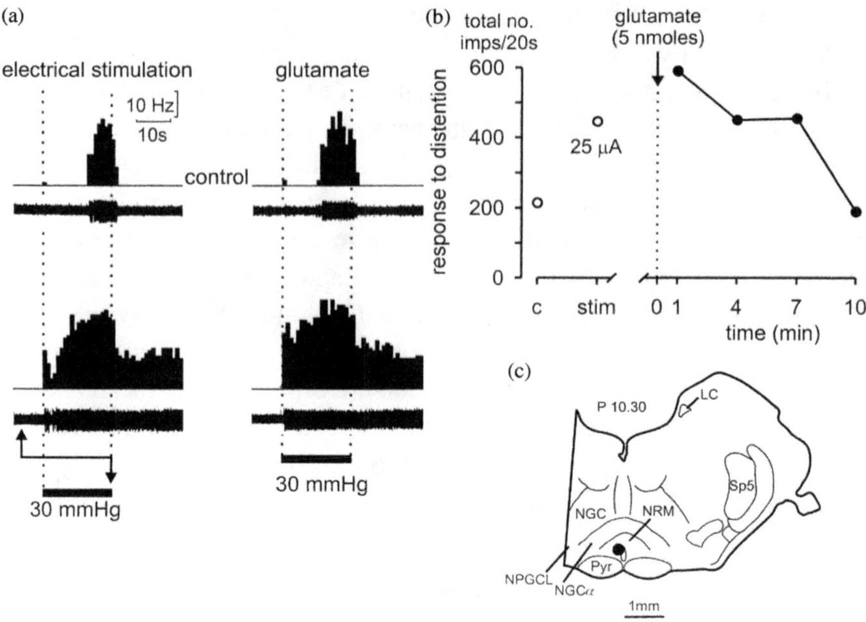

Figure 3.   Example of facilitation of spinal visceral transmission produced by electrical stimulation and glutamate in the NRM. (a) Peristimulus time histograms (1-s bin width) and corresponding oscillographic records in the absence (top histograms) and presence (bottom histograms) of electrical stimulation (25 µA) and glutamate (5 nmoles) given in the same site in NRM. The intensity and duration of colorectal distension are illustrated in the following; the period of electrical stimulation (25 s) is indicated by the arrows. (b) Summary of the data illustrated in A and time course of effect of glutamate given in NRM. The point above C represents the response to 30-mmHg colorectal distension; the point above stim represents the response to the same intensity of distension during stimulation in NRM. (c) Site of stimulation and injection of glutamate (adapted from Zhuo and Gebhart [33]).

conditions, such as allodynia. Enhanced responses may lead to the activation of cortical areas that receive ascending sensory inputs and thus the brain may interpret these as noxious stimuli or pain even though the intensity is non-noxious at periphery.

Masked by tonic descending inhibition

Facilitatory influences are observed only at lesser intensities of stimulation (or lesser concentrations of glutamate). In early studies, the presence

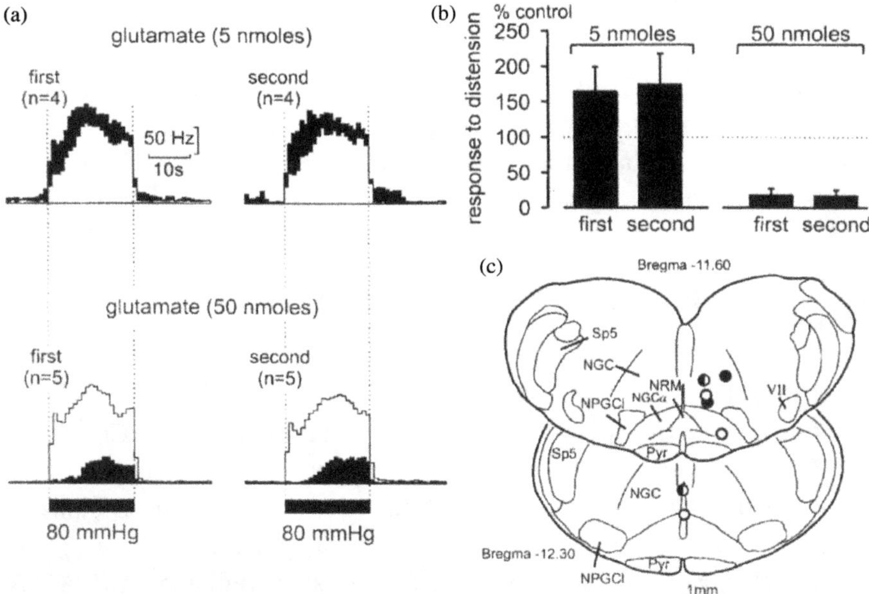

Figure 4. Summary of reproducibility of glutamate-produced facilitation and inhibition. (a) Mean peristimulus time histograms (PSTHs; 1-second bin width) representing the mean visceromotor responses before glutamate administration (unfilled PSTHs) and at 1 minute after glutamate administration (filled PSTHs). The period of distention (20 seconds) is indicated below by the horizontal bar. On the left, mean responses before and after the first glutamate administration at doses of 5 or 50 nmol are illustrated on top and bottom, respectively. Responses before and after the second glutamate administration at the same site are illustrated on the right top and bottom, respectively. (b) Graphic illustrations of the data in A expressed as a percentage of control responses to distention. (c) Summary of sites where glutamate at a low dose (5 nmol;○) or greater dose (50 nmol;●) was administered. At 2 sites (◑), both low and high doses of glutamate were tested. (adapted from Zhuo and Gebhart, [33].

of descending facilitation from the brainstem may have been missed because brain stimulation intensity-dependent functions were not performed. Descending inhibitory and facilitatory influences are likely to be simultaneously activated, and prepotent inhibitory effects masked the facilitatory effects. This idea has been confirmed in experiments investigating spinal pathways for descending modulation. Bilateral transections of the dorsolateral funiculi (DLFs) in the thoracic spinal cord not only abolished descending inhibition produced by electrical or chemical stimulation in the RVM but also unmasked descending facilitatory influences on spinal sensory neurons at the same high intensities of stimulation that only produced inhibition before the DLF transactions. These findings suggest that descending inhibitory and facilitatory influences can be simultaneously engaged by activation of sites in the RVM and that removal of the route conveying the inhibitory influences uncovers descending facilitatory effects on spinal sensory neurons.

Spinal mechanism for descending facilitation

It is important to show that these modulatory effects are due to changes in spinal sensory synaptic transmission and not due to modulation of pre-motor spinal interneurons or spinal local inhibitory synapses. Consistent with the biphasic modulatory effects of 5-HT on spinal nociceptive transmission and behavioral reflexes, we found that 5-HT produced biphasic modulation of excitatory synaptic responses in spinal cord slices [8–12] (see Chapter 4). 5-HT at high doses produces inhibition of AMPA/KA receptor-mediated excitatory postsynaptic currents (EPSCs), while a low dose of 5-HT or a selective 5-HT_2 receptor agonist induces facilitation of fast EPSCs in the lumbar spinal cord. 5-HT at low doses could facilitate fast EPSCs in the presence of an NMDA receptor antagonist, AP-5 (50 μM), indicating that the facilitatory effect is NMDA receptor-independent. Application of methysergide after administration of a serotonergic receptor agonist, DOI, failed to reverse the facilitatory effect of 5-HT. These results indicate that 5-HT triggers long-term plastic changes in spinal dorsal horn synapses and continuous activation of 5-HT receptors are not required for the expression of the facilitation.

5-HT may affect spinal sensory transmission by acting on presynaptic or postsynaptic receptors. Postsynaptic application of G protein inhibitors, introduced through the recording pipette, abolished the ability of 5-HT to facilitate synaptic transmission, suggesting that postsynaptic 5-HT receptors are critical for the effect. In support of this notion, we found postsynaptic Ca^{2+}-dependent processes to be required for 5-HT-induced facilitation. In experiments with chelating postsynaptic Ca^{2+} with BAPTA in the pipette solution, the facilitatory effect of 5-HT was abolished, indicating that an increase in postsynaptic Ca^{2+} is required. Additional evidence against a mechanism of 5-HT-induced synaptic facilitation involving modulation of presynaptic glutamate release comes from the observation that while 5-HT application clearly caused AMPA receptor-mediated EPSCs, NMDA receptor-mediated EPSCs were significantly decreased by 5-HT in the same neurons. This result suggests that postsynaptic enhancement of AMPA receptor-mediated currents by 5-HT is selective.

Spinal mechanism for descending inhibition

Electrophysiological studies using intracellular or whole-cell patch-clamp recordings of dorsal horn neurons allow for the investigation of the cellular mechanisms underlying antinociceptive, or analgesic, effects induced by these transmitters. In anesthetized whole animals, electrical stimulation applied to sites within the nucleus raphe magnus or PAG produced inhibitory postsynaptic potentials (IPSPs) in dorsal horn neurons including ascending projection spinothalamic tract cells. More detailed pharmacological analyses came from studies using an *in vitro* brain/spinal cord slice preparation. In trigeminal nuclei, all three major transmitters, acetylcholine, serotonin, and norepinephrine, are reported to inhibit glutamatergic transmission. In the lumbar spinal cord, less is known about the synaptic mechanisms underlying sensory inhibition by carbachol, clonidine, and serotonin [13,14].

Facilitation from the cortex

As mentioned above, most investigation of descending facilitation is focused on subcortical structures, such as the RVM. The possible central control of RVM-spinal facilitation has been less investigated. One possible structure is the PAG. The PAG-RVM is known to play a key analgesic effect in descending inhibition of pain. There are a few studies that report that PAG may exert descending facilitatory effects on spinal transmission. In addition, it has been known that cortical neurons could project to brainstem neurons and lead to the excitation of descending facilitation [15]. Activation of the ACC at high intensities (up to 500 μA) of electrical stimulation did not produce any antinociceptive effect. Instead, at most sites within the ACC, electrical stimulation produced significant facilitation of the TF reflex (i.e., decreases in TF latency) (Figure 5). Chemical

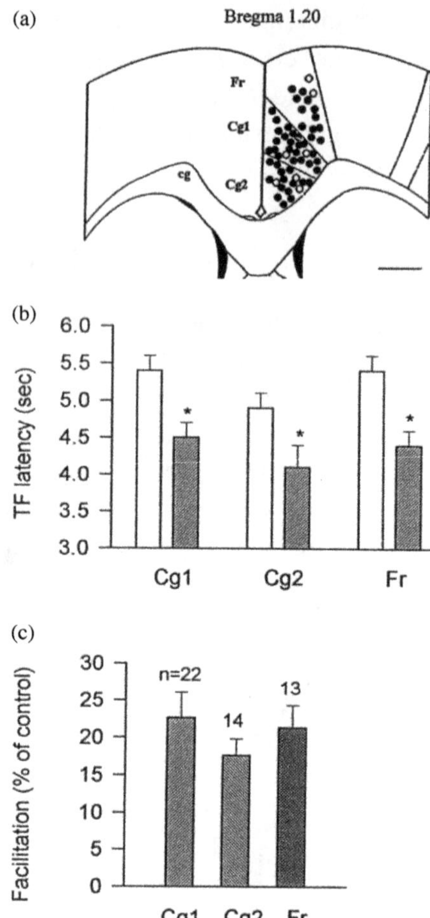

Figure 5. Facilitation of the TF reflex produced by electrical stimulation in the Cg1, Cg2, and Fr. (a) Brain sites in the ACC for electrical stimulation, illustrated on representative coronal brain sections. Filled circles show sites where electrical stimulation produced facilitation of the TF reflex. Open circles show sites where electrical stimulation at intensities tested (10–500 μA) produced no facilitation of the TF reflex. Cg1, cingulate cortex, area 1; Cg2, cingulate cortex, area 2; cg, cingulum; Fr, frontal cortex. Scale bar: 1 mm. (b) Electrical stimulation delivered to sites within the Cg1, Cg2, or Fr produced significant decreases in TF response latency. Open bars show TF response latencies measured without stimulation. Filled bars show TF latencies measured with stimulation. *$P<0.05$. (c) Percentage facilitation induced by electrical stimulation in three areas is similar (adapted from Calejesan *et al.* [15]).

activation of metabotropic glutamate receptors (mGluRs) within the ACC also produced a facilitatory effect. Interestingly, this descending facilitation from the ACC apparently relays on the RVM. In addition, the dorsal reticular nucleus (DRt) has also been proposed as another possible brainstem relay for descending facilitation from ACC [16]. However, descending facilitation from the DRt on spinal sensory transmission has not yet been characterized [16].

In addition to the ACC-RVM-spinal pathway, recent studies suggest that there may be ACC-spinal direct projection in parallel with descending facilitation modulation. Anatomic studies report that some prefrontal cortical areas, including part of the dorsal ACC, send descending projections to the spinal cord in rats, monkeys, and mice [17–19]. Electrical stimulation of ACC facilitated excitatory glutamate synaptic transmission in the spinal cord dorsal horn (Figure 6). This link provides a possible pathway for ACC neurons to directly regulate the spinal cord neurons. In this study, *in vivo* electrophysiological experiments find that activation of ACC enhances spinal sensory transmission, and this facilitation is independent of RVM activity.

Optogenetic activation of descending modulation *in vivo*

In support of descending facilitation from the ACC in freely moving animals, Kang *et al.* [20] used optogenetic methods to selectively activate ACC pyramidal vs inhibitory neurons in mice (Figure 7). They found that selective activation of pyramidal neurons rapidly and acutely reduced nociceptive thresholds and that this effect was occluded in animals made hypersensitive using Freund's Complete Adjuvant (CFA). Conversely, inhibition of ACC pyramidal neurons rapidly and acutely reduced hypersensitivity induced by CFA treatment. These results provide direct evidence of the pivotal role of ACC excitatory neurons, and their regulation by PV-expressing interneurons, in nociception [20]. Similarly, it has been reported that ACC pyramidal cells in the deep layers of the ACC send direct descending projecting terminals to the dorsal horn of the spinal cord

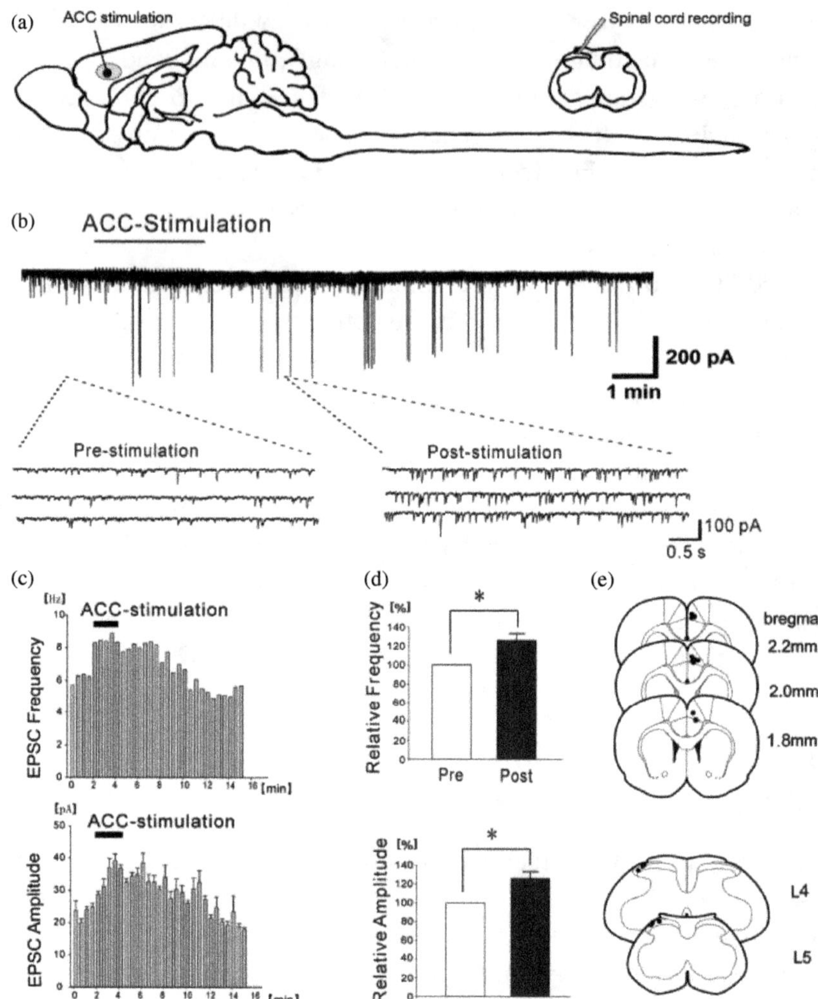

Figure 6.   ACC stimulation induced facilitation of the sEPSCs on the spinal cord SDH neurons. (a) A schematic diagram showing the experimental design for ACC electronic stimulation and *in vivo* spinal recording. (b, c) One sample (b) and histogram figure (c) showing ACC stimulation-induced facilitation of the frequency and amplitude of the sEPSC. (d) Summarized results from 12 SDH neurons. (E) Mapping of the stimulation sites in the ACC and locations of recorded SDH neurons in the lumbar spinal cord. *$P < 0.05$ (adapted from Chen *et al.* [21]).

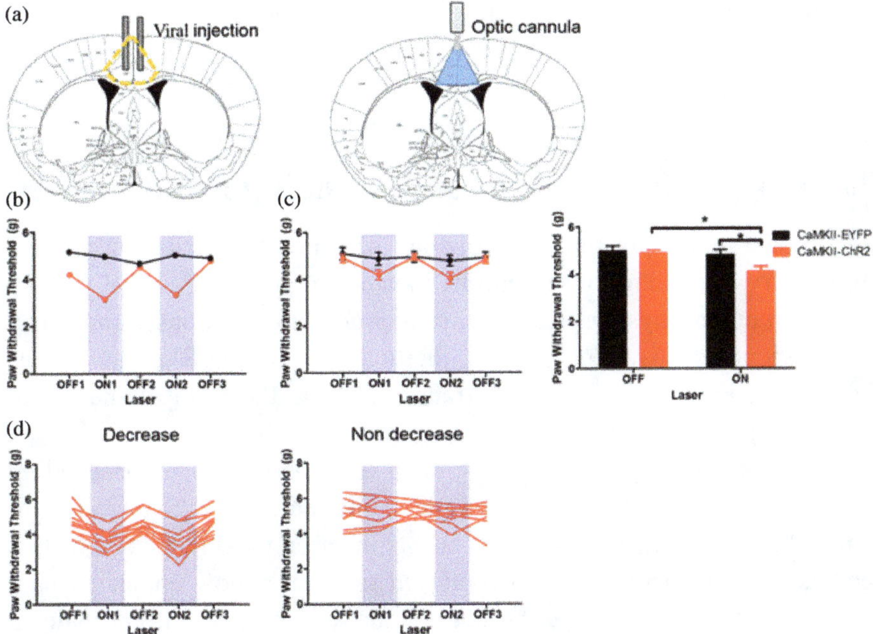

Figure 7. Descending facilitation of behavioral withdrawal by optogenetic activation of ACC pyramidal cells. (a) Schematic diagram of viral injection site (left) and optic cannula placement (right) in the ACC. (b) One example of the effects of blue light in a CaMKII-ChR2 expressing mouse (red) and an EYFP expressing mouse (black) in the von Frey test. (c) Pooled data for ChR2 and EYFP mice. On average, there was a reduction in the mechanical threshold in the ChR2 group. The graph on the right plots the combined data for the two ON and three OFF episodes. There was a significant effect of laser light in the ChR2 group but not in the EYFP group. (d) Not all animals showed a decrease in the mechanical threshold during light activation. Out of 18 animals, 10 showed a decrease while 8 did not. (e) The overlapped pattern of ChR2 expression in each group. Modified from Kang *et al.* (2016) (adapted from Chen *et al.* [21]).

(laminae I–III). After peripheral nerve injury, these projection cells are activated, and postsynaptic excitatory responses of these descending projecting neurons were significantly enhanced. Newly recruited AMPARs contribute to the potentiated synaptic transmission of cingulate neurons. PKA-dependent phosphorylation of GluA1 is important since enhanced synaptic transmission was abolished in GluA1 phosphorylation site serine-845 mutant mice. Our findings provide strong evidence that peripheral nerve injury induces long-term enhancement of cortical-spinal projecting cells in the ACC [14,19,21].

Pathological implications of descending facilitation

It is known that supraspinal neurons are activated by noxious stimuli or after injury, including those neurons located in the RVM and ACC [22–24]. Even under physiological conditions, such brief noxious stimuli may be sufficient to activate descending facilitation from the RVM as well as the ACC. Facilitated nociceptive transmission at the level of the spinal cord may assist animals to escape dangerous conditions.

Under pathological conditions, long-term activation of descending facilitation may contribute to chronic pain, including behavioral allodynia, hyperalgesia, and spontaneous pain after the injury. A positive feedback mechanism has been proposed to explain how enhanced pain transmission may directly contribute to the suffering of chronic pain in patients. Using various animal models, descending facilitation has been implicated in cancer pain conditions, chronic muscle pain, neuropathic pain, opioid-induced pain, headache, as well as inflammation-related pain [25–31]. Especially for the ACC-spinal top-down facilitation, it has been proposed that it may also contribute to emotional disorders caused by pain [19,32] (Figure 8).

Conclusions

A better understanding of the molecular and cellular mechanisms for central changes in various pain-related states gives us hope for improved treatments for chronic pain. At the synaptic level, it is important to understand molecular and cellular mechanisms for long-term plastic changes in the ACC, RVM, and spinal dorsal horn after peripheral tissue insult. At the network level, descending facilitation, as well as descending inhibition, provides a key mechanism to link pain-related neurons located in different regions of the brain. For facilitation, it will convey the cortical excitation

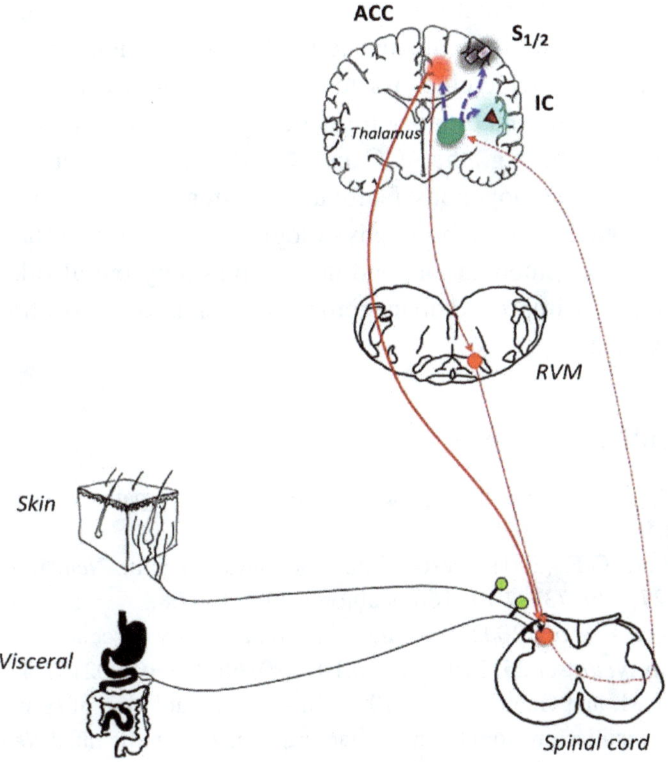

Figure 8. Top-down corticospinal network for emotional anxiety-related pain in the body. Proposed network that may explain anxiety-related somatic discomfort/pain among patients. Increased ACC activity due to elevated anxiety level trigger descending facilitatory system. Neurons in the deeper layers of the ACC send their axonal projections to neurons in the RVM of the brainstem that is known to send its modulatory projection to the spinal cord dorsal horn. Serotonin (5-HT) is the major neurotransmitter for triggering facilitatory effects on spinal nociceptive transmission. ACC neurons can also project to the spinal cord dorsal horn directly by corticospinal pathways, and glutamate is the likely candidate of transmitter. Both facilitatory systems can potentiate spinal excitatory sensory transmission, and enhance sensory inputs to the brain through spinothalamic pathways. Consequently, neurons in pain-related cortical regions are activated and contribute to unpleasantness or discomfort. Depending on the facilitation of sensory inputs from the region of the body, the activation of neurons in somatosensory cortices then contributes to the feeling of pain as if it comes from specific parts of the body (adapted from Zhuo [32]).

back to the level of spinal dorsal horn, a gate region for the entrance of painful information from the outside. The information on descending modulation will provide clues for testing new drug targets, for example, blocking descending facilitatory influences at different levels of the central nervous system (e.g., the ACC and RVM). It is clear that an improved understanding of endogenous facilitatory systems provides us not only with the knowledge of the basic physiological mechanisms related to sensory transmission, modulation, and neuron plasticity but also knowledge that can lead to improved management of persistent and chronic pain states in patients.

References

[1] Fields, H. (2004) State-dependent opioid control of pain. *Nat Rev Neurosci* 5, 565–575. 10.1038/nrn1431.

[2] Gebhart, G.F. (2004) Descending modulation of pain. *Neurosci Biobehav Rev* 27, 729–737. 10.1016/j.neubiorev.2003.11.008.

[3] Porreca, F. *et al.* (2002) Chronic pain and medullary descending facilitation. *Trends Neurosci* 25, 319–325. 10.1016/s0166-2236(02)02157-4.

[4] Zhuo, M. and Gebhart, G.F. (1997) Biphasic modulation of spinal nociceptive transmission from the medullary raphe nuclei in the rat. *J Neurophysiol* 78, 746–758. 10.1152/jn.1997.78.2.746.

[5] Zhuo, M. and Gebhart, G.F. (1992) Characterization of descending facilitation and inhibition of spinal nociceptive transmission from the nuclei reticularis gigantocellularis and gigantocellularis pars alpha in the rat. *J Neurophysiol* 67, 1599–1614. 10.1152/jn.1992.67.6.1599.

[6] Zhuo, M. and Gebhart, G.F. (2002) Modulation of noxious and non-noxious spinal mechanical transmission from the rostral medial medulla in the rat. *J Neurophysiol* 88, 2928–2941. 10.1152/jn.00005.2002.

[7] Zhuo, M. *et al.* (2002) Biphasic modulation of spinal visceral nociceptive transmission from the rostroventral medial medulla in the rat. *J Neurophysiol* 87, 2225–2236. 10.1152/jn.2002.87.5.2225.

[8] Zhuo, M. and Gebhart, G.F. (1990) Characterization of descending inhibition and facilitation from the nuclei reticularis gigantocellularis and gigantocellularis pars alpha in the rat. *Pain* 42, 337–350. 10.1016/0304-3959(90)91147-B.

[9] Zhuo, M. and Gebhart, G.F. (1991) Spinal serotonin receptors mediate descending facilitation of a nociceptive reflex from the nuclei reticularis

gigantocellularis and gigantocellularis pars alpha in the rat. *Brain Res* 550, 35–48. 10.1016/0006-8993(91)90402-h.

[10] Zhuo, M. (2000) Silent glutamatergic synapses and long-term facilitation in spinal dorsal horn neurons. *Prog Brain Res* 129, 101–113. 10.1016/S0079-6123(00)29008-0.

[11] Hori, Y. *et al.* (1996) Long-lasting synaptic facilitation induced by serotonin in superficial dorsal horn neurones of the rat spinal cord. *J Physiol* 492 (Pt 3), 867–876. 10.1113/jphysiol.1996.sp021352.

[12] Wang, G.D. and Zhuo, M. (2002) Synergistic enhancement of glutamate-mediated responses by serotonin and forskolin in adult mouse spinal dorsal horn neurons. *J Neurophysiol* 87, 732–739. 10.1152/jn.00423.2001.

[13] Zhuo, M. and Gebhart, G.F. (1990) Spinal cholinergic and monoaminergic receptors mediate descending inhibition from the nuclei reticularis giganto-cellularis and gigantocellularis pars alpha in the rat. *Brain Res* 535, 67–78. 10.1016/0006-8993(90)91825-2.

[14] Zhuo, M. (2017) Descending facilitation. *Mol Pain* 13, 1744806917699212. 10.1177/1744806917699212.

[15] Calejesan, A.A. *et al.* (2000) Descending facilitatory modulation of a behavioral nociceptive response by stimulation in the adult rat anterior cingulate cortex. *Eur J Pain* 4, 83-96. 10.1053/eujp.1999.0158.

[16] Zhang, L. *et al.* (2005) Anterior cingulate cortex contributes to the descending facilitatory modulation of pain via dorsal reticular nucleus. *Eur J Neurosci* 22, 1141–1148. 10.1111/j.1460-9568.2005.04302.x.

[17] Galea, M.P. and Darian-Smith, I. (1994) Multiple corticospinal neuron populations in the macaque monkey are specified by their unique cortical origins, spinal terminations, and connections. *Cereb Cortex* 4, 166–194. 10.1093/cercor/4.2.166.

[18] Gabbott, P.L. *et al.* (2005) Prefrontal cortex in the rat: Projections to sub-cortical autonomic, motor, and limbic centers. *J Comp Neurol* 492, 145–177. 10.1002/cne.20738.

[19] Chen, T. *et al.* (2014) Postsynaptic potentiation of corticospinal projecting neurons in the anterior cingulate cortex after nerve injury. *Mol Pain* 10, 33. 10.1186/1744-8069-10-33.

[20] Kang, S.J. *et al.* (2015) Bidirectional modulation of hyperalgesia via the specific control of excitatory and inhibitory neuronal activity in the ACC. *Mol Brain* 8, 81. 10.1186/s13041-015-0170-6.

[21] Chen, T. *et al.* (2018) Top-down descending facilitation of spinal sensory excitatory transmission from the anterior cingulate cortex. *Nat Commun* 9, 1886. 10.1038/s41467-018-04309-2.

[22] Zhuo, M. (2008) Cortical excitation and chronic pain. *Trends Neurosci* 31, 199–207. 10.1016/j.tins.2008.01.003.

[23] Robinson, D. *et al.* (2002) Long-lasting changes in rostral ventral medulla neuronal activity after inflammation. *J Pain* 3, 292–300. 10.1054/jpai.2002.125183.

[24] Guo, W. *et al.* (2006) Supraspinal brain-derived neurotrophic factor signaling: A novel mechanism for descending pain facilitation. *J Neurosci* 26, 126–137. 10.1523/JNEUROSCI.3686-05.2006.

[25] Edelmayer, R.M. *et al.* (2009) Medullary pain facilitating neurons mediate allodynia in headache-related pain. *Ann Neurol* 65, 184–193. 10.1002/ana.21537.

[26] Wang, R. *et al.* (2013) Descending facilitation maintains long-term spontaneous neuropathic pain. *J Pain* 14, 845–853. 10.1016/j.jpain.2013.02.011.

[27] Huang, Z.X. *et al.* (2014) Involvement of RVM-expressed P2X7 receptor in bone cancer pain: Mechanism of descending facilitation. *Pain* 155, 783–791. 10.1016/j.pain.2014.01.011.

[28] Donovan-Rodriguez, T. *et al.* (2006) Evidence of a role for descending serotonergic facilitation in a rat model of cancer-induced bone pain. *Neurosci Lett* 393, 237–242. 10.1016/j.neulet.2005.09.073.

[29] Suzuki, R. *et al.* (2004) Descending facilitatory control of mechanically evoked responses is enhanced in deep dorsal horn neurones following peripheral nerve injury. *Brain Res* 1019, 68–76. 10.1016/j.brainres.2004.05.108.

[30] Tillu, D.V. *et al.* (2008) Descending facilitatory pathways from the RVM initiate and maintain bilateral hyperalgesia after muscle insult. *Pain* 136, 331–339. 10.1016/j.pain.2007.07.011.

[31] Vanderah, T.W. *et al.* (2001) Tonic descending facilitation from the rostral ventromedial medulla mediates opioid-induced abnormal pain and antinociceptive tolerance. *J Neurosci* 21, 279–286

[32] Zhuo, M. (2016) Neural mechanisms underlying anxiety-chronic pain interactions. *Trends Neurosci* 39, 136–145. 10.1016/j.tins.2016.01.006.

[33] Zhuo, M. and Gebhart, G.F. (2002) Facilitation and attenuation of a visceral nociceptive reflex from the rostroventral medulla in the rat. *Gastroenterology* 122, 1007–1019. 10.1053/gast.2002.32389.

Chapter 12

Current Drug and Novel Drug Target for Treating Chronic Pain

News and Views

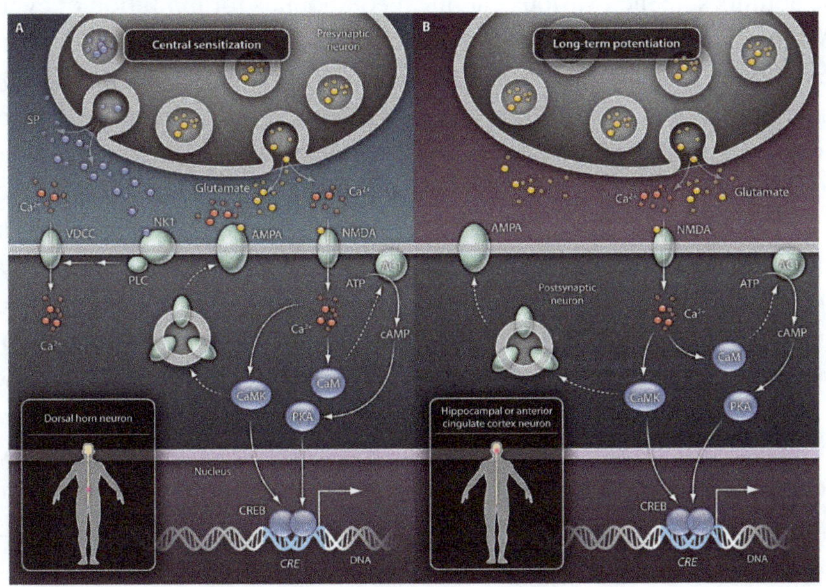

No Gain from Pain

Pain from a hot stove or an injury can be positive, as it can help prevent more serious damage from occurring. Neuropathic pain, on the other hand — which includes burning or aching pain —seemingly has no purpose. Analgesics that block only neuropathic pain are desirable but scarce. Wang *et al.* recently identified a promising new drug candidate by screening for drugs that selectively block a type of calcium-activated adenylyl cyclase that participates in neuropathic pain. NB001 is a novel compound, which can block neuropathic pain in rodents, without apparent side effects.

Adenylyl cyclase 1 has many characteristics of a good drug target for neuropathic pain. Primarily adenylyl cyclase 1 is an activity-dependent enzyme — expressed selectively in neurons — that is critical for pain-related neural plasticity which is believed to underlie this kind of pain. The authors screened various chemical compounds for inhibition of cyclic AMP production and the transcription factor CREB in human cells transfected with adenylyl cyclase 1. One of these compounds, named NB001, was found to be most effective and also inhibited cyclic AMP production in mouse brain slices and human neurons. NB001 prevented allodynia (a condition in which an innocuous stimulus causes pain) in mice in which certain nerves were ligated, and in mice with chronic inflammatory pain, produced by an injection of an irritant into a paw. When NB001 was injected directly into the anterior cingulate cortex (a brain region involved in neuropathic pain generation), it also prevented allodynia, although to a lesser extent, suggesting that NB001 acts on multiple sites in the body.

Just as important as the positive effects of NB001 on chronic pain are the effects that it does not have. NB001 does not interfere with normal nociception — the sensation that allows the animal to escape dangerous heat. It does not affect neurotransmission of the critical hormone glutamate or the size of glutamate-induced currents. Tests of anxiety, motor function, and fear all showed that NB001 had no effects on these endpoints, a good sign for the potential safety profile of this drug.

A clue as to how NB001 works can be gleaned from recent results that show that it prevents the synapses in the dorsal horn of the spinal cord and the anterior cingulate cortex from "learning" — a process triggered in neuropathic pain conditions. This effect may underlie its analgesic ability, a conclusion consistent with the fact that it does not alter plasticity in the hippocampus, a non-pain-related brain region. If the selective action of NB001 on neuropathic pain, along with its lack of serious side effects, also holds true in humans, it may prove to be a valuable treatment for pain.

Summary

Basic investigations of nociception and pain have provided us with a better understanding of the mechanisms of pain transmission, modulation, and plasticity. However, the clinical treatment of chronic pain has not kept pace with these discoveries. This failure to discover new and improved treatments for chronic pain is at least in part due to the business model of pharmaceutical industries and an oversensitive legal system. Many early-stage drug trials are terminated for reasons other than their scientific value. Considering that the majority of cells in the body use most of the common signaling proteins to form key signaling pathways, it is the next great challenge for researchers to develop selective, yet effective, painkillers without dangerous side effects.

Keywords: Analgesia; LTP; AC1; chronic pain; NB001; inflammatory pain

Introduction

Revealing a novel mechanism and identifying a new target for pain are the main focuses for most pain researchers and neuroscientists. The most direct, and simple, idea for controlling pain is to identify selective proteins that are preferentially involved in the pain process and discover/design selective inhibitors for those proteins. However, the history of pain medicine research proves that this is easier said than done. Recent studies using modern molecular and cellular neurobiology approaches found that pain was not, in fact, transducted by a few signaling proteins. Even more bad news is that many of the key proteins involved in pain processes, such as chronic pain, also play key roles in physiological conditions. One good example is NMDA NR2B receptor, while critical for cognitive functions, its activity also plays a critical role in chronic pain. This chapter will review current drugs by focusing on their basic neurobiological mechanisms and introduce several new signaling proteins involved in chronic pain, in order to explore potential future targets and treatments.

Distinguish physiological pain and pathological pain

Physiological pain is imperative for human survival. Physiological pain is the first sign that tells an individual that something is wrong. The position, quality, and duration of pain are crucial for a doctor to locate and diagnose disease and also to appreciate the efficacy of treatment. Pain also serves as a protective function. Pain, or the threat of pain, warns an individual to avoid dangerous situations and to protect against self-harm. This protective function of pain is well demonstrated in patients with diabetes or congenital insensitivity to pain. In diabetes cases, patients often lose pain sensation due to a degeneration of peripheral nerve fibers. This degeneration prevents patients from feeling pain, often resulting in damage or injury without awareness. Patients with congenital pain insensitivities similarly struggle with injuring themselves inadvertently, and they often die young as a result. Thus, understanding the physiological mechanism(s) for pain, including how injury-producing stimuli transmit from the peripheral site to the CNS and how this nociceptive process could be modulated by the CNS, is crucial for treatment of those patients with both acute and chronic pain.

Injury most often leads to pain which lasts for an extended period of time after the injury (persistent pain or chronic pain). Long-lasting pain, or pathological pain, often leads to a wide range of dysfunctions in the brain including anxiety, depression, and even suicide. While physiological pain is likely to be conducted through precise physiological pain pathways — and under fine modulation of descending biphasic modulation — pathological pain is more likely to be caused by activity-dependent changes in a variety of brain regions. An understanding of the molecular and cellular mechanisms of plastic changes in sensory-related brain areas is essential to the development of clinical strategies aimed at alleviating chronic pain. Because physiological pain is such an important function for daily living, the next generation of drugs will need to selectively reduce chronic pain while keeping acute pain intact. Studies designed to treat chronic pain need to consider how to maintain physiological pain while reducing or abolishing pain-induced unpleasantness or discomfort selectively.

Inhibiting pain transmission as drug action mechanism: Traditional approach to develop analgesic

Based on what we have discussed so far, it is possible to think of different methods for controlling pain. Most of the clinically effective drugs can be explained by using the animal model of chronic pain. It must be noted that there are possible differences between the species, and the side effects of drugs on humans cannot be predicted with animal models. More importantly, the human body is made of many common signaling proteins from the testis to the central neurons. It may be unavoidable that even the most effective pain drugs come with significant side effects. Based on basic mechanisms, this chapter will discuss pain drugs in the following order:

1. drugs that targeted peripheral transmission and sensitization,
2. drugs that affect sensory transmission in the spinal cord or supraspinal structures,
3. drugs that act to activate descending modulation,
4. drugs that act to block pain plasticity.

Table 1 summarizes most of the current drugs or obvious drug candidates for the treatment of chronic pain.

Peripheral inhibition

For targeting peripheral transmission, there are several drugs available for topical use. The first is lidocaine. Lidocaine is an amide-type local anesthetic that produces its effect by blocking voltage-dependent sodium channels. It reduced excitability in peripheral nociceptors. The second drug is capsaicin. Capsaicin mediates its effect by activating the transient receptor potential type 1 (TRPV1) receptor. TRPV1 is preferentially

Table 1. Pain medicines currently used and their synaptic mechanisms

Classes	Clinical drugs	Types of drugs	Major synaptic mechanisms	Clinical pain conditions
Type I	SNX-111	N-type calcium channel blocker	Inhibiting transmitter release	Cancer pain
	LY293558	AMPA/kainate receptor antagonist	Inhibiting postsynaptic excitatory currents	Hyperalgesia and/or allodynia
Type II	Ketamine, CPP	NMDA receptor antagonist	Blocking glutamatergic synaptic plasticity and transmission	Neuropathic pain, post-operative pain
Type III	Clonidine	α2 adrenergic receptor agonist	Inhibiting sensory synapses	Cancer pain, neuropathic pain, postoperative pain
	Neostigmine	Cholinesterase inhibitor	Inhibiting sensory synapses	Cancer pain, postoperative pain
	Morphine and other similar drugs	μ-opioid receptor agonist	Inhibiting sensory synapses	Cancer pain, postoperative pain
	Celebrex	Cyclooxygenase subtype 2 inhibitor	Inhibiting sensory synapses	Cancer pain, postoperative pain
Type IV	GABApentin		Enhancing inhibitory transmission	Neuropathic pain

expressed in nociceptive primary afferent neurons [1,2]. It likely produced analgesic effects by altering the sensitivity of peripheral sensory fibers. One lead drug target for pain is a novel receptor antagonist for TRPV1 receptors. Based on its selectively involvement in nociceptive transmission, it is likely that it may contribute to injury-related hyperalgesia but not allodynia. All these drugs are less, or not, effective in treating centrally related chronic pain. Furthermore, recent studies indicate that TRPV1 receptors are also distributed in other parts of the brain such as the cortex and hippocampus — areas that are also important for memory and other key brain functions. The TRPV1 receptor is found to play a key role in body temperature maintenance and the TRPV1 antagonist is reported to produce dramatic changes in body temperature. In the ACC, a selective TRPV1 antagonist failed to produce any inhibition of chronic pain-related LTP (see Figure 1), suggesting that it is unlikely to affect cortical sanitization in chronic pain.

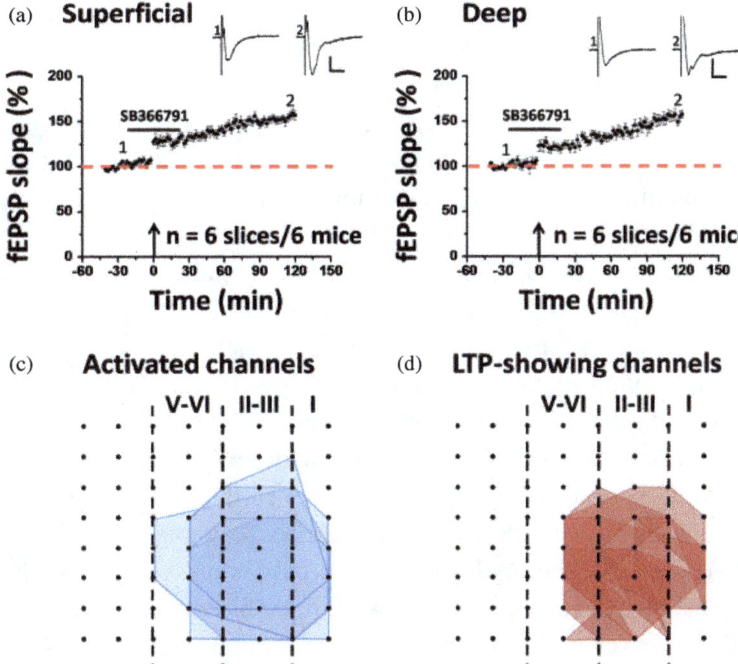

Figure 1. SB366791 cannot block the induction of LTP in the adult mice ACC. (a) Grouped data from 6 slices of 6 mice for SB366791 (20 μM), showing the normal induction of LTP in the superficial layer. (b) Summarized data in the deep layer (n = 6 slices/6 mice). Sample fEPSP recordings taken at the times indicated by the corresponding numbers are shown at the top of each plot. Arrows in (a and b) show the starting point of TBS application. Horizontal bars denote the period of drug delivery. Calibration: 100 μV, 10 ms. Error bars represent SEM. (c, d) Spatial analysis of the effect of SB366791 on LTP distribution in the ACC. Shown are polygonal graphs of the channels that were activated (blue, c) and that exhibited LTP (red, d) when TBS was delivered in the presence of SB366791 (n = 6 slices/6 mice). Vertical lines denote the specific layers in the ACC slice. SB366791 has no effect on the LTP distribution map in the ACC (adapted from Liu and Zhuo [66]).

Opioids

Drugs targeting sensory synaptic transmission and modulation are more well-known, such as opioids, cannabinoids, calcium channel blockers, Cox-2 inhibitors, as well as anticonvulsants, and antidepressants. Opioid analgesics include morphine and other opioid agonists. These drugs are used to treat moderate to severe pain. Morphine and other opioid drugs produce analgesia by binding to opioid receptors expressed in the central nervous system, including the cortex, brainstem, spinal cord, and peripheral tissues. Activation of these receptors affects both excitatory and inhibitory transmission in the central synapses. There are three major subtypes of opioid receptors: the mu-, kappa-, and delta-opioid receptors. Most of the opioid drugs act by activation of μ receptor, such as morphine, hydromorphone, oxycodone, fentanyl, and meperidine. Activation of opioid receptors triggers G_i/G_o protein-coupled signaling pathways. At the synaptic level, it will inhibit voltage-gated calcium channels that are required for neurotransmitter release presynaptically and induce an inwardly rectifying potassium conductance postsynaptically. In addition, it may also trigger receptor internalization which explains the relatively short analgesic effect of opioids. It has also been reported that opioids affect inhibitory transmission in the spinal cord, thus not selective for excitatory transmission. Since injury-related pain plasticity is NMDA receptor-dependent, the application of opioids is unlikely to block pain-related plasticity. A recent study showed that a brief application of a high opioid dose can reverse injury-related LTP in spinal cord C fiber synapses, effectively erasing a spinal memory trace of pain.

The action of opioids likely varies at the different levels of the CNS. Peripheral effects have been reported, although the major analgesic effect is likely from its action in the CNS. In the cortex, opioids are highly expressed and inhibited excitatory transmission in the pain-related cortical areas, such as the ACC. Its cortical action is likely reducing pain perception by affecting excitatory synapses. In addition, the PAG and the RVM are important for opioid analgesia. Opioids activate descending inhibitory pathways to the spinal cord which are mediated by multiple neurotransmitters including serotonin, noradrenaline, acetylcholine, and neuropeptides in the dorsal horn. At the spinal cord level, mu opioids act presynaptically to inhibit neurotransmitter release and postsynaptically to inhibit dorsal horn neurons firing to peripheral stimulation.

In addition to the relatively rapid loss of analgesic effects over time, opioid usage for the treatment of chronic pain is also limited by its very powerful side effects. Side effects can include sedation, mental clouding or confusion, respiratory depression, nausea, vomiting, constipation, pruritus (itch), and urinary retention.

Cannabinoids

Cannabinoids are compounds that activate the cannabinoid receptors. They include endocannabinoids (produced endogenously in the body), phytocannabinoids (produced by Cannabis plants), and cannabimimetics (produced synthetically and by various plant species). One of the most notable cannabinoids is Δ9-tetrahydrocannabinol (THC), the primary psychoactive compound of cannabis. However, there are numerous other cannabinoids with various effects. There are two major types of cannabinoid receptors: CB1 and CB2. CB1 receptors are found primarily in the brain, including the basal ganglia, limbic system, cerebellum, and reproductive systems. CB1 receptors are mainly responsible for the euphoric and anticonvulsive effects. CB2 receptors are predominantly found in the immune system and maybe be responsible for the possible anti-inflammatory effects.

Due to its central distribution, activation of CB1 receptors has been shown to inhibit nociceptive responses in central sensory synapses, which suggests that CB1 receptor agonists may be beneficial for controlling pain. However, the effect of CB1 receptor activation is not selective. It also affects synaptic regulation and learning-related plasticity (LTP and LTD) in the cortex, hippocampus, and cerebellum [3]. This explains the side effects of these drugs on cognitive and emotional functions. Endocannabinoids are known to serve as intercellular 'diffusible messengers' in key brain regions, including the hippocampus. They also contribute to synaptic LTP that is known to be critical for memory. Although some animal behavioral studies suggest that CB2 agonists may be analgesic for chronic pain, the basic mechanisms for such analgesics are still lacking. It is worthwhile to point out that conclusions from pure animal models of pain often fail to translate in human patients.

Calcium channel blockers

Voltage-gated calcium channels, especially N-type calcium channels, are important for sensory synaptic transmission. The calcium channel blocker ziconotide is a synthetic version of the naturally occurring omega-conotoxin MVIIA and inhibits voltage-gated N-type calcium channels. N-type calcium channels are widely distributed in the nervous system and are required for normal physiological synapse functions. They are also widely distributed in the postsynaptic structures of neurons. Due to the negative side effects ziconotide has on the CNS (nausea, vomiting, and dizziness), it is limited in intrathecal applications in pain patients.

Sodium channel blockers

Inhibition of local sensory sodium channels is thought to be the ideal and safe way to reduce chronic pain, especially neuropathic pain. Different chemical inhibitors have been developed to target certain sodium channel subtypes, however the administration of sodium channel blockers is also limited by central side effects. Some of the similar subtype channels are also expressed in the heart, as well as other key central brain areas. The ineffectiveness of inhibitors is also reported in certain forms of chronic pain.

Cox-2 inhibitor

Mild to moderate pain is often treated with non-steroidal anti-inflammatory drugs (NSAIDs). NSAIDs are effective in reducing the hyperalgesia associated with tissue injury and inflammation but are less effective for treating neuropathic pain. NSAID drugs produce their effects by inhibiting cyclooxygenase (COX). COX is the rate-limiting enzyme in the production of prostaglandins, which are derived from C-20 unsaturated fatty acids, mainly arachidonic acid. Arachidonic acid is formed from membrane phospholipids by activation of phospholipase A2. COX converts arachidonic acid to prostaglandin H2 (PGH2), which is subsequently catalyzed to PGD2, PGE2, PF2a, and PGI2 by specific synthases for the different prostaglandins. In addition, COX catalyzes the generation of thromboxanes, leukotrienes, and lipoxins. Together, these substances are referred to as eicosanoids. The prostaglandins mediate their action by binding to specific membrane receptors (DP, EP, FP, and IP) that are G-protein-coupled receptors. Tissue injury and inflammation result in prostanoid production at the site of tissue injury and inflammation. Once released at the site of injury, prostaglandins act on prostaglandin receptors on primary afferent terminals to produce sensitization and behavioral hyperalgesia. In addition, tissue injury and inflammation are associated with elevated prostaglandin levels and induction of COX-2 at the spinal cord level. Thus, prostaglandins may not only produce sensitization in the periphery but also contribute to sensitization mechanisms at the spinal cord level. The major benefit of COX-2, over Cox inhibitors, is the reduced gastrointestinal side effects. It is unclear whether Cox-2 inhibitors are any better than non-selective Cox inhibitors in terms of the neurobiological effects, however, the recent discovery of negative cardiovascular side effects of COX-2 has led to a diminished use of Cox-2 drugs for pain control.

Anticonvulsants

The anticonvulsants represent a group of compounds with diverse chemical structures and different mechanisms of action. Several anticonvulsants are potent blockers of voltage-gated sodium channels, including carbamazepine, phenytoin, lamotrigine, topiramate, and felbamate. Gabapentin is the only anticonvulsant approved for the treatment of neuropathic pain (postherpetic neuralgia), and its activity is thought to be a result of the inhibition of voltage-sensitive calcium channels and modulation of the GABA system. In a recent study, it was found that gabapentin reduced excitatory synaptic transmission in the ACC while LTP was not blocked [24] (Figure 2). Pregabalin is a similar compound to gabapentin.

Antidepressants

Tricyclic antidepressant drugs (TCAs), such as amitriptyline, may be effective in treating painful conditions, such as neuropathic pain. There are different mechanisms that may contribute to the pain-relieving effects of TCAs, such as the reuptake inhibition of noradrenaline and serotonin, blockade of Na ion channels, and the inhibition of NMDA receptors. Changes in NE and 5-HT levels may affect descending pain modulatory systems, especially descending inhibitory systems. The exact mechanisms at the cellular and molecular levels remain to be investigated, but this family of drugs will affect physiological pain.

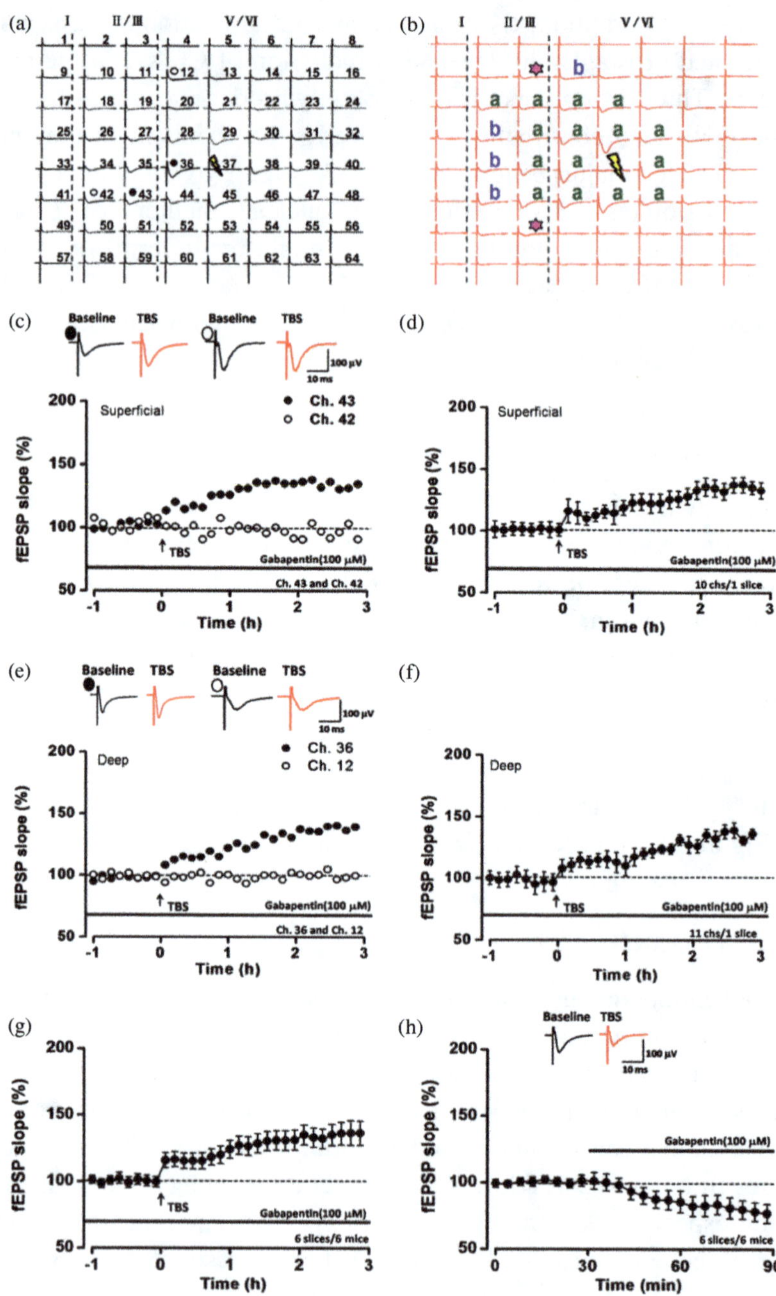

Figure 2.   Basic synaptic transmission and LTP in the ACC with gabapentin application. (a, b) Two mapped figures show the LTP of the fEPSP slope which was induced in the presence of gabapentin (100 μM). The baseline responses (a) were potentiated 3 h after TBS was applied on one channel (yellow lightning bolt) (b). Channels with L-LTP and no L-LTP were marked as "a" (green) and "b" (blue), and the TBS-recruited synaptic responses were marked as asterisks. (c) The sample traces and the plotted slope results of one channel showing L-LTP (Ch. 43) and the other without showing L-LTP (Ch. 42) in the superficial layer with gabapentin application. (d) The plotted slope shows the L-LTP of the fEPSP of 8 channels in the superficial layer with gabapentin application. (e) The sample traces and the plotted slope results of one channel showing L-LTP (Ch. 36) and the other without showing L-LTP (Ch. 12) in the deep layer of the same slice with gabapentin application. (f) The plotted slope shows the L-LTP of 11 channels in the deep layer with gabapentin application. (g) The summarized fEPSP slopes of 6 slices/6 mice with gabapentin application. (h) The summarized fEPSP slopes of 6 slices/6 mice show gabapentin decreases the basic synaptic transmission (adapted from Chen *et al.* [48]).

Substance P receptor NK1 antagonist

Substance P (SP) is a key pain transmitter in the spinal cord. The search for SP receptor NK1 antagonists is mainly for pain treatment. However, both animal studies using gene knockout mice, or pharmacological inhibitors, show that inhibiting SP receptors can have biphasic effects in various animal models of chronic pain. The activation of NK1 receptors can contribute to pain transmission and is also important for descending pain inhibition. Clinical studies in patients with neuropathic pain did not see any significant analgesic effects of NK1 antagonists. One good outcome from this 'negative' discovery is that NK1 antagonists produced unique antidepressant, anxiolytic, and antiemetic properties, which have been a great help to patients receiving cancer chemotherapy.

KA receptors: Diversity, structure, and synaptic function

Among the glutamate receptors, the AMPA and NMDA receptors are the most commonly studied [4]. KA receptors are one of the three subtypes of ionotropic receptors for glutamate and are composed of five different subunits: GluR5, GluR6, GluR7, KA1, and KA2. GluR5 is involved in both *in vitro* and *in vivo* nociceptive responses. The involvement of spinal KA receptors in acute behavioral nociception was first demonstrated by Li and colleagues. They found that intrathecal injection of the AMPA and KA receptor antagonist CNQX produced greater antinociceptive effects than an antagonist selective for AMPA receptors alone, in the HP and TF but not cold plate tests. Moreover, they showed that the non-selective KA receptor antagonist, SYM2081, alone could have analgesic effects. SYM2081 had a similar effect on acute pain induced by mechanical or thermal stimulation. In contrast, intrathecal injection of ATPA was reported to have antinociceptive effects in mechanical or thermal stimulation-induced acute pain. By using spinal cord slices or cultured neurons, it has been reported that various KA receptor subtypes play different roles in spinal transmission and modulation of pain transmission.

A number of recent studies have focused on the role of KA receptors in chronic pain. Behavioral experiments using systemic drug administration showed that GluR5-containing KA receptors may play a role in formalin-induced inflammatory pain while SYM 2081 was shown to attenuate mechanical allodynia and thermal hyperalgesia in a rat model of nerve injury. Similar analgesic effects were found by several groups using different chronic pain models and pharmacological tools. These effects were found in intraperitoneal injection of SYM2081 in a freeze injury neuropathic pain model, intrathecal injection of NS-102 (KA receptor antagonist) or LY382884 in the CFA inflammation model, intrathecal or intraperitoneal injection of SYM2081 in capsaicin-induced hyperalgesia or carrageenan-induced inflammation, intraperitoneal injection of NS1209 (AMPA/GluR5 antagonist) in the formalin test or chronic constriction nerve injury model, and oral administration of LY467711 or LY525327 (GluR5 antagonist) in capsaicin-induced hyperalgesia, formalin, or carrageenan inflammation. Most recently, Ko *et al.* found that GluR5, but not GluR6, deficient mice showed reduced responses to capsaicin or formalin-induced inflammatory pain. All these animal studies point to the importance of KA receptors in nociception. Clinical studies also provide some preliminary evidence for the role of KA receptors in pain. Sang *et al.* reported that intravenous infusion of LY293558 reduced capsaicin-evoked hyperalgesia in humans, reducing pain intensity, pain unpleasantness, and the receptive field. Moreover, a recent study reported that intravenous administration of LY293558 was efficacious in the treatment of acute migraines.

Collectively, these behavioral studies indicate that pharmacological or genetic manipulation of KA receptors has the potential to affect acute or chronic nociceptive transmission. Although the exact contribution of each KA receptor subunit to acute nociception versus chronic pain remains unexplored, selectively targeting different KA receptor subunits may provide a useful strategy for treating persistent pain, considering the potentially different distribution and function of KA receptor subunits in the pain pathway.

Basic neuronal mechanism for chronic pain

Cumulative evidence using integrative neurobiological approaches suggests that physiological pain (or acute pain) is distinguished from pathological pain (chronic pain) at the cellular and molecular levels. For example, in excitatory synapses, physiological pain is likely mediated by normal sensory synaptic transmission without activating the activity-dependent signaling pathways. By contrast, pathological pain is likely triggered by transcriptional and translational events and may also contain structural changes and reorganization in the central synapses [5–11]. Most conventional analgesics often unselectively affect synaptic transmission (excitatory and inhibitory transmission), yet there is not a single clinical drug for chronic pain that selectively targets pathological pain.

Recent molecular studies of different levels of the CNS have underscored the fact that the mechanisms leading to chronic neuropathic pain are far more complicated than previously assumed. At peripheral sites, injuries trigger sensitization and induce prolonged abnormal neural activity along primary afferent fibers [12–14]. In the spinal cord, LTP of sensory synaptic transmission in the spinal dorsal horn projecting neurons often occurs [5,15]. Recent work demonstrates that long-term synaptic plasticity also takes place in cortical areas that participate in pain perception, such as the ACC, PFC, and IC [6,7,16–20]. Similar changes can be also found in the neurons located in subcortical areas, such as the amygdala [21,22]. Furthermore, behavioral evidence indicates that changes in cortical excitation may regulate subsequent pain transmission in the spinal cord by top-down descending facilitation modulation [23–26].

Targeting synaptic potentiation for the treatment of chronic pain

Considering that spinal and cortical potentiation contributes significantly to chronic pain, it is conceivable that proteins and ion channels that are involved in the induction and expression of plastic changes in somatosensory pathways are potential drug targets for treating chronic pain. Among many candidates that play roles in central plastic changes, the NMDA receptor, a major glutamate receptor that is important for synaptic LTP [27,28], has been intensively investigated [29]. The NMDA receptor NR2B subunit has been selected as a potential drug target since it has less side effects than other NMDA receptor subunits [29–32].

One possible solution to avoid NMDA receptor-related side effects is to target NMDA receptor downstream proteins or messengers. Many intracellular proteins have been reported to act downstream from NMDA receptors, such as AC1, PKC, CaMKII, and tyrosine kinases Among them, calcium-stimulated AC1 plays a critical role in pain-related LTP in both the spinal cord dorsal horn and the ACC. Behaviorally, AC1 and AC8 knockout (KO) mice have shown reduced inflammatory, deep muscle pain, and neuropathic pain, whereas other physiological functions remain intact, including acute pain [33], hippocampal LTP, and related Morris water maze performance, as well as anxiety-like behaviors and motor functions [34]. Considering that AC1 is mainly expressed in the CNS, AC1 may be a suitable neuron-specific drug target for treating neuropathic pain (Figure 3).

Consideration of cortical proteins as drug targets for treating chronic pain

The majority of pain research focuses on the spinal cord dorsal horn as a major target for treating chronic pain. One obvious consideration is the

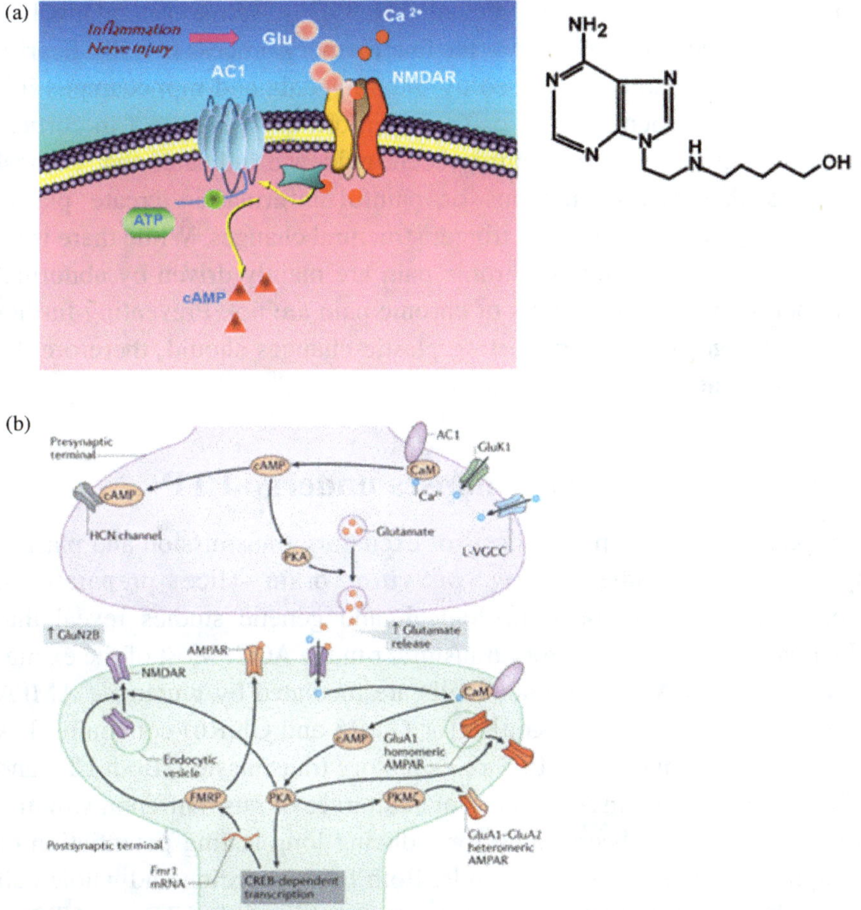

Figure 3. Identification of a selective inhibitor for AC1. (a) The model shows that AC1 acts downstream from the glutamate NMDA receptors and is activated in a calcium-dependent manner. On the right is the chemical structure of NB001. (b) Synaptic mechanism for ACC post- and pre-LTPs. NB001 produces analgesic effects by inhibiting both forms of LTPs (adapted from Wang *et al.* [52]; Bliss *et al.*, 2016).

lack of CNS side effects if drugs are acting locally only. However, in reality, most analgesic drugs act on CNS, including gabapentin, opioids, and calcium channel blockers. Cumulative evidence from animal and human studies suggests that the cortex and its related areas are important not only for processing pain perception but also for undergoing plastic changes and may play an active role in chronic pain [16,35,36]. Active cortical involve-ment in chronic pain is similar to the fear memory process [37,38]. Fearful information is mainly processed by the amygdala and hippocampus ini-tially, and permanent painful or fearful information is stored in cortical areas. For tissue or nerve injury, recent evidence indicates that cortical synapses that receive noxious or painful information create plastic changes and, in the long term, trigger structural changes. While there is no doubt that certain forms of chronic pain are mainly driven by abnormal peripheral inputs, many forms of chronic pain are not. Preventing further plastic changes or erasing existing plastic changes should, therefore, be the focus of new drug development.

Pain-related cortical synapses undergo LTP

Synaptic and cellular mechanisms of excitatory transmission and plastic-ity are best studied using *in vitro* brain slice preparations. Electrophysiological, pharmacological, and genetic studies reveal that glutamate is the fast excitatory transmitter in the ACC. Most of the excita-tory currents in ACC pyramidal cells are mediated by glutamate AMPA receptors [39]. GluK5 and GluK6 (or GluR5 and GluR6)-containing KA receptors contribute to 10–15% of excitatory transmission. Both LTP and LTD can be induced in ACC slices of adult mice or rats. Different stimula-tion protocols have been found for inducing long-lasting potentiation of synaptic responses in the ACC cells. Both field recording and whole-cell patch-clamp recording techniques can be used to study LTP and LTD in the ACC slices [40,41]. For field recording from adult rat or mouse ACC (*in vitro* slices or *in vivo* recording), glutamatergic synapses in the ACC can undergo long-lasting potentiation in response to TBS — a paradigm more closely related to the activity of ACC neurons. LTP is also observed after peripheral injury. The potentiation lasted for at least 40–120 min [42]. Whole-cell patch-clamp recordings allow for better investigation of synaptic mechanisms for LTP in the ACC [41]. LTP can be induced using three different protocols, including the pairing training, spike-EPSP pair-ing, and TBS protocol [41]. Unlike the field recordings, LTP induced by the pairing protocol is mainly triggered by the activation of NMDA recep-tors but not through L-VDCCs [41]. These studies provide clear evidence that LTP induced by different stimulation protocols may share some common signaling pathways. Similar LTP has been reported in other pain-related excitatory synapses, such as PFC, somatosensory cortices, and IC [16,42].

LTP can be divided into at least two major phases. Most of the electrophysiological studies of LTP focused on early LTP or E-LTP. Late-phase LTP, or L-LTP that lasts for at least three hours after the induction, is reported in the hippocampus [43–46] but is less investigated in pain-related cortex, such as ACC. Using multiple channel field recording techniques, strong TBS is able to induce L-LTP that lasted for 3–6 hrs in ACC slices [47,48]. This key mechanism shows that chronic pain and ACC LTP share similar mechanisms as seen in the occlusion experiments. If ACC LTP is already induced in mice with chronic pain, one would expect that TBS will fail to induce any further potentiation. In fact, nerve injury has been reported to occlude LTP in ACC slices, especially L-LTP, while early potentiation still can be induced in slices of animals with nerve injury. These experiments provide direct evidence that L-LTP and chronic pain share similar mechanisms in the ACC.

NMDA NR2B receptors

As described in Chapter 5, NMDA receptors play critical roles in activity-dependent synaptic plasticity in the CNS. Along somatosensory pathways, NMDA receptors are located in DRG, spinal cord, and supraspinal structures that are critical for pain perception. Activation of NMDA receptors expressed on the peripheral terminals of primary sensory afferents by endogenously released glutamate during injury or inflammation causes pain behavior. Peripheral administration of MK801, a non-competitive NMDA receptor antagonist, produces local anesthetic-like effects and inhibits formalin-induced inflammatory pain. However, due to imitated electrophysiological studies, it is unclear if the activation of NMDA receptors at peripheral terminals triggers any obvious neuronal plasticity.

Although spinal NMDA receptors receive a great deal of attention, evidence is accumulating that NMDA receptors located in supraspinal structures such as the brain stem play an important role in chronic pain. It has been shown that there was an upregulation of mRNAs encoding

NMDA receptor subunits, such as NR1, NR2A, and NR2B, in the RVM after inflammation. However, electrophysiological data were lacking for the functional upregulation of NMDA receptor-mediated responses in these studies. Microinjection of selective NMDA receptor antagonists prevented the inflammation-induced increase in RVM excitability. Thus, the enhanced descending modulation appears to be mediated by changes in the activation of the NMDA receptor in the RVM (see the following). However, for clinic consideration, the application of drugs to the RVM is not practical in human patients.

Recent studies from our group demonstrate the important role of cortical NMDA receptors, particularly NR2B-containing receptors in chronic pain [31]. Evidence for the involvement of cortical NR2B receptors in pathological pain is primarily coming from so-called Doogie smart mice. Doogie smart mice are transgenic mice with selective forebrain NMDA NR2B overexpression. Electrophysiological and behavioral studies with Doogie mice found that NR2B transgenic mice are super performers in memory tests with significantly enhanced hippocampal CA1 LTP. In the case of tissue injury and inflammation, inflammatory pain and allodynia were significantly enhanced without any effect on acute pain, providing the first genetic evidence that NR2B-containing NMDA receptors in forebrain neurons may encode information related to pathological pain. Consistently, a recent study using cortex-specific NR1 knockout mice showed reduced formalin-induced inflammatory pain. NR2B subunit-containing NMDA receptors are localized predominantly in pain-relevant structures, such as in dorsal root ganglion cells, superficial layers of the dorsal spinal horn, thalamus, hippocampus, and cortex. In particular, NR2B has very low expression in cerebellum. The restricted distribution of NR2B makes them promising candidates as targets of side-effect-free analgesic drugs. [49] Indeed, NR2B antagonists, such as ifenprodil and related compounds, are effective in neuropathic pain in animals, and in patients, and show better separation between efficacy and side effects than non-selective NMDA receptor blockers. For example, antinociceptive doses of ifenprodil and eliprodil appear to be devoid of psychotomimetic effects as well as motor deficits. Importantly, these drugs have minimal potential for abuse and may otherwise have effects on the development of morphine-induced conditioned place preference. After the disclosure of ifenprodil, the second generation of compounds, such as Ro25-6981 and CI-1041, were developed and have demonstrated efficacy in a number of animal pain models and appear to possess superior side effect profiles compared to earlier analogs. One compound, traxoprodil (CP-101,606), has progressed to phase II clinical trials. A preliminary report indicates that intravenous administration of the compound is effective in patients suffering from central pain, such as in spinal cord injuries, but show no typical psychotomimetic effects. Therefore, although there is debate about whether spinal NR2B or brain NR2B is responsible for mediating antino-ciceptive effects, it is clear that the NR2B-containing NMDA receptor is one of the best potential targets for neuropathic pain.

AC1 as a key messenger in neuron

Cyclic adenosine monophosphate (cAMP) is a key intracellular second messenger and plays a critical role in many physiological functions, such as learning and memory, chronic pain, emotional fear, and drug abuse [16,46][68]. AC is the enzyme that catalyses ATP to cAMP. Two major families of AC have been reported: nine membrane-bound (AC1-9) and one soluble form AC (sAC). The unique organ and cellular distributions and activation mechanisms support distinct physiological functions of each AC isoform in biological systems.

Among the more than 10 subunits, AC1 and AC8 are two of the AC subtypes that respond positively to calcium-CaM [50]. As compared with AC8, AC1 is more sensitive to a calcium increase. In the ACC, AC1 is highly expressed in cingulate neurons located in most of the layers [33]. AC1 is selective for plastic changes and gene deletion of AC1 does not affect basal glutamate transmission in the ACC. By contrast, LTP induced by TBS or pairing stimulation is abolished in cingulate pyramidal cells [51]. Whole-cell patch-clamp recordings also revealed that AC1 activity is required for the induction of LTP in ACC pyramidal cells. By using chemical design and biochemical screening, several selective inhibitors of AC1 have been identified. It has been consistently demonstrated that the pharmacological inhibition of AC1 in ACC neurons abolishes LTP induced by pairing training [52].

Our recent data also found that AC1 is essential for the induction of L-LTP in the ACC synapses. In wild-type mice, TBS-induced L-LTP lasted for at least 3–6 hrs. TBS failed to induce any significant potentiation in ACC slices of AC1 KO mice [48] (Figure 4). In addition to the

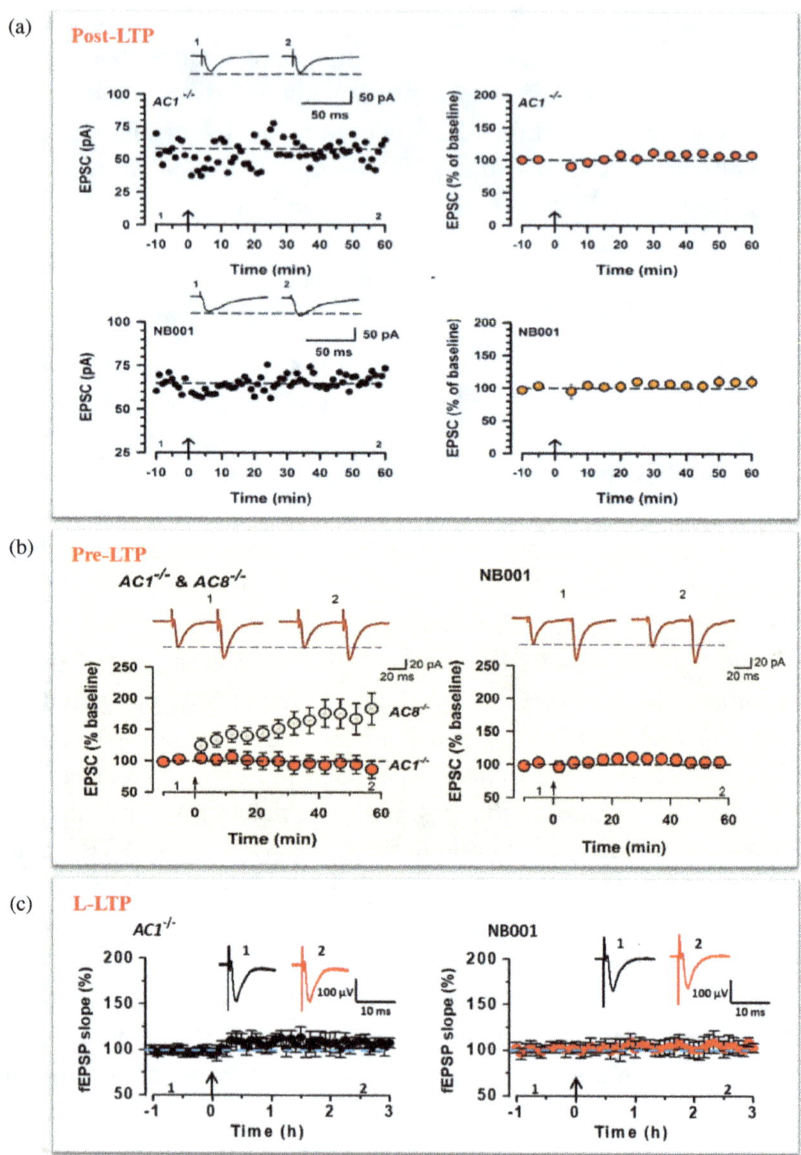

contribution to the ACC, AC1 activity is likely to contribute to other pain-related cortical areas, such as PFC, insular cortex, and somatosensory cortex. It has been reported that AC1 activity is required for injury-activated immediate early gene activity in these areas. Similar LTP induction protocols also induce LTP in PFC, somatosensory cortex, and IC areas.

AC1 contributes to behavioral sensitization and spinal facilitation/potentiation

Behavioral contribution of AC1 to chronic pain was investigated using AC1 KO mice [33]. While wild-type, and AC1, as well as AC8, KO mice were indistinguishable in tests of acute pain, behavioral responses to peripheral injection of two inflammatory stimuli, formalin and complete Freund's adjuvant, were reduced or abolished in AC1 KO mice. AC1 also contributes to inflammation-induced activation of CREB. Using an acute persistent inflammatory muscle pain model, we found that the behavioral nociceptive responses of both the late phase of acute muscle pain and the chronic muscle inflammatory pain were significantly reduced in AC1 KO mice [53].

The roles of AC1 in chronic pain-related plasticity changes are not just limited to the cortex. AC1 activity has also been shown to be critical for spinal cord facilitation and plasticity. In the spinal cord dorsal horn, application of low dose 5-HT in slices of young animals or co-application

---

Figure 4. (*Continued*)   AC1 is required for different forms of cortical LTPs. (a) AC1 KO and NB001 blocked the induction of pairing stimulation-induced post-LTP in the IC of mice. Top: Data of one neuron (left) and pooled data (right) show that post-LTP failed to induce LTP in the IC of AC1 KO mice. Bottom: NB001 blocked the induction of LTP in one neuron (left) and pooled data of NB001 (right). (b) AC1 KO (left) and NB001 (right) blocked the induction of pre-LTP in the ACC of mice. Left: Sample traces of EPSCs with paired-pulse stimulation at 50 ms and pooled data show that pre-LTP was absent in AC1 KO mice (red). AC8 KO mice showed normal pre-LTP (gray). Right: A specific AC1 inhibitor, NB001, blocked pre-LTP (red circle). (c) The summarized L-LTP of the fEPSP slopes shows that AC1 KO (left) and NB001 application (right) blocked the L-LTP induction (adapted from Yamanaka *et al.*, 2017; Koga *et al.* [67]; Chen *et al.* [48]) (adapted from Koga *et al.* [67]).

of forskolin and 5-HT in slices of adult animals produced long-lasting facilitation of excitatory synapses transmission between primary afferent fibers and dorsal horn neurons. This enhancement requires the recruitment or trafficking of functional AMPA receptors. Studies using AC1 KO mice found that calcium-sensitive, CaM-regulated AC1 is required for the enhancement [54].

In addition, AC1 also contributes to the activation of the extracellular signal-regulated kinase (Erk) either after peripheral tissue inflammation *in vivo* or by glutamate or SP *in vitro* spinal cord slices. It also contributes to spinal LTP induced by pairing protocol in spinal dorsal horn neurons [55].

Cortical amplification of synaptic transmission after injury

Unlike spinal cord dorsal horn neurons or neurons in somatosensory cortex, neurons in the ACC show a wide-spread and diffuse receptive field. Often, it can cover the whole body of animals or humans [56]. The unique property of ACC neurons allows us to detect long-term plastic changes in ACC neurons without using selective markers. By recording AMPA receptor-mediated EPSCs in pyramidal neurons in the layer II/III of the ACC in mice with peripheral nerve ligation [57], we found that input (stimulation intensity)–output (EPSC amplitude) curve of AMPA receptor-mediated current was significantly shifted to the left after peripheral nerve injury, compared with that in control group. These results suggest that excitatory synaptic transmission was increased in the ACC after peripheral nerve injury [19].

Two major synaptic mechanisms may contribute to immediate synaptic potentiation [58]. For presynaptic amplification, data collected using three different experimental measurements consistently suggested that presynaptic release of glutamate was significantly enhanced. Electrophysiological measurements include PPF, miniature EPSC analyses, and the use of MK801. PPF is a transient form of plasticity commonly used as a measure of presynaptic function, in which the response to the second stimulus is enhanced as a result of residual calcium in the presynaptic terminal after the first stimulus. After nerve ligation, there was a significant reduction in

PPF in ACC neurons compared with those from control mice. In the case of AMPA receptor-mediated miniature EPSCs (mEPSCs), following peripheral nerve injury, there was an obvious increase in mEPSC frequency in ACC neurons as compared to that of control group. Finally, the blocking rate of an irreversible NMDA receptor blocker, MK-801, was measured in both control mice and mice with neuropathic pain. The blocking rate of NMDA receptor-mediated synaptic current by MK801 is used to estimate the transmitter release probability. Significantly faster decay time by MK801 was observed in mice with nerve ligation than in control mice. Taken together, these results indicate that the enhanced excitatory synaptic transmission in the ACC after nerve injury is due to an increase in the probability of presynaptic neurotransmitter release.

For postsynaptic changes, we have detected changes in the amplitude of mEPSCs, suggesting that there is an increase in postsynaptic responsiveness after nerve injury. By using western blot, we found that there was no difference in the expression levels of both GluA1 (or GluR1) and Glu A2&3 (GluR2&3) receptors in ACC between the control mice and mice with nerve ligation. However, we found that the induction of neuropathic pain by nerve ligation was associated with an increase in the abundance of the GluA1 subunits in the membrane fraction and a corresponding decrease in the levels in the cytosolic fraction. By contrast, nerve ligation had no effect on the intracellular distribution of GluA2&3 subunits in ACC neurons. The data show that AMPAR GluA1 subunit is redistributed in ACC neurons after nerve injury.

AC1 contributes to injury-induced cortical changes

Is AC1 activity required for injury-triggered presynaptic and postsynaptic changes? In AC1 KO mice, behavioral sensitization caused by tissue inflammation or nerve injury was significantly reduced or inhibited [33]. It is therefore expected that ACC presynaptic and postsynaptic changes triggered by the injury may be prevented if these cortical changes are indeed important for behavioral sensitization in animal models of chronic pain. Indeed, we found that following nerve injury, the AC1 KO mice did not exhibit reduction in PPF of AMPA receptor-mediated EPSCs as shown by the wild-type mice after the nerve injury [19]. No increase in mEPSC frequency or amplitude was observed in AC1 KO mice. Moreover, we compared the evoked EPSCs in wild-type and AC1 KO mice and there is no significant difference between the two groups. These results indicate that both presynaptic and postsynaptic enhancements of excitatory synaptic transmission in the ACC depend on AC1 in neuropathic pain. Figure 5B summarizes possible synaptic mechanisms for ACC LTP and periphery injury-triggered plastic changes within the ACC.

Biochemical studies also confirmed the importance of AC1. The expressions of GluA1 and GluA2&3 were not altered in the ACC of AC1 KO mice compared to wild-type mice. Furthermore, the phosphorylation levels of GluA1 subunit in the ACC were unchanged in AC1 KO mice with neuropathic pain. The phosphorylation of GluA1 subunits of AMPA receptors is critical for synaptic expression of the receptors, their channel properties, and synaptic plasticity [59]. Interestingly, we found that the phosphorylation levels of GluA1 (at Ser 845) were significantly increased in the ACC after nerve injury. These data indicate that the nerve injury can increase the phosphorylation levels of GluA1 through the PKA signaling pathway [19]. cAMP may also contribute to injury-triggered enhancement of presynaptic glutamate release. Using a novel transgenic mouse model, heterologously expressing an Aplysia octopamine receptor (Ap oa1), we found that activation of Ap oa1 by octopamine enhanced glutamatergic synaptic transmission in the ACC by increasing presynaptic glutamate release *in vitro* [60]. Bilateral microinjection of octopamine into the ACC significantly facilitated behavioral responses to inflammatory pain but not acute pain.

PKMζ as a key mechanism for maintaining pain-related cortical L-LTP

Multiple protein kinases are thought to contribute to the induction of LTP and initial consolidation of information storage [61][68]. PKMζ, an atypical isoform of PKC, has been detected in many regions of the brain including the frontal cortex [62]. In animal models of neuropathic pain, we found that the activity of PKMζ was upregulated. This upregulation can either increase protein synthesis or increase phosphorylation. The

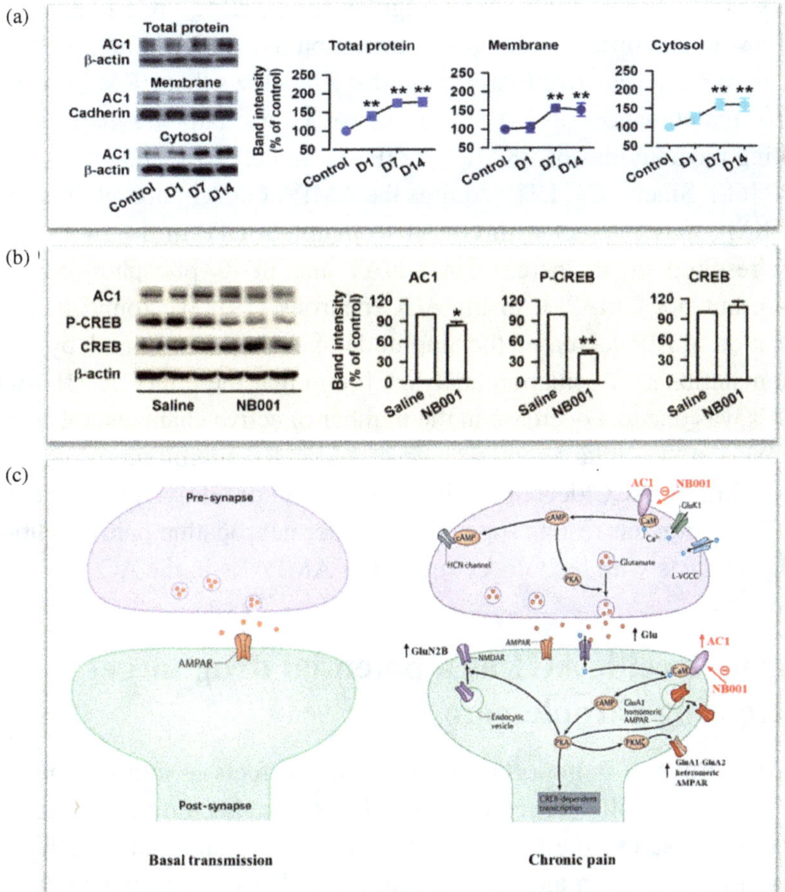

Figure 5. Upregulation of AC1 in chronic visceral pain. (a) Representative western blots for total, membrane, and cytosol AC1 in the ACC of control and zymosan-induced chronic visceral pain mice. Total, membrane, and cytosol AC1 levels in the ACC were markedly increased on day 7 and 14 after zymosan injection. (b) Representative western blots for AC1, p-CREB, and CREB in the ACC of chronic visceral pain mice treated with saline or NB001. NB001 injection reduced the levels of AC1 and p-CREB in the ACC but not the total CREB of model mice. (c) Signalling pathways of the AC1-dependent LTP in the ACC of chronic pain. Synapses in the ACC undergo long-term presynaptic and postsynaptic changes after chronic pain. The presynaptic glutamate release and expression of GluA1-GluA2 containing AMPARs and GluN2B-containing NMDARs are upregulated in models of chronic pain. AC1 is essential for both the long-term presynaptic and postsynaptic potentiation. NB001 inhibited the AC1-dependent LTP as a drug target for the treatment of chronic pain (adapted from Liu *et al.* [68]).

activity of AC1 is critical for the upregulation of PKMζ. AC1 KO mice did not show any significant changes in the amounts of PKMζ and p-PKMζ.

The use of a selective PKMζ inhibitor ZIP shows that PKMζ is critical for the maintenance of L-LTP in the ACC excitatory synapses [68]. Blocking the activities of PKMζ by ZIP at 3 hrs after the induction erased L-LTP [68]. Since ACC LTP requires the AMPA GluA1 subunit, it is likely that PKMζ may interact with GluA1 to maintain LTP in the ACC. Nerve injury resulted in an increase in GluA1 and in its phosphorylation on Ser845, but not GluA2/3, in the ACC neurons [19]. We found that bath application of ZIP decreased the amplitude of evoked EPSCs and, by applying non-stationary fluctuation analysis, found that the effect of ZIP on the eEPSCs was due to a decrease in the number of active channels, rather than a decrease in the unitary conductance of AMPARs. Inhibiting the activities of PKMζ in the ACC decreased the protein level of GluA1 in the synapses [68]. Therefore, our results suggest that under neuropathic pain conditions, PKMζ interacts with the GluA1 subunit of AMPARs in the ACC.

Neuron-specific AC1 as a potential drug target for treating chronic pain

For CNS drugs, it is important to avoid side effects in non-neuronal tissues. Many clinically proven effective drugs have been removed from the market as a result of their side effects in non-neuronal organs, such as the cardiovascular system and liver. There are at least three major strategies to avoid side effects of CNS analgesic drugs which are as follows:

(1) drug targets that are mainly expressed in neurons,
(2) target proteins that are recruited or activated in an activity-dependent manner with an almost "silent" status in normal physiological status,
(3) choosing target proteins that are critical for chronic pain-related neuronal plasticity but not for other cognitive functions.

Few target proteins meet all three criteria. Based on the data reviewed so far, AC1 would be a top target for treating chronic pain [63,64] [see also 62 and 63 for the consideration of ACs as drug targets]. First, AC1 is primarily expressed in neurons, and no AC1 gene expression was found in heart, liver, or kidney cells. Second, AC1 is activated in a calcium-CaM-dependent manner. Third, it acts downstream from the glutamate NMDA receptors and contributes to chronic pain-related neuronal plasticity in the cortex and spinal cord. Finally, gene KO mice lacking AC1 showed reduced or blocked behavioral sensitization to non-noxious mechanical stimuli in animal models of inflammatory and neuropathic pain.

NB001 as the first generation of selective inhibitor for AC1

Using a heterologous expression system, chemical screening experiments have been performed to identify AC1 inhibitors. NB001 has been identified as a selective inhibitor for AC1 [52] (see Figure 5). In both human embryonic kidney (HEK) 293 cells in which AC1 was stably expressed and adult mouse neurons, NB001 produced a dose-dependent inhibition of AC1 activity with an estimated IC50 5–8 μM. Furthermore, NB001 produces a dose-dependent inhibition of cAMP production triggered by excitatory transmitter glutamate in the ACC slices, indicating its effectiveness in inhibiting activity-dependent cAMP at physiological or pathological conditions in adult animals. Furthermore, it produced significant inhibition of glutamate-induced cAMP production in the human neuroblastoma SH-SY5Y cells with an estimated IC50 of 8.3 μM. This is consistent with the fact that mouse AC1 gene and human AC1 share up to 90% homology [52]. The inhibitory effect of NB001 is subtype selective. NB001 did not significantly affect AC5-8 activity at effective inhibiting doses for AC1, and the efficacy differences between AC1 and AC5-8 are more than 10 fold.

Similar to results in AC1 KO mice, NB001 inhibited sensory-related LTP in two key areas (Figure 6). In mouse ACC slices, postsynaptic application of NB001 completely blocked the induction of LTP in ACC pyramidal neurons. In the spinal cord, NB001 prevented the induction of LTP induced by pairing protocol. These results provide strong evidence for the critical roles of AC1 in both spinal and ACC LTP — one possible cellular mechanism contributing to neuropathic pain. Recent studies using a multiple-channel recording system found that bath application of NB001 produced inhibition of L-LTP that lasted at least 3 hr after the induction [48].

Analgesic effects of NB001 in animal models of neuropathic and inflammatory pain

Previous studies with genetic KO mice lacking AC1 showed that behavioral allodynia (pain experienced by a usually innocuous stimulus) in animal models of neuropathic pain and inflammatory pain was significantly reduced [33]. However, it is difficult to rule out the possible contribution of developmental defects in AC1 KO mice that may contribute to behavioral results. With the identification of NB001 as an inhibitor for AC1, we examined the effects of NB001 on behavioral allodynia in animal models of neuropathic pain induced by nerve ligation [57]. Consistent with genetic studies using AC1 KO mice, administration of NB001 (0.1 mg/kg, i.p.) given 30 min before behavioral allodynia testing produced a significant analgesic effect. NB001 at a higher dose of 1 mg/kg produced a greater inhibition of behavioral allodynia [52,55].

In general, application of NB001 at different dosages did not cause any abnormal behaviors in animals. Animals injected with NB001 were calm and less responsive to behavioral allodynic measurement than the control animals. To compare our results with NB001 with the analgesic effect of currently available drugs for neuropathic pain, we also performed experiments with gabapentin. Intraperitoneal injection of gabapentin (30 mg/kg) produced significant analgesic effects in animals with neuropathic pain. The inhibitory effects were comparable to those produced by NB001 at 0.1 mg/kg. Behavioral allodynia induced by the hind paw injection of complete Freund's adjuvant (CFA) (50%) has been commonly used for evaluating a drug's analgesic effects in chronic inflammatory pain. One day after the CFA injection, NB001 (1 and 3 mg/kg, ip) produced significant analgesic effects. The inhibitory effect was dose-related: Greater inhibition was found with a higher dose of NB001. Similar analgesic effects were also obtained 3 days after CFA injection.

New cellular mechanism: AC1 positive feedback mechanisms

AC1 is thought to be an activity-dependent signaling protein that is essential to produce the second messenger cAMP. However, there is less knowledge about whether it is also regulated in disease conditions. A recent study demonstrates that AC1 itself is also activity-dependent for regulation in cortical neurons. Liu *et al.* (2019) reported that AC1 significantly increased in the ACC in an animal model of IBS and persisted for at least a few weeks (Figure 3). By contrast, acute pain did not affect the protein level of AC1. Inhibiting AC1 activity by NB001 significantly reduced the upregulation of AC1 protein in the ACC, suggesting that AC1 activity itself is critical for AC1 protein upregulation [64]. This finding indicates that AC1 may form a positive regulation in the cortex during chronic visceral pain.

HCN2 ion channel blockers

It has been recently reported that HCN2-driven action potential firings in NaV1.8 expressing nociceptors are critical for neuropathic pain in animal models, suggesting that HCN2 could be a potential drug target for neuropathic pain, as well as inflammatory pain. Although HCN2 may be specifically related to chronic pain in sensory DRG cells, it is also widely distributed in other parts of the brain, such as the cortex and hippocampus as well as other parts of the body. Any attempt at a systemic drug application, for an extended period of time, would be considered risky.

Targeting "Top-down" descending endogenous facilitatory systems

In addition to the neuronal interactions between different types of sensory afferent fibers, spinal sensory transmission, including pain transmission, receives strong descending modulation from the brain. Spinal sensory transmission receives numerous descending projections from central nuclei directly or indirectly, including the ACC, amygdala, PAG, and brainstem. Among them, a major descending pathway consists of the PAG-RVM and spinal cord connections. Many other central nuclei interact with this so-called endogenous analgesia system and produce antinociceptive/analgesic effects. As the last step of relay nuclei, neurons in several nuclei in the brainstem play an important role in descending inhibition of spinal sensory transmission. In addition to descending inhibition, descending facilitatory systems from the brainstem or forebrains have also been characterized. Biphasic modulation of spinal nociceptive transmission from the RVM offers fine regulation of spinal sensory thresholds and responses. While descending inhibition is primarily involved in regulating suprathreshold responses to noxious stimuli, descending facilitation reduces the neuronal threshold to nociceptive stimulation.

In the spinal cord, serotonin receptors are involved in the facilitation of synaptic responses as well as behavioral reflexes. One synaptic mechanism for serotonin-mediated facilitation is the recruitment of functional AMPA receptors at pure NMDA receptor-containing synapses. Postsynaptic G protein-coupled activation of PKC is important for serotonin-produced facilitation. Furthermore, the interaction between AMPA receptors and the PDZ protein glutamate receptor-interacting protein (GRIP) likely contributes to the recruitment of functional AMPA responses. Descending facilitation can be activated under physiological conditions, and a possible physiological reason for descending facilitation is to enhance animals' ability to detect potential dangerous signals in the environment. Indeed, neurons in the RVM not only respond to noxious stimuli but also show 'learning'-type changes during repetitive noxious stimuli. More importantly, RVM neurons can undergo plastic changes during, and after, tissue injury and inflammation. Descending facilitation is likely activated after the injury, contributing to secondary hyperalgesia. Blocking descending facilitation, by lesion of the RVM, or spinal blockade of serotonin receptors, is antinociceptive. The descending facilitatory system therefore serves as a double-edged sword in the central nervous system. On one hand, it allows neurons in different parts of the brain to communicate with each other and enhance sensitivity to potentially dangerous signals; on the other hand, prolonged facilitation of spinal nociceptive transmission after injury speeds up central plastic changes related to chronic pain.

Alternative treatment of chronic pain

There are several alternative treatments for chronic pain. These include acupuncture, electrical stimulation, physical therapy, and rehabilitation [65]. Most of these alternative treatments employ the techniques of stimulation of the endogenous analgesic systems or shifting the attention or focus of pain. They are often effective in controlling acute pain. However, these techniques do not work well for chronic pain in general. Considering the plastic changes in endogenous descending modulatory systems, it is expected that electrical stimulation will not produce any powerful analgesic effects in patients with chronic pain. Future studies into the molecular mechanisms of alternative medicines for pain control are needed.

Potential new drugs targeted at central plasticity and descending facilitation

Based on recent progress in the understanding of central synaptic modulation and plasticity, the following potential drug targets for chronic pain are suggested:

(1) only or mostly activated during high-frequency neuronal firing or injury-related pattern of activity but not resting or physiological activity,
(2) inhibition of the target or function selectively during chronic but not acute pain or threshold for physiological pain, e.g., AC1 and AC8, NMDA NR2B,
(3) inhibition will unlikely affect the established memory and with less effects on the formation of new experiences in animal models.
(4) target proteins showing unique expression in neurons, especially in areas related to the pain process,

(5) enhancing the functions of target proteins should selectively enhance persistent pain with less or no effect on acute pain.

The following is a list of possible candidates suggested based on recent findings:

Glutamate receptors
NMDA NR2B: Inhibiting NMDA NR2B receptors may affect chronic pain with fewer side effects.
KA receptors: Blocking function of GluR5 or other KA receptors.
AMPA receptors: Blocking the postsynaptic AMPA traffic triggered by activity.

Second and third messengers
Adenylyl cyclase AC1 and AC8: Inhibiting calcium-stimulated ACs.
Protein kinases/phosphatases: Blocking calcium-stimulating protein kinases, such PKA, PKC, and ERK.
Protein–protein interaction: Inhibitors for PDZ93 and PDZ95, AMPA receptor-related GRIP1/2.
Gene expression: Preventing the injury-related gene expression, such as Egr1 and CREB.
Neurotrophic factors: BDNF and other trophic factors.

Presynaptic regulation
Regulation of sensory transmitter releases: Inhibiting the plasticity-related enhancement of transmitter release.

Inhibitory mechanism
GABA/Glycine receptors: Drugs to enhance central inhibitory mechanisms or prevention of central disinhibition due to injury.

Descending facilitation
Serotonin subtype receptors: Drugs to block 5-HT receptor mediating descending facilitation as well as those activate 5-HT subtype receptors mediating descending inhibition

Conclusions and future directions

In conclusion, it is clear that we are just beginning to explore the molecular and cellular mechanisms of chronic pain. Due to rapid progress in the areas of genetics, neuroscience, imaging, and molecular biology, we no longer need to treat the neuronal circuits involved in the pain as a black (or gray) box. Instead, future studies will allow us to open the box and investigate the structures of the boxes at different levels. We will almost certainly be able to identify new molecular targets for controlling pain in this exciting exploration.

References

[1]   Szallasi, A. *et al.* (2006) TRPV1: A therapeutic target for novel analgesic drugs? *Trends Mol Med* 12, 545–554. 10.1016/j.molmed.2006.09.001.

[2]   Gunthorpe, M.J. and Chizh, B.A. (2009) Clinical development of TRPV1 antagonists: Targeting a pivotal point in the pain pathway. *Drug Discov Today* 14, 56–67. 10.1016/j.drudis.2008.11.005.

[3]   Di Marzo, V. (2008) Targeting the endocannabinoid system: To enhance or reduce? *Nat Rev Drug Discov* 7, 438–455. 10.1038/nrd2553.

[4]   Wu, L.J. *et al.* (2007) Kainate receptors and pain: From dorsal root ganglion to the anterior cingulate cortex. *Curr Pharm Des* 13, 1597–1605. 10.2174/138161207780765864.

[5]   Sandkuhler, J. (2007) Understanding LTP in pain pathways. *Mol Pain* 3, 9. 10.1186/1744-8069-3-9.

[6]   Zhuo, M. (2004) Central plasticity in pathological pain. *Novartis Found Symp* 261, 132–145; discussion 145–154.

[7]   Zhuo, M. (2007) A synaptic model for pain: Long-term potentiation in the anterior cingulate cortex. *Mol Cells* 23, 259–271.

[8]   Zhuo, M. (2007) Neuronal mechanism for neuropathic pain. *Mol Pain* 3, 14. 10.1186/1744-8069-3-14.

[9]   Costigan, M. *et al.* (2009) Neuropathic pain: A maladaptive response of the nervous system to damage. *Annu Rev Neurosci* 32, 1–32. 10.1146/annurev.neuro.051508.135531.

[10] Zhuo, M. (2011) Cortical plasticity as a new endpoint measurement for chronic pain. *Mol Pain* 7, 54. 10.1186/1744-8069-7-54.

[11] Zhuo, M. *et al.* (2011) Neuronal and microglial mechanisms of neuropathic pain. *Mol Brain* 4, 31. 10.1186/1756-6606-4-31.

[12] Ueda, H. (2008) Peripheral mechanisms of neuropathic pain — involvement of lysophosphatidic acid receptor-mediated demyelination. *Mol Pain* 4, 11. 10.1186/1744-8069-4-11.

[13] Reichling, D.B. and Levine, J.D. (2009) Critical role of nociceptor plasticity in chronic pain. *Trends Neurosci* 32, 611–618. 10.1016/j.tins.2009.07.007.

[14] Hucho, T. and Levine, J.D. (2007) Signaling pathways in sensitization: Toward a nociceptor cell biology. *Neuron* 55, 365–376. 10.1016/j.neuron.2007.07.008.

[15] Ruscheweyh, R. *et al.* (2011) Long-term potentiation in spinal nociceptive pathways as a novel target for pain therapy. *Mol Pain* 7, 20. 10.1186/1744-8069-7-20.

[16] Zhuo, M. (2008) Cortical excitation and chronic pain. *Trends Neurosci* 31, 199–207. 10.1016/j.tins.2008.01.003.

[17] Wu, L.J. *et al.* (2005) Kainate receptor-mediated synaptic transmission in the adult anterior cingulate cortex. *J Neurophysiol* 94, 1805–1813. 10.1152/jn.00091.2005.

[18] Zhao, M.G. *et al.* (2006) Enhanced presynaptic neurotransmitter release in the anterior cingulate cortex of mice with chronic pain. *J Neurosci* 26, 8923–8930. 10.1523/JNEUROSCI.2103-06.2006.

[19] Xu, H. *et al.* (2008) Presynaptic and postsynaptic amplifications of neuropathic pain in the anterior cingulate cortex. *J Neurosci* 28, 7445–7453. 10.1523/JNEUROSCI.1812-08.2008.

[20] Metz, A.E. *et al.* (2009) Morphological and functional reorganization of rat medial prefrontal cortex in neuropathic pain. *Proc Natl Acad Sci U S A* 106, 2423–2428. 10.1073/pnas.0809897106.

[21] Bird, G.C. *et al.* (2005) Protein kinase A-dependent enhanced NMDA receptor function in pain-related synaptic plasticity in rat amygdala neurones. *J Physiol* 564, 907–921. 10.1113/jphysiol.2005.084780.

[22] Neugebauer, V. and Li, W. (2003) Differential sensitization of amygdala neurons to afferent inputs in a model of arthritic pain. *J Neurophysiol* 89, 716–727. 10.1152/jn.00799.2002.

[23] Calejesan, A.A. *et al.* (2000) Descending facilitatory modulation of a behavioral nociceptive response by stimulation in the adult rat anterior cingulate cortex. *Eur J Pain* 4, 83–96. 10.1053/eujp.1999.0158.

[24] Zhuo, M. and Gebhart, G.F. (2002) Modulation of noxious and non-noxious spinal mechanical transmission from the rostral medial medulla in the rat. *J Neurophysiol* 88, 2928–2941. 10.1152/jn.00005.2002.

[25] Zhuo, M. *et al.* (2002) Biphasic modulation of spinal visceral nociceptive transmission from the rostroventral medial medulla in the rat. *J Neurophysiol* 87, 2225–2236. 10.1152/jn.2002.87.5.2225.

[26] Gebhart, G.F. (2004) Descending modulation of pain. *Neurosci Biobehav Rev* 27, 729–737. 10.1016/j.neubiorev.2003.11.008.

[27] Bliss, T.V. and Collingridge, G.L. (1993) A synaptic model of memory: Long-term potentiation in the hippocampus. *Nature* 361, 31–39. 10.1038/361031a0.

[28] Collingridge, G.L. and Bliss, T.V. (1995) Memories of NMDA receptors and LTP. *Trends Neurosci* 18, 54–56.

[29] Wei, F. *et al.* (2001) Genetic enhancement of inflammatory pain by forebrain NR2B overexpression. *Nat Neurosci* 4, 164–169. 10.1038/83993.

[30] Zhuo, M. (2002) Glutamate receptors and persistent pain: Targeting forebrain NR2B subunits. *Drug Discov Today* 7, 259–267. 10.1016/s1359-6446(01)02138-9.

[31] Wu, L.J. and Zhuo, M. (2009) Targeting the NMDA receptor subunit NR2B for the treatment of neuropathic pain. *Neurotherapeutics* 6, 693–702. 10.1016/j.nurt.2009.07.008.

[32] Zhuo, M. (2009) Plasticity of NMDA receptor NR2B subunit in memory and chronic pain. *Mol Brain* 2, 4. 10.1186/1756-6606-2-4.

[33] Wei, F. *et al.* (2002) Genetic elimination of behavioral sensitization in mice lacking calmodulin-stimulated adenylyl cyclases. *Neuron* 36, 713–726. 10.1016/s0896-6273(02)01019-x.

[34] Wong, S.T. *et al.* (1999) Calcium-stimulated adenylyl cyclase activity is critical for hippocampus-dependent long-term memory and late phase LTP. *Neuron* 23, 787–798. 10.1016/s0896-6273(01)80036-2.

[35] Han, J.S. *et al.* (2010) Facilitation of synaptic transmission and pain responses by CGRP in the amygdala of normal rats. *Mol Pain* 6, 10. 10.1186/1744-8069-6-10.

[36] Sharif-Naeini, R. and Basbaum, A.I. (2011) Targeting pain where it resides ... In the brain. *Sci Transl Med* 3, 65ps61. 10.1126/scitranslmed.3002077.

[37] Tang, J. *et al.* (2005) Pavlovian fear memory induced by activation in the anterior cingulate cortex. *Mol Pain* 1, 6. 10.1186/1744-8069-1-6.

[38] Zhang, Y. *et al.* (2011) Induction and requirement of gene expression in the anterior cingulate cortex and medial prefrontal cortex for the consolidation of inhibitory avoidance memory. *Mol Brain* 4, 4. 10.1186/1756-6606-4-4.

[39] Wu, L.J. *et al.* (2005) Upregulation of forebrain NMDA NR2B receptors contributes to behavioral sensitization after inflammation. *J Neurosci* 25, 11107–11116. 10.1523/JNEUROSCI.1678-05.2005.

[40] Wei, F. *et al.* (1999) Loss of synaptic depression in mammalian anterior cingulate cortex after amputation. *J Neurosci* 19, 9346–9354.

[41] Zhao, M.G. *et al.* (2005) Roles of NMDA NR2B subtype receptor in prefrontal long-term potentiation and contextual fear memory. *Neuron* 47, 859–872. 10.1016/j.neuron.2005.08.014.

[42] Wei, F. *et al.* (2002) Calcium calmodulin-dependent protein kinase IV is required for fear memory. *Nat Neurosci* 5, 573–579. 10.1038/nn0602-855.

[43] Frey, U. *et al.* (1993) Effects of cAMP simulate a late stage of LTP in hippocampal CA1 neurons. *Science* 260, 1661–1664. 10.1126/science. 8389057.

[44] Nguyen, P.V. *et al.* (1994) Requirement of a critical period of transcription for induction of a late phase of LTP. *Science* 265, 1104–1107. 10.1126/science.8066450.

[45] Huang, Y.Y. *et al.* (1996) Mice lacking the gene encoding tissue-type plasminogen activator show a selective interference with late-phase long-term potentiation in both Schaffer collateral and mossy fiber pathways. *Proc Natl Acad Sci USA* 93, 8699–8704. 10.1073/pnas.93.16.8699.

[46] Kandel, E.R. (2001) The molecular biology of memory storage: A dialogue between genes and synapses. *Science* 294, 1030–1038. 10.1126/science. 1067020.

[47] Li, X.Y. *et al.* (2010) Alleviating neuropathic pain hypersensitivity by inhibiting PKMzeta in the anterior cingulate cortex. *Science* 330, 1400–1404. 10.1126/science.1191792.

[48] Chen, T. *et al.* (2014) Adenylyl cyclase subtype 1 is essential for late-phase long term potentiation and spatial propagation of synaptic responses in the anterior cingulate cortex of adult mice. *Mol Pain* 10, 65. 10.1186/1744-8069-10-65.

[49] Gurwitz, D. and Weizman, A. (2002) The NR2B subunit of glutamate receptors as a potential target for relieving chronic pain: Prospects and concerns. *Drug Discov Today* 7, 403–406. 10.1016/s1359-6446(02) 02242-0.

[50] Xia, Z. and Storm, D.R. (1997) Calmodulin-regulated adenylyl cyclases and neuromodulation. *Curr Opin Neurobiol* 7, 391–396. 10.1016/ s0959-4388(97)80068-2.

[51] Liauw, J. *et al.* (2005) Calcium-stimulated adenylyl cyclases required for long-term potentiation in the anterior cingulate cortex. *J Neurophysiol* 94, 878–882. 10.1152/jn.01205.2004.

[52] Wang, H. *et al.* (2011) Identification of an adenylyl cyclase inhibitor for treating neuropathic and inflammatory pain. *Sci Transl Med* 3, 65ra63. 10.1126/scitranslmed.3001269.

[53] Vadakkan, K.I. *et al.* (2006) Genetic reduction of chronic muscle pain in mice lacking calcium/calmodulin-stimulated adenylyl cyclases. *Mol Pain* 2, 7. 10.1186/1744-8069-2-7.

[54] Wang, G.D. and Zhuo, M. (2002) Synergistic enhancement of glutamate-mediated responses by serotonin and forskolin in adult mouse spinal dorsal horn neurons. *J Neurophysiol* 87, 732–739. 10.1152/jn.00423.2001.

[55] Wei, F. *et al.* (2006) Calcium calmodulin-stimulated adenylyl cyclases contribute to activation of extracellular signal-regulated kinase in spinal dorsal horn neurons in adult rats and mice. *J Neurosci* 26, 851–861. 10.1523/JNEUROSCI.3292-05.2006.

[56] Koga, K. *et al.* (2010) In vivo whole-cell patch-clamp recording of sensory synaptic responses of cingulate pyramidal neurons to noxious mechanical stimuli in adult mice. *Mol Pain* 6, 62. 10.1186/1744-8069-6-62.

[57] Vadakkan, K.I. *et al.* (2005) A behavioral model of neuropathic pain induced by ligation of the common peroneal nerve in mice. *J Pain* 6, 747–756. 10.1016/j.jpain.2005.07.005.

[58] Descalzi, G. *et al.* (2009) Presynaptic and postsynaptic cortical mechanisms of chronic pain. *Mol Neurobiol* 40, 253–259. 10.1007/s12035-009-8085-9.

[59] Esteban, J.A. *et al.* (2003) PKA phosphorylation of AMPA receptor subunits controls synaptic trafficking underlying plasticity. *Nat Neurosci* 6, 136–143. 10.1038/nn997.

[60] Wu, L.J. *et al.* (2008) Enhancement of presynaptic glutamate release and persistent inflammatory pain by increasing neuronal cAMP in the anterior cingulate cortex. *Mol Pain* 4, 40. 10.1186/1744-8069-4-40.

[61] Sacktor, T.C. (2008) PKMzeta, LTP maintenance, and the dynamic molecular biology of memory storage. *Prog Brain Res* 169, 27–40. 10.1016/S0079-6123(07)00002-7.

[62] Naik, M.U. *et al.* (2000) Distribution of protein kinase Mzeta and the complete protein kinase C isoform family in rat brain. *J Comp Neurol* 426, 243–258. 10.1002/1096-9861(20001016)426:2<243::aid-cne6>3.0.co;2-8.

[63] Pierre, S. *et al.* (2009) Capturing adenylyl cyclases as potential drug targets. *Nat Rev Drug Discov* 8, 321–335. 10.1038/nrd2827.

[64] Pavan, B. *et al.* (2009) Adenylyl cyclases as innovative therapeutic goals. *Drug Discov Today* 14, 982–991. 10.1016/j.drudis.2009.07.007.

[65] Han, J.S. (2011) Acupuncture analgesia: Areas of consensus and controversy. *Pain* 152, S41–S48. 10.1016/j.pain.2010.10.012.

[66] Liu, M.G. and Zhuo, M. (2014) No requirement of TRPV1 in long-term potentiation or long-term depression in the anterior cingulate cortex. *Mol Brain* 7, 27. 10.1186/1756-6606-7-27.

[67] Koga, K. *et al.* (2015) Coexistence of two forms of LTP in ACC provides a synaptic mechanism for the interactions between anxiety and chronic pain. *Neuron* 86, 1109. 10.1016/j.neuron.2015.05.016.

[68] Liu, S.B. *et al.* (2020) Cyclic AMP-dependent positive feedback signaling pathways in the cortex contributes to visceral pain. *J Neurochem* 153, 252–263. 10.1111/jnc.14903.

[69] Xia, Z. and Storm, D.R. (2005) The role of calmodulin as a signal integrator for synaptic plasticity. *Nat Rev Neurosci* 6(4), 267–276.

Index

Printed in the USA
CPSIA information can be obtained
at www.ICGtesting.com
LVHW020240121123
763440LV00001B/5